Diamane

Fabrication, properties and new advances in 2D diamond

Online at: https://doi.org/10.1088/978-0-7503-5939-9

Diamane

Fabrication, properties and new advances in 2D diamond

Edited by
Santosh K Tiwari
Department of Chemistry, NMAM Institute of Technology Nitte (Deemed to be University), Mangaluru, Karnataka 547110, India

Arpan Kumar Nayak
Department of Education in Science and Mathematics, Regional Institute of Education, National Council of Educational Research and Training (NCERT) Mysuru, Mysuru – 570006, Karnataka, India

IOP Publishing, Bristol, UK

ISBN 978-0-7503-5939-9 (ebook)
ISBN 978-0-7503-5937-5 (print)
ISBN 978-0-7503-5940-5 (myPrint)
ISBN 978-0-7503-5938-2 (mobi)

DOI 10.1088/978-0-7503-5939-9

Version: 20240401

IOP ebooks

British Library Cataloguing-in-Publication Data: A catalogue record for this book is available from the British Library.

Published by IOP Publishing, wholly owned by The Institute of Physics, London

IOP Publishing, No.2 The Distillery, Glassfields, Avon Street, Bristol, BS2 0GR, UK

US Office: IOP Publishing, Inc., 190 North Independence Mall West, Suite 601, Philadelphia, PA 19106, USA

Dedicated to Lord Jagannath

Contents

Preface

Diamane, its derivatives, and two-dimensional (2D) diamond stand at the forefront of materials innovation, poised to reshape a multitude of industries. While diamane remains in its nascent stages of development, it has already exhibited remarkable potential across a diverse spectrum of applications. From automotive engineering to textiles, from flexible electronics to cutting-edge optoelectronics, from the realm of quantum electronics to solar cells, supercapacitors, batteries, and the next generation of energy storage devices—diamane holds the promise of transformative progress. Herein, we have added a comprehensive exploration of diamanes and diamanoids, encompassing their physical and chemical properties and delving into their myriad applications.

A striking prediction from research reports suggests that carbon-based materials are poised to supplant nearly 55%–60% of non-carbonaceous semiconductor materials by 2040, thereby becoming the cornerstone of advanced technologies, such as supercomputers and next-generation communication devices. This meteoric rise can be attributed to the unique tunability of carbon materials and their exceptional bonding capabilities. Interestingly, data from the Scopus database shows that a remarkable 30%–35% of total scientific publications are directly or indirectly linked to carbon-based materials, underscoring the overarching influence of carbon materials research in the scientific community. This dominance extends across a wide array of industrial applications, making carbon materials research the vanguard of scientific exploration. In essence, this book focuses on unraveling the physical and chemical properties of sp^3 carbon, as found in diamane and its derivatives. It also addresses various facets of diamane and diamanoid synthesis, delving into their fundamental properties and showcasing their advanced applications. In this endeavor, we have endeavored to spotlight the progress made in these newly developed materials, while also acknowledging the challenges and the path to commercialization.

As we embark on this journey through the world of diamane and diamanoids, we invite readers to explore the limitless possibilities and transformative potential of these remarkable carbon-based materials. It is our hope that this book will serve as a valuable resource for researchers, scientists, engineers, and enthusiasts, as we collectively navigate the exciting frontier of materials science and technology.

Santosh K Tiwari and Arpan Kumar Nayak

Acknowledgements

We are very grateful to all the contributors (authors and co-authors of the chapters), in particular for their timely completion and revision as per the reviewers' comments. We are also very grateful to IOP Publishing, in particular Dr Caroline Mitchell, for allowing enough time to prepare the manuscript. We very sincerely thank the researchers whose work is directly or indirectly cited in this book in the form of text, research data, figures, and tables.

Editor biographies

Santosh K Tiwari

Dr. Santosh K. Tiwari is an Assistant Professor and faculty in-charge of "Centre for New Materials & Surface Engineering", at NMAMIT, Nitte, India (Visiting faculty, GIFT, GU, Nanning, China). He received a PhD in 2D materials-based polymer nanocomposites from IIT Dhanbad, India. He then worked for over five years with several highly ranked research labs and universities around the globe, including HSCL, Hanynag University in Seoul, Korea, GIFT, GU in Nanning, China, and the University of Warsaw in Poland. He has published more than 70 articles (h-index 27) and five books with reputed international publishers. His research interests are in the areas of new materials, 2D nanomaterials, materials for extreme environment and mechanical properties.

Arpan Kumar Nayak

Dr. Arpan Kumar Nayak is a Research Professor at Konkuk University, Seoul, South Korea. His research mainly focuses on the synthesis of various nanostructured materials towards environment and energy applications. He has published more than 80 articles in various international journals (h-index 25, i-index 46), Six books, and is the editor-in-chief of the International Research Journal of Multidisciplinary Technovation and Bulletin of Scientific Research.

List of contributors

S K Rozi Unnesa Alam
Department of Chemical Engineering, University of Calcutta, 92 A.P.C. Road, Kolkata 700009, India

Blessy Thomas Baby
Centre for Industrial Material Development and Lambton Manufacturing Innovation Centre, Lambton College Research and Innovation, Sarnia, N7S 6K4, Canada

Ramachandran Balaji
Department of Electronics and Communication Engineering, Koneru Lakshmaiah Educational Foundation, Andhra Pradesh 522302, India

Sagar Ban
Department of Chemical Science and Engineering, Kathmandu University, PO Box 6250, Dhulikhel, Kavre, Nepal

Narendhar Chandrasekarb
Department of Chemistry, Gachon University, 1342 Seongnam-daero, Sujeong-gu, Seongnam-si, Gyeonggi-do 13120, Republic of Korea

Paritosh Chaudhuri
Institute for Plasma Research, Bhat, Gandhinagar 382428, India

Maya Devi
Department of Physics, School of Applied Sciences, KIIT Deemed to be University, Bhubaneswar 24, Odisha, India

Muqaddas Farooq
Polymer and Process Engineering (PPE) Department, University of Engineering and Technology (UET) Lahore, Pakistan

Ankita Subhrasmita Gadtya
School of Applied Sciences, Centurion University of Technology and Management, Odisha, India

Pradeep Kumar Gautam
Civil and Environmental Engineering, Colorado School of Mines, Golden, CO, USA

Tapas Ghatak
Vellore Institute of Technology, Vellore, India

Radha Gupta
Department of Chemistry, SRM Institute Science and Technology, Delhi-NCR Campus, Modinagar, Ghaziabad, Uttar Pradesh 201204, India

Rajendra Joshi
Department of Pharmacy, School of Science, Kathmandu University, PO Box 6250, Dhulikhel, Kavre, Nepal

Myung Jong Kim
Department of Chemistry, Gachon University, 1342 Seongnam-daero, Sujeong-gu, Seongnam-si, Gyeonggi-do 13120, Republic of Korea

Chandan Kumar Maity
Department of Chemistry, Gachon University, 1342 Seongnam-daero, Sujeong-gu, Seongnam-si, Gyeonggi-do 13120, Republic of Korea

Umer Mehmood
Polymer and Process Engineering (PPE) Department, University of Engineering and Technology (UET) Lahore, Pakistan

Soumyadeep Mitra
Centre for Research in Nanoscience and Nanotechnology, University of Calcutta, JD-2, Sector 3, Salt Lake City, Kolkata 700098, India

Srikanta Moharana
School of Applied Sciences, Centurion University of Technology and Management, Odisha, India

Sanglap Mondal
Department of Chemical Engineering, University of Calcutta, 92 A.P.C. Road, Kolkata 700009, India

Abdul Mumin
Centre for Industrial Material Development and Lambton Manufacturing Innovation Centre, Lambton College Research and Innovation, Sarnia, N7S 6K4, Canada

Arpan Kumar Nayak
Department of Education in Science and Mathematics, Regional Institute of Education, National Council of Educational Research and Training (NCERT) Mysuru, Mysuru – 570006, Karnataka, India

Rabia Nazar
Polymer and Process Engineering (PPE) Department, University of Engineering and Technology (UET) Lahore, Pakistan

Sunday J Olusegun
Department of Environmental Biotechnology, Faculty of Energy and Environmental Engineering, Silesian University of Technology, Gliwice, Poland

Swapnadipa Patra
Department of Chemical Engineering, University of Calcutta, 92 A.P.C. Road, Kolkata 700009, India

Swetapadma Praharaj
Department of Physics, School of Applied Sciences, KIIT Deemed to be University, Bhubaneswar 24, Odisha, India

Dibyaranjan Rout
Department of Physics, School of Applied Sciences, KIIT Deemed to be University, Bhubaneswar 24, Odisha, India

Subhasis Roy
Department of Chemical Engineering, University of Calcutta, 92 A.P.C. Road, Kolkata 700009, India

Suresh Sagadevan
Nanotechnology and Catalysis Research Centre, University of Malaya, Kuala Lumpur 50603, Malaysia

Rajashree Sahoo
Institute for Plasma Research, Bhat, Gandhinagar 382428, India

Bibhuti B Sahu
Department of Basic Science and Humanities, Nalanda Institute of Technology (NIT), Bhubaneswar, Odisha

Aditya L Shinde
Vellore Institute of Technology, Vellore, India

Rakesh Shrestha
Department of Chemical Science and Engineering, Kathmandu University, PO Box 6250, Dhulikhel, Kavre, Nepal
Centre for Industrial Material Development and Lambton Manufacturing Innovation Centre, Lambton College Research and Innovation, Sarnia, N7S 6K4, Canada

Santosh K Tiwari
Department of Chemistry, NMAM Institute of Technology Nitte (Deemed to be University), Mangaluru, Karnataka 547110, India

Tarun Yadav
Department of Basic Sciences, IES University, Bhopal (M.P.), India

IOP Publishing

Diamane

Fabrication, properties and new advances in 2D diamond

Santosh K Tiwari and Arpan Kumar Nayak

Chapter 1

Introduction to different types of 2D carbon and nanodiamond

Rajashree Sahoo, Paritosh Chaudhuri and Arpan Kumar Nayak

This chapter provides an in-depth exploration of different types of two-dimensional (2D) carbon materials, including graphene, graphene oxide (GO), carbon nanotubes (CNTs), and nanodiamond. These materials have garnered significant attention due to their unique structures and properties, offering immense potential for various applications. This chapter begins by discussing the definition, structure, and properties of graphene, followed by an examination of synthesis methods and applications in electronics, optoelectronics, energy storage, sensors, and composite materials. GO, derived from graphene, is then introduced, highlighting its structure, synthesis approaches, and applications in membranes, biomedical devices, and catalysis. The subsequent section focuses on carbon nanotubes, exploring their structure, types, synthesis methods, and applications in electronics, conductive films, nanocomposites, and energy storage. Finally, nanodiamond is introduced, elucidating its structure, synthesis methods, surface functionalization, and applications in biomedical devices, quantum sensing and imaging, and tribology. This chapter concludes by discussing the future perspectives and emerging trends in the field of 2D carbon materials, such as hybrid materials, scalable synthesis, integration with other nanomaterials, and considerations for environmental sustainability. Overall, this chapter provides a comprehensive overview of the diverse properties, synthesis methods, and applications of 2D carbon materials and nanodiamond, setting the stage for further research and technological advancements in this rapidly evolving field.

1.1 Introduction

Carbon, a versatile and abundant element, has captivated the scientific community with its remarkable properties and potential for innovation. In recent years, the discovery and exploration of 2D carbon materials have revolutionized the field of materials science and opened up exciting possibilities for a wide range of applications. This chapter delves into the fascinating world of different types of 2D carbon

doi:10.1088/978-0-7503-5939-9ch1

materials, including graphene, GO, CNTs, and nanodiamond, as well as their unique properties and applications.

Graphene, the most renowned 2D carbon material, has attracted immense attention due to its extraordinary electrical conductivity, mechanical strength, and thermal stability [1, 2]. Graphene is composed of a single layer of carbon atoms arranged in a honeycomb lattice and exhibits remarkable properties that have the potential to revolutionize fields such as optoelectronics, electronics [3], energy storage [4], sensors [5], and composite materials. The synthesis methods for graphene, ranging from mechanical exfoliation [6] to chemical vapor deposition (CVD) [7] and epitaxial growth [8], allow for the fabrication of high-quality graphene films and structures with precise control over their properties.

GO, derived from graphene through oxidation processes, offers an exciting avenue for the development of new materials. With the introduction of oxygen functional groups, GO becomes soluble and amenable to various processing techniques. This characteristic enables its applications in membranes for filtration and separation, biomedical devices for tissue engineering, and drug delivery, and catalysis for chemical transformations [9].

Carbon nanotubes exhibit a cylindrical structure composed of rolled graphene sheets [10]. These tubular structures possess extraordinary electrical conductivity, mechanical strength, and thermal stability, making them promising candidates for applications in electronics, conductive films, nanocomposites, and energy storage devices. Carbon nanotubes can be synthesized through various methods, including arc discharge [11, 12] and CVD [13, 14], offering control over their properties and scalability.

Nanodiamond has a diamond crystal structure at the nanoscale and presents intriguing properties that differ from its bulk counterpart. Nanodiamonds exhibit exceptional hardness, chemical stability, and biocompatibility, making them attractive for diverse applications. The synthesis of nanodiamonds can be achieved through the detonation method [15] or CVD [16], leading to particles with sizes ranging from a few to a few hundred nanometers. Surface functionalization of nanodiamonds further expands their potential by allowing the attachment of specific functional groups and biomolecules. Nanodiamonds find applications in biomedical devices, quantum sensing and imaging, and tribology and lubrication.

As we delve deeper into the realm of 2D carbon materials and nanodiamond, we gain a deeper understanding of their unique structures, properties, synthesis methods, and diverse applications. This chapter aims to provide a comprehensive overview of these materials, their synthesis approaches, and their applications in various fields. Additionally, we will explore the future perspectives and emerging trends in the field of 2D carbon materials and nanodiamond, as we continue to push the boundaries of scientific exploration and technological innovation.

1.2 Graphene

1.2.1 Definition and structure

Graphene, often referred to as the 'wonder material,' is a 2D allotrope of carbon consisting of a single layer of carbon atoms arranged in a hexagonal lattice. Each

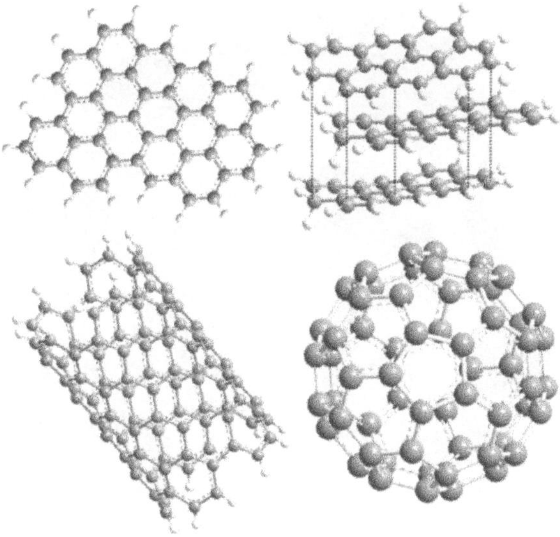

Figure 1.1. Various graphene structures: graphene sheet, stacked graphene, wrapped graphene, and rolled graphene.

carbon atom is bonded to three neighboring carbon atoms, forming strong sp^2 hybridized bonds [17, 18]. The resulting structure gives graphene its exceptional mechanical, electrical, and thermal properties. Some of the different structures of graphene (i.e., graphene sheet, wrapped graphene, stacked graphene, and rolled graphene) are shown in figure 1.1.

1.2.2 Properties of graphene

Graphene exhibits a remarkable combination of properties, including high electrical conductivity, excellent thermal conductivity, exceptional mechanical strength, and optical transparency [19]. It is also highly flexible, lightweight, and impermeable to gases and liquids. Graphene's unique electronic properties, such as quantum confinement effects and high electron mobility, make it a promising material for various applications.

1.2.3 Synthesis methods

The different synthesis methods (figure 1.2) and applications (figure 1.3) of graphene are explained in detail in this subsection.

1.2.3.1 Mechanical exfoliation

Mechanical exfoliation, also known as the 'Scotch tape' method, was the first technique used to isolate graphene. It involves repeatedly peeling off thin flakes from a graphite crystal using adhesive tape, resulting in the formation of single-layer graphene sheets [20–22]. Although it is a time-consuming and low-yield method, it produces high-quality graphene with exceptional properties.

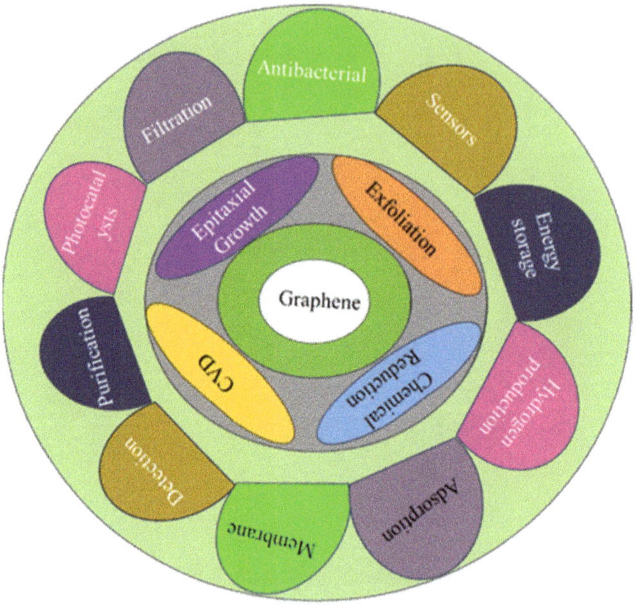

Figure 1.2. Synthesis methods and applications of graphene.

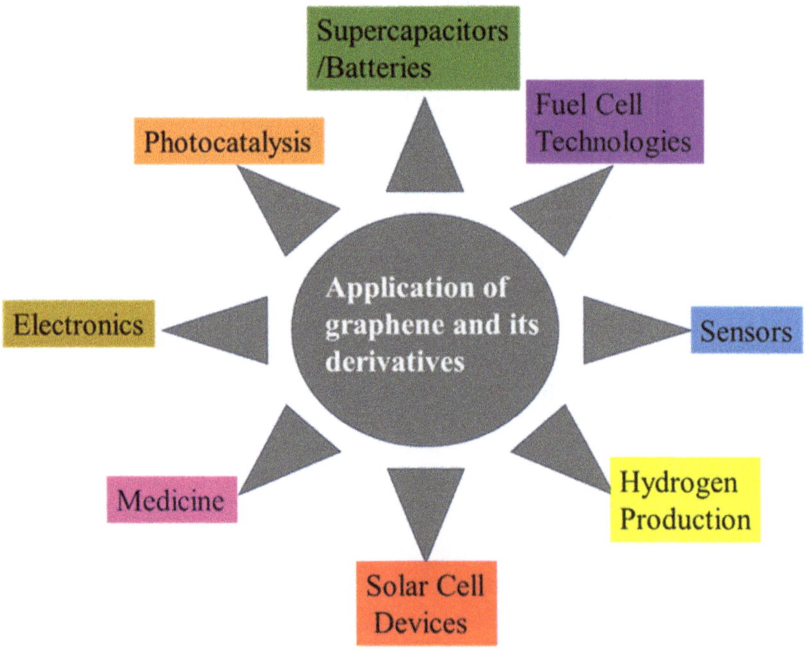

Figure 1.3. Schematic diagram of applications of graphene and its derivatives.

1.2.3.2 Chemical vapor deposition

CVD is a highly utilized method for the large-scale synthesis of graphene. It involves the deposition of C atoms onto a substrate in the presence of a carbon-containing precursor gas, such as methane or ethylene. The carbon atoms arrange themselves into a graphene layer on the substrate's surface. CVD allows control over the graphene's size, number of layers, and doping, making it suitable for various applications [23–25].

1.2.3.3 Epitaxial growth

Epitaxial growth involves the deposition of graphene on a substrate that has a similar crystal structure. Common substrates include silicon carbide (SiC) and metal surfaces. Epitaxial graphene is typically grown by heating the substrate in a controlled environment, causing the sublimation of silicon or other elements from the substrate and the subsequent formation of a graphene layer [24]. This method produces high-quality graphene but is limited to specific substrates.

1.2.4 Applications of graphene

1.2.4.1 Electronics and optoelectronics

Graphene's exceptional electrical properties make it suitable for various electronic devices, including transistors, flexible displays, and photodetectors. Its high electron mobility, low resistivity, and transparency enable faster and more efficient electronic components.

1.2.4.2 Energy storage

Graphene-based materials are extensively studied for energy storage applications such as batteries and supercapacitors. Graphene's chemical stability, high surface area, and electrical conductivity enhance the performance of energy storage devices, enabling faster charging, higher energy density, and longer cycle life.

1.2.4.3 Sensors and biosensors

Graphene's sensitivity to changes in its electrical conductivity makes it an ideal material for sensors and biosensors. Graphene-based sensors can detect gases, chemicals, and biological molecules with high sensitivity, selectivity, and speed, offering potential applications in environmental monitoring, healthcare diagnostics, and food safety.

1.2.4.4 Composite materials

Graphene can be incorporated into various composite materials to enhance their electrical, mechanical, and thermal properties. Graphene composites show improved strength, stiffness, and toughness, making them valuable for aerospace, automotive, and construction industries. Additionally, graphene-based composites have shown potential in thermal management and electromagnetic shielding applications.

1.2.5 Introduction to graphene oxide

GO is a derivative of graphene that contains oxygen-containing functional groups, such as hydroxyl, epoxy, and carboxyl groups, attached to its basal plane and edges (figure 1.4). These functional groups give GO distinct properties and make it highly soluble in water and other solvents [26, 27]. GO exhibits unique characteristics which makes it suitable for a wide range of applications.

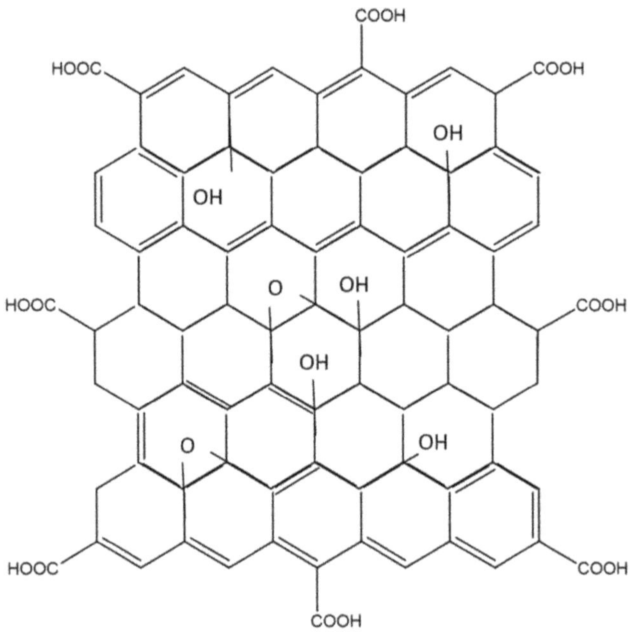

Figure 1.4. Structure of GO.

1.2.6 Structure and properties

GO consists of a layered structure similar to graphene, with a carbon lattice interspersed with oxygen functional groups. The presence of these functional groups introduces sp^3 hybridization and disrupts the delocalization of π-electrons, resulting in reduced electrical conductivity compared to graphene. However, GO still maintains excellent mechanical strength and flexibility.

1.2.7 Synthesis approaches

1.2.7.1 Hummers' method

Hummers' method is one of the most commonly used techniques to synthesize GO. It involves the oxidation of graphite using a mixture of $KMnO_4$, H_2SO_4, and $NaNO_3$. This method yields GO with a relatively high degree of oxidation and a significant number of oxygen functional groups [28].

1.2.7.2 Brodie's method

Brodie's method is another popular approach to produce GO, which involves the oxidation of graphite using a mixture of concentrated nitric acid (HNO_3) and potassium chlorate ($KClO_3$). The resulting GO obtained from Brodie's method typically has a lower degree of oxidation than that produced by Hummers' method [29, 30].

1.2.8 Functionalization of graphene oxide

GO's surface can be further functionalized by modifying the oxygen functional groups with various molecules or polymers. This functionalization enhances the dispersibility, stability, and compatibility of GO in specific solvents or matrices, enabling its tailored integration into desired applications. Functionalized GO offers improved properties and opens up opportunities for targeted applications [26].

1.2.9 Applications of graphene oxide

1.2.9.1 Membranes and filtration

GO membranes have shown exceptional properties for water purification and gas separation. The layered structure of GO provides selective permeability, allowing the filtration of small molecules while blocking larger ones. GO membranes have the potential to revolutionize desalination processes and offer energy-efficient solutions for water treatment [31].

1.2.9.2 Biomedical applications

GO has gained significant interest in biomedical research due to its unique physicochemical properties. It exhibits excellent biocompatibility, high surface area, and the ability to interact with biomolecules. GO-based materials find applications in drug delivery, tissue engineering, bioimaging, and biosensing. GO's versatility and potential for functionalization make it a promising candidate for targeted therapy and diagnostics [32].

1.2.9.3 Catalysis

GO and its derivatives have shown promising catalytic activity in various chemical reactions. The oxygen functional groups on GO's surface provide active sites for catalytic reactions. GO-based catalysts have been explored in applications such as fuel cells, water splitting, and organic synthesis. GO's stability and tunable surface chemistry enable the design of efficient and selective catalysts [33].

1.3 Carbon nanotubes

1.3.1 Introduction to carbon nanotubes

CNTs are cylindrical nanostructures composed of carbon atoms arranged in a hexagonal lattice. They possess unique structural, electrical, mechanical, and thermal properties, making them highly sought-after materials in various fields of science and technology. CNTs can be categorized into single-walled carbon nanotubes (SWCNTs) and multi-walled carbon nanotubes (MWCNTs), depending on the number of concentric tubes present [34].

1.3.2 Structure and types of CNTs

CNTs have a tubular structure with a high aspect ratio. SWCNTs consist of a single graphene sheet rolled into a seamless cylinder, while MWCNTs consist of multiple concentric graphene layers nested within each other. The arrangement of carbon atoms and the tube diameter and chirality influence the properties and behavior of CNTs [35].

1.3.3 Properties of CNTs

CNTs possess remarkable properties due to their unique structure. They exhibit high electrical conductivity, excellent mechanical strength, and thermal stability. CNTs also have a large surface area, high aspect ratio, and exceptional flexibility [36]. The combination of these properties makes CNTs suitable for a wide range of applications.

1.3.4 Synthesis methods

1.3.4.1 Arc discharge

Arc discharge is one of the earliest and most common methods for producing CNTs. It involves passing a high current between two graphite electrodes in an inert gas atmosphere. The high temperature generated vaporizes the carbon, leading to the formation of CNTs. Arc discharge synthesis typically produces a mixture of SWCNTs and MWCNTs [37].

1.3.4.2 Chemical vapor deposition

CVD is a widely used method for controlled synthesis of CNTs. It involves the decomposition of carbon-containing precursor gases, such as methane or ethylene, in the presence of a catalyst. CVD allows the growth of CNTs on a substrate, enabling the control of diameter, chirality, and alignment. It offers a scalable and versatile approach for large-scale production of CNTs [38].

1.3.5 Applications of carbon nanotubes

1.3.5.1 Electronics and conductive films

CNTs possess exceptional electrical conductivity, making them ideal candidates for electronic applications. They can be integrated into transistors, interconnects, and other electronic components to enhance performance. CNT-based conductive films also find applications in touchscreens, flexible electronics, and transparent conductive coatings [39].

1.3.5.2 Nanocomposites

CNTs serve as reinforcing agents in nanocomposites, enhancing their mechanical strength, stiffness, and electrical properties. The incorporation of CNTs into polymers, metals, and ceramics improves the overall performance and enables the development of lightweight and high-performance materials. CNT-based nanocomposites have applications in automotive, aerospace, and structural engineering [40].

1.3.5.3 Energy storage

CNTs have demonstrated potential in energy storage devices, supercapacitors, including lithium-ion batteries, and fuel cells. Their high surface area, electrical conductivity, and mechanical robustness enhance the charge storage and transport properties of electrodes, leading to improved energy storage and conversion capabilities [41].

1.3.6 Introduction to nanodiamond

Nanodiamond refers to diamond nanoparticles with sizes typically in the range of a few nanometers to a few hundred nanometers (figure 1.5). These nanoparticles exhibit unique properties derived from their diamond structure and have gained significant interest in various fields of research and application. Nanodiamonds can be produced through different synthesis methods and offer a wide range of potential applications.

Many of the characteristics of bulk diamond are shared by nanodiamonds. They share traits such being a wide-band semiconductor with the biggest optical bandgap of any known materials (5.4–5.6 eV at ambient temperature) [42–44]. In addition, nanodiamonds have high refractive index, strength, hardness (high Young's modulus), electrical resistivity, and thermal conductivity. They have optical transparency throughout a wide spectrum, from deep ultraviolet to the infrared region, exhibit minimum thermal expansion, and have good breakdown voltage and carrier mobility [42–44]. Nanodiamond particles exhibit progressively distinctive properties driven by finite-size (quantum) phenomena as their size decreases. Additionally, the proportion of sp^3 to sp^2 carbon atoms in nanodiamonds significantly affects their characteristics. The breadth of the curve increases as the number of sp^2 carbon atoms decreases. The bandgap widens with a reduction in the proportion of sp^2 carbon atoms. The bandgap widths of nanodiamonds can be very different from

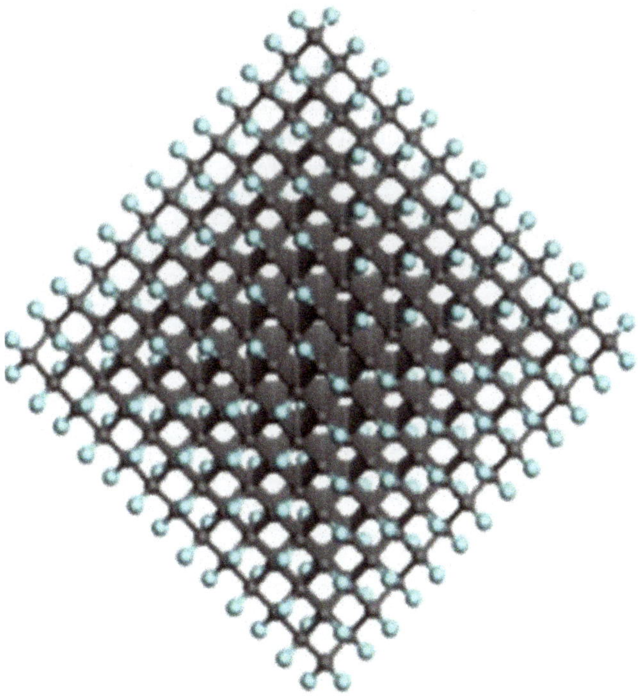

Figure 1.5. Representative atomic structures of the nanodiamond $C_{455}H_{196}$. Reprinted from [47]. Copyright (2007), with permission from Elsevier.

those of bulk diamond, and they can show substantial visual absorption. These findings imply that the carbon atoms on their surfaces are sp^2-hybridized. The existence and type of impurities within or on the surfaces of nanodiamonds can also drastically alter their physical characteristics [45].

Excellent chemical stability and corrosion or deterioration resistance can be found in nanodiamonds. Additionally, they have a substantial surface area that makes it possible for them to successfully bind materials like proteins [46]. Nanodiamonds are readily functionalized, making them suitable as solid delivery systems for biomolecules and pharmaceuticals. They are more biocompatible and eco-friendlier than other carbon nanostructures [46, 47], which makes them potentially appropriate for a range of biomedical applications. Nanodiamond particles have unpaired electrons on their surfaces, which allows them to act as multi-charged radical donors and effectively absorb the free radicals.

1.3.7 Structure and properties

Nanodiamonds possess a diamond crystal structure, consisting of a three-dimensional (3D) network of carbon atoms arranged in a face-centered cubic lattice. The small size of nanodiamonds results in a high surface area-to-volume ratio, which can lead to enhanced reactivity and interactions with other materials. They exhibit exceptional hardness, chemical stability, and biocompatibility, making them attractive for numerous applications.

1.3.8 Synthesis methods

Figure 1.6 provides an overview of the significant characteristics of monodispersed single-digit nanodiamonds and pertinent characteristics [48]. Detonation-produced nanodiamond particles typically range in size from 3 to 6 nm and have a spherical

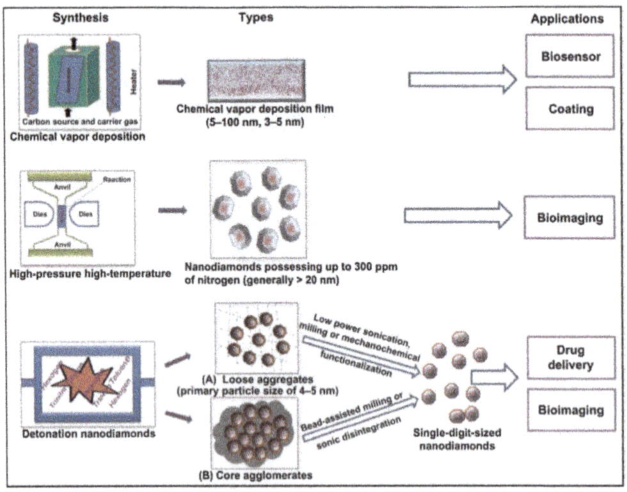

Figure 1.6. Diverse morphologies of nanodiamonds with their model applications and various synthesis processes for nanodiamonds. Reproduced from [48]. CC BY 4.0. Adapted from [49].

shape with 3D symmetry. However, particular nanodiamond aspect ratios (about 300–450 $m^2\ g^{-1}$) are needed in some applications [50–52], such as sorbent molecular systems or building a strong network within a polymer matrix. These uses demand a precise form and size for the nanoparticles. Spherical nanodiamond particles serve as nanoscale ball bearings in lubricating and polishing applications, which is helpful. In contrast, regulated morphology in pharmaceutical applications might increase the loading capacity for drug transportation and offer more routes for protein adsorption.

Detonation-synthesized nanodiamonds have been found to provide considerable advantages over materials of comparable sizes in applications requiring a minimum amount of cytotoxicity and related actions, often involving particles with a size between 5 and 10 nm. Although detonated nanodiamonds are not hazardous by nature, the addition of sp^2 carbon and metals during the purifying process can affect their toxicity [51–53]. The particular cell type used in the evaluation affects the toxicity levels of these nanodiamonds (table 1.1).

Table 1.1. Historical information about nanodiamonds research.

Year	Remarks
1956	Efforts at dynamic synthesis are not fruitful.
1960	After evidence supporting the possibility of creating diamonds by compressing graphite under shock was discovered, experimental research at VNIITF was started.
1961	The preparation of diamonds while preserving shock-compressed graphite.
1962	Using spherical and cylindrical ampoules with shock-compressed graphite diamond formation occurs, resulting in a diamond yield of approximately 2%.
1963	Diamonds were created by compressing combinations of graphite and Me (metallic element), and carbon black and Me, with a yield of 20% or so.
1963	It was possible to create ultra-nano-crystalline diamond (UDD) by employing carbon that was derived from explosive waste. The yield of diamonds obtained ranged between 8% and 12% of the mass of the TG40 charge.
1965	Graphite was used for the synthesis of diamond.
1976	Compressing a graphite–copper combination with a charge mass of 5 t in cylindrical ampoules results in the commercial manufacturing of diamond micropowder, known as Mypolex.
1982	UDD formation.
1983	Diamond micropowders DAG and DAS were produced at a pilot scale by explosively detonating 100 g charges of a mixture containing graphite or carbon black with hexogen within an explosion chamber.
1984–90	Large-scale production of ultrafine-dispersed diamonds (UDDs).
1992	Pilot-scale production of UDDs involved using TG40 charges weighing 10 kg and implementing water cooling for diamond particles within a 100 m^3 chamber.
1994	Initiation of research on sintering UDDs in static conditions.
2003	Controversial hypothesis of nanodiamonds in Younger Dryas boundary layer.
2006	Widespread nanodiamonds found in multiple geological layers.
2009	Confirmation of nanodiamonds in the Younger Dryas boundary layer.
2010	Potential role of nanodiamonds in extraterrestrial events explored.
2012	Evidence suggesting nanodiamonds from cosmic impact.
2014	Development of advanced techniques to study nanodiamonds.
2017	Discovery of nanodiamonds in prehistoric human teeth.
Present	Nanodiamonds find applications in diverse fields, including composites, biomedical, tribology, electronics, and more.

1.3.8.1 Precursors

Nanodiamonds can be produced using various precursors, including gaseous mixtures of CH_4 and H_2 with inert gases. However, there have been reports of utilizing oxygen and nitrogen as well, but this approach is not considered to be economical or efficient due to the introduction of multiple flaws [54, 55]. Studies have demonstrated that adjusting the amount of oxygen and nitrogen inclusion can affect the shape, microstructure, grain size, crystalline grade, and growth rate of nanodiamonds. Increasing the percentage of inclusions has been found to improve the crystal properties of nanodiamonds.

In addition to traditional precursors, alternative materials such as carbon nanotubes have been used as precursors for nanodiamond synthesis. For example, diamond deposition in a hydrogen microwave discharge has been achieved by employing multiwall carbon nanotubes as precursors [56]. Silicon (Si) has also been utilized in the production of nanodiamonds by incorporating it into a pure carbon film. The growth of synthesized nanodiamonds can be controlled by tuning the Si content amount and heating temperature. In addition to Si, other precursors can be used for nanodiamond synthesis, including graphite, fullerenes, SiC, carbide-organic systems, trotyl/hexogen, and 1,1,1-trichloroethane polycarbonate [54, 55].

1.3.8.2 Detonation method

Detonation synthesis utilizes the energy released from an explosion to drive the production of diamond phases. The process involves detonating a combination of explosives, usually consisting of 60% TNT and 40% hexogen, within a sealed metal chamber (figure 1.7) [1]. The precursor graphite that is placed inside the explosion chamber or the explosive molecules themselves provide the carbon atoms needed for the production of nanodiamonds. The 'dry' approach, in which the chamber is filled with a gas (N_2, Ar, or CO_2) functioning as a coolant, and the 'wet' method, in which

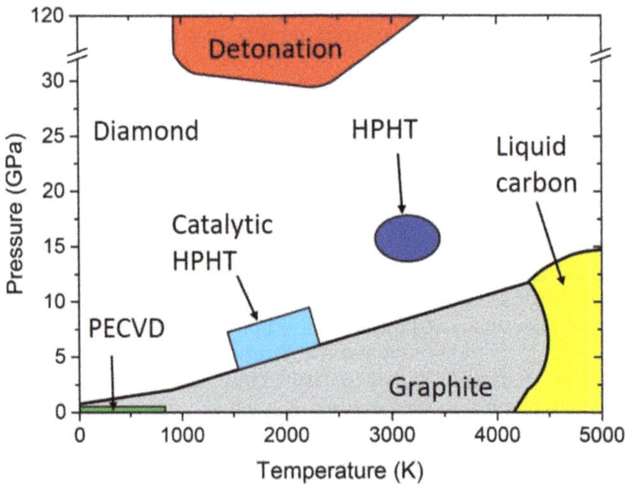

Figure 1.7. The carbon phase diagram illustrates the presence of graphite (shown in gray), diamond (represented in white), and the liquid state (highlighted in yellow). Reproduced from [15]. CC BY 4.0.

water (ice) serves as the coolant, are two versions of this procedure. The carbon yield is greatly impacted by the cooling media, which typically accounts about 10% of the explosive weight [15, 16, 57].

When explosive molecules detonate, the carbon atoms that were liberated during the process condense and crystallize into nanoclusters [16].

The detonation-induced pressure and temperature increase (figure 1.7) leads to the crystallization of the diamond phase from the carbon nanoclusters. The newly produced nanodiamonds then expand and group together to create nanodiamond clusters that measure 4–5 nm in size. This method does, however, have significant shortcomings, such as detonation soot in the resultant sample, which includes contaminants including non-diamond carbon (25–85 wt%) and metals from the chamber wall (1–8 wt%) [31]. To get rid of these impurities, purification is necessary, which entails a risky and expensive procedure of cleaning the detonation soot in a harsh acid environment (usually a mixture of $HNO_3/H_2SO_4/HClO_4$) [15].

The agglomeration of nanodiamonds during synthesis, which results in nanodiamond clusters of several hundred nanometers, is another problem. Standard deaggregation techniques for isolating individual 4–5 nm nanodiamonds include ultrasonic disintegration or milling with ceramic microbeads, which introduces impurities. These techniques are required to produce nanodiamonds that have the right size and purity, and to increase the scope of their potential applications (figure 1.8).

The synthesis process of nanodiamonds during shock wave propagation can be described as follows (as shown in figure 1.9):

(i) The shock wave resulting from detonation causes the chemical dissociation of the carbon precursor.
(ii) The carbon precursor can either be the molecules from the explosives or existing graphite.

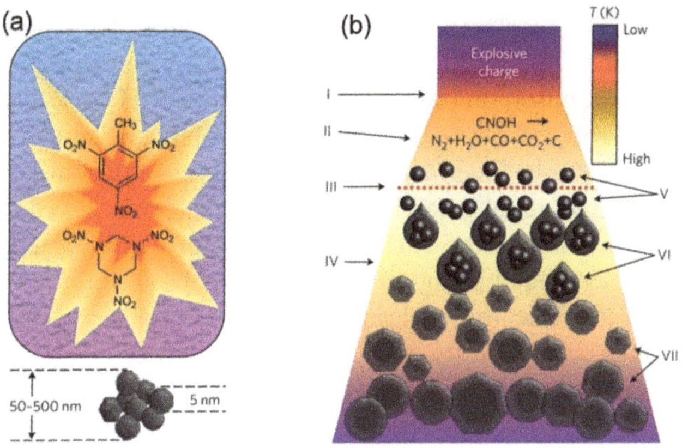

Figure 1.8. When the mixture containing 60% TNT ($C_6H_2(NO_2)CH_3$) and 40% hexogen ($C_3H_6N_6O_6$) detonates inside a metallic chamber and is exposed to water (in liquid or ice form) or a gas atmosphere (composed of N_2, Ar, or CO_2) it leads to the formation of nanodiamonds (nanodiamonds) with a size of 4–5 nm. These nanodiamonds agglomerate into submicrometer clusters. Reprinted from [42], Copyright (2012), with permission from Springer Nature.

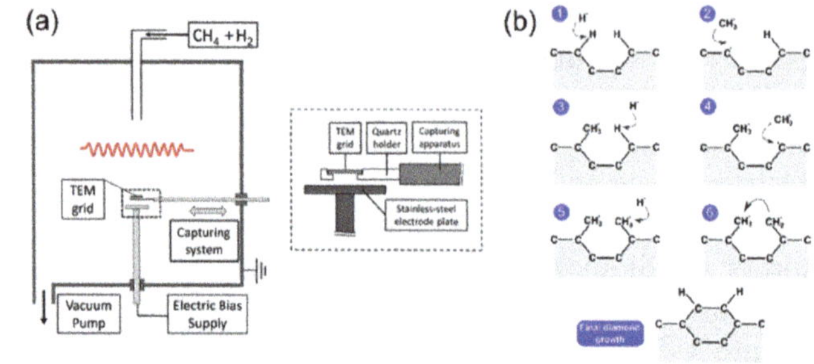

Figure 1.9. (a) The experimental setup employed in the CVD synthesis process. Reprinted from [58], Copyright (2016), with permission from Elsevier. (b) Schematic of the standard growth mechanism of nanocrystalline diamond film during CVD synthesis. Reproduced from [15]. CC BY 4.0.

(iii) The shock wave reaches a critical pressure (represented by the dashed line, III) necessary for diamond formation.
(iv) The products of detonation rapidly expand within the chamber (IV).
(v) During this expansion, carbon atoms in the products start to condense and crystallize (V), forming nanoclusters (VI) [15].

1.3.8.3 Chemical vapor deposition

CVD is a popular technique for forming thin films, such as nanocrystalline diamond sheets [15]. Carbon atoms are deposited in CVD by decomposing a gas mixture containing a carbon-containing molecule, commonly methane (CH_4), in the presence of too much hydrogen. The gas combination is broken down using a heated filament or microwave plasma, producing vital radicals such as H• and CH_3• that promote diamond formation.

The procedure is carried out on a substrate, typically a silicon wafer coated in diamond powder that is smaller than a micrometer and serves as a seed for nucleation [15]. On the substrate, the diamond film develops progressively until it is a continuous layer. The CVD synthesis setup is displayed in figure 1.9(a) [15].

It is to be noted that based on the relative proportion of H_2 and CH_4, the resulting film's grain size can range from tens of microns to a few nanometers. In contrast, a greater CH_4/H_2 ratio causes the grain size to decrease to tens of nanometers for 15% CH_4/H_2, while a low CH_4 concentration promotes the formation of microcrystalline diamonds. Carbon atoms on the surface of diamond seeds are left with dangling bonds as a result of hydrogen being removed by H• radicals in the typical diamond formation mechanism. The molecules of CH_3• then fill these bonds. The newly bound carbon atoms can integrate into the diamond lattice when this process takes place at two nearby places, as shown in figure 1.9(b) [15].

1.3.8.4 Hydrothermal synthesis

This strategy utilizes hydrothermal techniques, such as those described in references [59], to grow crystals of compounds from water-based solutions under high temperatures and vapor pressures. The process typically takes place in an autoclave, a steel pressure tank, where all the reactants are combined with distilled water. This synthesis method is considered effective for scaling up the production of nanodiamonds. In the hydrothermal preparation of diamond from a SiC organic compound, chlorinated hydrocarbons are commonly used as precursors. Several studies have extensively investigated the use of dichloromethane and 1,1,1-trichloroethane in this context [59].

1.3.8.5 Laser ablation

The advent of the ruby laser marked the beginning of the laser ablation (LA) era. In 1987, LA was employed to mix iron oxide nanoparticles by using pulsed laser beams to irradiate water interfaces. Pulsed laser ablation in liquids (PLAL) has gained popularity as a method for synthesizing pure liquids. Nanodiamonds can be produced physically by using carbon powder in contact with liquids or solids. Moreover, the PLAL technique combines the benefits of pulsed laser deposition and soft-chemical routes to create nanoparticles in the form of stable colloidal suspensions. This method involves subjecting an object immersed in a liquid, like ether, ethanol, acetone, or their compounds, to laser treatment. The atomic level liquids and fluids are utilized by combining plasma contraction and expansion under specific conditions, such as temperature and pressure. The laser strength on the target surface is typically higher than 108 W cm^{-2}, leading to target evaporation and release. Microcrystalline graphite is favored over carbon wafers or crystal graphite wafers as a target. The production of crystalline nanodiamonds can be faciltated using shock wave and extinguishing operations. Several factors, including laser power, laser intensity, solvent, excitation wavelength, system temperature, and pressure, can influence the properties of the produced nanodiamonds [60]. PLAL holds the potential to be economical. The capillary effect of the curvature in a nanoscopic crystalline nucleus on the nucleation and growth of nanocrystals can be explained using a kinetic theory approach [61–63]. Table 1.2 demonstrates the synthesis process of various nanodiamonds using LA methods.

1.3.8.6 Ultrasound

The utilization of ultrasonic treatment, either as a pre-treatment or during the production steps, has shown potential benefits in the process of nanodiamond production [69]. Research has indicated that carbon materials containing an alkyne system can undergo structural changes, such as the formation of diamond thin films and fine-grain diamond, when exposed to x-rays or ultrasonic waves [69, 70]. By subjecting suspensions of hexagonal crystal graphite in various organic solvents to ultrasonic cavitation at approximately 120°C and atmospheric pressure, diamond microcrystals measuring 6–9 μm in size and accounting for approximately 10% of the total weight have been successfully produced [64]. Additionally, Iijima and colleagues patented a method to generate nanoscale diamond particles as a powder through ultrasonic agitation [71, 72].

Purification and deaggregation of nanodiamonds

Detonation soot contains non-diamond carbon, metallic, and incombustible impurities that are typically located outside the aggregated nanodiamonds formed during detonation. To prevent these contaminants from being exposed to the oxidizing medium, they can be effectively filtered out or confined within the compact

Table 1.2. Various nanodiamond synthesis processes using the LA method.

Sl. No.	Wavelength (nm)	Power density (W cm^{-2})	Pulse width	Graphite target	Liquid	nanodiamond size (nm)	Reference
1	532	10^{11}	10 nc	Polycrystalline	Acetone	30–40	[64]
2	532	66 J cm^{-2}	10 nc	Graphite disk	Cyclo hexane		[65]
3	532	28 J cm^{-2}	15 nc	Graphite	Water	~100	[66]
4	532	10^{11}	10 nc	Polycrystalline	Water	300–400	[67]
5	532	66 J cm^{-2}	10 nc	Graphite disk	Water		[68]
6	1064	4×10^6 W cm^{-2}	1.2 ms	Graphite particles < 2.0 μm	Water	2–13	[68]
7	355	40–1000Jcm^{-2}	5 nc	Pyrolytic graphite	Water	5–15	[69]
8	1064	4×10^6 W cm^{-2}	0.4 ms	Graphite particles < 2.0 μm	Water	2–7	[68]
9	1064	4 × 106 W cm^{-2}	1.2 ms	Graphite particles < 2.0 μm	Water	3–6	[70]

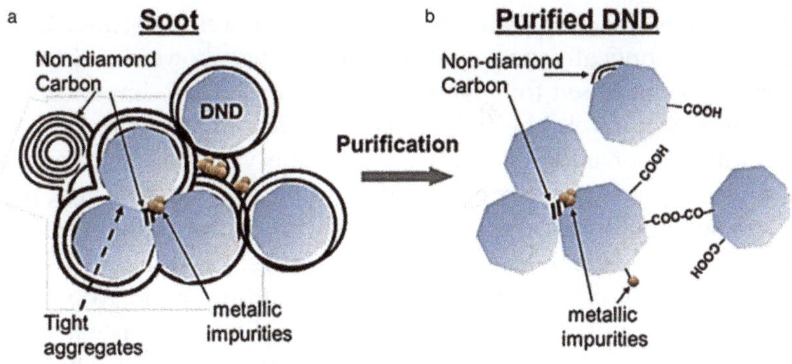

Figure 1.10. A schematic of the key physical and structural elements of detonation soot, and (b) ultrapure commercially available explosive nanodiamonds. Reproduced from [48]. CC BY 4.0. Adapted from [78].

aggregates, as shown in figure 1.10. Various filtration methods have been developed to address this issue in diamonds and nanodiamonds produced using different techniques. These methods include magnetic, electrolysis, ultrasonic, laser, electric field, and chemical treatments with strong oxidants [73, 74]. Additionally, a gas-phase method for purifying and modifying nanodiamonds has also been discussed. This method involves exposing a gas mixture to an original substrate while applying pressure up to 0.8 MPa and simultaneously heating it from 20°C to 550°C [74–77].

1.4 Properties of nanodiamond

Diamond possesses fundamental properties characterized by its arrangement of sp^3 hybridized carbon atoms in a strictly cubic configuration. The tightly packed carbon atoms and strong sp^3 hybridization through covalent bonding contribute to the chemical inertness and durability of diamond in various conditions [79–81]. These exceptional stability and inertness properties of nanodiamonds make them highly advantageous in extreme environments, including their use in microelectronic products where they are currently deposited using CVD techniques [82, 83].

Luminescent color centers have been identified within the diamond lattice. The performance of these color centers is influenced by the lattice's predominantly strain-free nature, which enhances the stability and durability of optical signals. Diamond is widely utilized as a polishing material, particularly in the field of metallographic research, due to its exceptional hardness and high resistance to wear [82, 83]. The robust diamond core of nanodiamonds makes them valuable as additives in lubricants, with several studies demonstrating improvements in tribological properties such as reduced wear and friction. Moreover, the fine polishing effects achieved through the use of nanodiamonds have led to increased fuel efficiency in both gasoline and diesel vehicles by reducing friction [82, 83].

1.4.1 Dopants and defects

The unique crystal lattice and atomic arrangement of nanodiamonds provide them with long-lasting properties and distinctive features, which can be harnessed for various applications [84, 85]. Naturally occurring as well as artificially produced diamonds possess optically active imperfection sites; however, their quantities usually need to be increased for practical applications. These imperfections can be introduced at higher levels through processes, such as ion with high-energy or electron irradiation, or additional doping to create vacancies [84, 85]. Currently, there is significant practical interest in creating stable imperfection sites, particularly nitrogen-vacancy (NV) centers, which consist of substitutional nitrogen and vacancies. NV centers can occur in two main forms: NV0 and NV−. The NV− state is created when a nearby nitrogen (N) atom substitutes a carbon (C) atom in the diamond lattice and donates an electron, resulting in a negatively charged imperfection site. The NV− center's attractive qualities, such as its ability to exhibit optically detected magnetic resonance and stable photoluminescence, make it highly suitable for various applications like nanoscale sensing, quantum computing, and bioimaging, both in vivo and in vitro [85]. Diamond has more than five hundred optically active imperfection sites identified, suggesting a vast potential for numerous yet undiscovered functions [85].

Fluorescent nanodiamonds have become a focal point in biological applications because of their biocompatibility and photostability. When it comes to optical thermometry, both NV color centers and silicon-vacancy color centers in diamonds, especially those with NV centers, display remarkable sensitivity, as demonstrated in the studies conducted by Alkahtani *et al* [84, 85].

Incorporating boron (B) into diamond to create electrically conductive nanodiamonds holds significant scientific value. Producing B-doped nanodiamond seed crystals is crucial for advancing conductive diamond thin films because conventional diamond seeds have high resistivity in devices, posing a risk [85, 86]. Additionally, B-doped nanodiamonds can be utilized as electrodes in fuel cells, offering improved resistance to oxidation. The introduction of boron into diamonds also leads to phosphorescence, presenting exciting possibilities for imaging applications [85]. Another promising application involves doping the nanodiamond core with tritium, which could produce highly stable radiolabeled nanodiamonds for bioimaging purposes [53].

1.4.2 Morphology

This distinct shape, combined with their small size, makes nanodiamonds highly suitable for applications requiring a large specific surface area (around 300–400 m^2 g^{-1}), such as loading medicines and bioligands. Meanwhile, arranging nanodiamond nanofillers in a dense network within a polymer matrix opens up possibilities for different applications. Additionally, circular-shaped nanodiamond particles are particularly efficient for purposes such as lubrication and polishing. Overall, the morphology of nanodiamonds offers versatility and advantageous characteristics for a range of applications in various fields.

1.4.3 Optical properties

Natural diamonds can be classified into Type I or Type II based on the presence or absence of nitrogen (N) defects in their crystal lattice. Type I diamonds are further divided into Type Ia or Type Ib, depending on the arrangement of the N flaws. Type Ib diamonds have isolated substitutional N defects (C-centers) that are randomly distributed throughout the lattice. Meanwhile, Type Ia diamonds have N atoms aggregated in the lattice. Type Ia diamonds can be subcategorized as Type IaA, which contains two aggregated N atoms (A-center), and Type IaB, where four aggregated N atoms surround the nearest lattice vacancy (B-center). Finally, Type IaA/B exhibits characteristics of both IaA and IaB [87].

1.4.4 Toxicity and biocompatibility of nanodiamonds

The developmental toxicity of nanodiamonds has received a lot of attention [88]. A number of studies have provided Information about the hazardous and non-toxic properties of nanodiamonds to organs. This argument is due to dose-dependent problems or variations in nanodiamond quality. For instance, because researchers noted that non-purified nanodiamonds might cause the membranes of white and red blood cells to burst, the toxicity of nanodiamonds in the direction of blood has been explained in terms of contaminants and inadequate purification [89]. Alternatively, in the case of acid-purified drugs, no evidence of hematological toxicity has been discovered [90]. Concern is raised about what happens to nanodiamonds in other creatures, particularly in mammals. Purified nanodiamonds are hazardous because of particle aggregation and greater nanodiamond concentration [91]. Like metal nanoparticles, nanodiamonds are unable to be digested and dissolved in the human body [92]. Table 1.3 provides a summary of the toxicity profile of nanodiamonds.

Table 1.3. Summary of toxicity profile for nanodiamonds.

Nanodiamond type	Groups	Remarks
Detonated nanodiamond	OH, COOH	Dose-dependent toxicity, impurities <5 wt% (Fe, Al, Ba, Mn, Ni).
Detonated nanodiamond	COOH	There was zero change of the mitochondrial membrane, no oxidative stress, 0.8% no contaminants, and liquid oxidation was used to purify it.
F-HPTNd-fluorescent HPHT nanodiamond	COOH	No cytotoxicity, no contaminants, and liquid oxidation purified.
Detonated nanodiamond	OH	urine retention in the lungs, cleaned by fernocide, with no impurities.
Fluorescent high-pressure–high-temperature nanodiamond	COOH	There was no change in the ability to reproduce, stress, liquid oxygen purification, or impurity.
Chemical vapor deposited nanodiamond		No impurities, toxic.

1.5 Applications of nanodiamond

Diamond soot is commonly produced by controlled explosive detonation of carbon-containing explosives. However, manufacturers of nanodiamond preludes are exploring cost-effective approaches by repurposing outdated equipment. Various methods are employed to generate nanodiamond particles possessing a substantial sp^3 composition, including the application of acid to specific sp^2 carbon, the occurrence of oxidation throughout the exploding diamond procedure, and the liberation of nanodiamond sediment due to detonation in an environment rich in oxygen. Recently, nanodiamonds have become increasingly popular because of their capability to create appealing composites when combined with other substances such as graphene, polymers, and metals. This results in enhanced strength, durability, and heat resistance.

1.5.1 Photovoltaic devices

Nanodiamond compounds have captivated interest in the field of dye-sensitive solar cells (DSSCs). Researchers have reported that incorporating natural dyes with diamond areas can result in a photocurrent of approximately 120 nA cm^{-2} under visible light [93]. This achievement was made possible by employing oligothiophene and Suzuki to establish a connection between the levels of electron donation and reception. Moreover, B-doped nanodiamonds demonstrated even higher performance, generating a photocurrent of around 4–6 mA cm^{-2} using the same setup. When polymer dye-functionalized polycrystalline B-doped nanodiamonds were employed as electrodes in DSSCs with a liquid electrolyte solution, a high current density was observed [94]. These findings indicate that nanodiamonds have the potential to be utilized as photo-electrodes under illumination. Furthermore, employing nanodiamonds in DSSCs as optical scatter has been shown to significantly enhance performance and increase power conversion efficiency, paving the way for diamond-based nanocomposites in solar driven devices and flat panel displays.

1.5.2 Thin film electronics

Polymer nanohybrid films reinforced with nanodiamonds have garnered significant attention due to their exceptional physical and mechanical properties. An organo-modified nanodiamond filler has been utilized to create a high heat resistance assisted radiant and flexible fluorinated polymer nanohybrid films with improved mechanical characteristics. Structural stability is a crucial parameter in nanodiamonds, which can be observed from the outer layer of nanodiamonds. Electrostatic interactions between nanodiamonds and the membrane play a critical role in achieving replication and similarity, irrespective of the substrate's shape and structure. Detonation nanodiamond seeds offer an alternative to glass or silicon (Si) substrates for CVD diamond films [95].

The incorporation of nanodiamonds into polymer nanofibers and microfibers enhances their mechanical properties. However, the presence of defects in nanodiamond films can lead to structural degradation. Numerous approaches have been investigated to create nanodiamond films without any pinhole defects. These methods include substrate scratch, the formation of carbide/carbon interlayers, enhanced nucleation bias, and the utilization of electrostatic seed-carrying nanodiamond colloid. Furthermore, the implantation method has shown significant benefits in exhibiting homogeneous high-density nucleation sites. Nanodiamond-based polymer composites are suitable for 3D printing, enabling the production of materials in diverse shapes.

Ultra-nanocrystalline diamond films with a high level of nitrogen incorporation (nanodiamond films) can function as field output elements. Nevertheless, a significant drawback of these films is their dependence on a high substrate growth temperature (>800°C) to facilitate the n-doping process. Despite this limitation, nanodiamond films are preferred for applications where a distinctive set of dielectric, thermal, and mechanical properties, along with UV protection and scratch resistance, are essential [96].

1.5.3 Energy storage devices

The increasing demand for energy storage solutions with excellent power density, energy density, and long durability is driven by the rapid growth of laptop, mobile and portable devices; e-mobility; machine tools; and power grid applications. Nanodiamonds are highly sought-after for energy storage systems because they offer versatility, conductivity, large surface area, high electrochemical potential windows, and excellent stability. These characteristics make nanodiamonds a preferred option for meeting the diverse and demanding requirements of modern energy storage applications. Effective methods for growing nanodiamond on conducting layers in subatmospheric environments have accelerated the development of nanodiamond-based electrode materials. Diamond electrodes doped with boron (B-doped diamond electrodes) have demonstrated remarkable stability across a wide range of operating potentials, even in aqueous environments. Compared to carbon-containing electrodes or metals, heavily boron-doped nanodiamond electrodes exhibit intriguing properties [97]. Surface functionalization of nanodiamonds can further enhance the ion-storage capacity of these electrodes by introducing additional pseudo-capacitive processes. Techniques such as electrochemical and plasma treatments can be employed to modify the surface of nanodiamonds with oxygen (O_2), hydrogen (H_2), hydroxyl (OH) groups, thereby enabling adjustable aqueous and hydrophobic properties. By manipulating the energy levels of interaction at the nanodiamond/electrolyte interfaces, nanodiamonds have been effectively utilized as electrodes in energy-related applications [98].

1.5.4 Electrochemical sensors

Nanodiamonds have received significant attention in the development of novel electrochemical sensors for diverse applications, including drug detection, gas monitoring, and genetic sequence diagnostics. The key parameters for any sensor system are photostability, accuracy, and resolution of electrochemical reactions.

Nanodiamonds offer distinctive properties such as chemical resistance, durability, biocompatibility, fluorescence, catalytic structures, and spin break, which make them an excellent option for improving the sensitivity of electrochemical sensors [99]. Several techniques have been created to enhance the sensory capabilities of nanodiamonds, such as metal catalysis, self-assembled monolayers, liquid chemical process, and oxygen plasmas etc. Through these modifications, nanodiamonds exhibit improved properties, such as optimized melting behavior and specific binding to targeted analytes for sensing purposes. For example, a PZA sensor utilized scattered nanodiamonds on a glassy carbon electrode and employed square-wave voltammetry for its development [100]. Another application involved the use of carbon black nanoparticle modified nanodiamonds in an electrochemical transmitter for dopamine monitoring. This approach achieved a remarkable detection limit of 0.06 mM using cyclic voltammetry [101].

1.5.5 Drug delivery

To create an effective drug delivery system, certain requirements such as biocompatibility, versatility in accommodating various treatments, water dissipation, and reliability need to be considered. Personalized therapy that can be used in conjunction with imaging is also important. Researchers have combined nanodiamonds with the cancer drug duxorubin to treat lung and breast cancer, and to reduce the circulation half-life of the nanodiamond–duxorubin complex [102]. Nanodiamonds have been coated with polyethylenimine 800 (PEI800) to deliver nucleic acids and small molecules with an improvement of 70-fold in GFP plasmid transfection efficiency [103]. In addition to drug delivery, ND-PEI800 has been shown to be more efficient in suppressing GFP expression with siRNA than the widely used delivery platform lipofectamine. Nanodiamonds have also been used to deliver various drugs and proteins, as well as cancer-specific siRNA in acidic tumor conditions. Furthermore, nanodiamonds have been extensively researched as injectable therapeutic agents for widespread drug delivery. In addition, the combination of parylene with nanodiamonds has shown promise in enabling localized drug release over specific periods, ranging from 2 days to 1 month.

1.5.6 Other applications

Nanodiamonds possess high biocompatibility and can serve as a promising platform for enhancing drug delivery to treat skin cancer. In addition, incorporating nanodiamonds in beauty products can enhance the performance of active compounds by improving their adsorption rate and allowing them to penetrate deeper into skin layers. Nanodiamonds have high moisture retention capacity, which keeps the skin hydrated for a longer period, and makes them quickly and fully absorbed. Nanodiamonds can also aid in wound healing and act as a dressing. The versatile surface chemistry of nanodiamonds allows for modifications that can facilitate gene delivery and influence cell entry, making nanodiamond-based gene delivery systems a promising choice for wound healing applications. Gene transfer is being explored in the context of wound healing to address the challenges associated with limited availability of growth factors at the site of injury. An example of this is the use of bovine collagen, which acts as a biocompatible matrix supporting tissue repair and functioning as a carrier for gene therapy in wound healing [104].

Research is growing rapidly on developing nanodiamond nanocomposite polymers with enhanced mechanical and thermal properties. Different techniques have been developed to fabricate nanocomposites with uniformly dispersed nanodiamonds, which hold promise for applications in areas such as aeronautics, automotive, membranes, coatings, and lubricants. Nitrogen-doped detonation nanodiamond nanocomposites have been used in quantum computing, microelectromechanical systems (MEMS), and quantum coherent devices. These materials, including B-doped nanodiamonds, have been characterized using techniques such as scanning electron microscope, Raman spectroscopy, cyclic voltammetry, and electrochemical response. Moreover, nanocomposites have been investigated for potential use in drug delivery, owing to their biocompatibility and bioresorbability. In any drug delivery system, the capability to transport a diverse array of drugs is crucial, along with ensuring the dispersion and scalability of the system in the intended medium. Furthermore, these materials hold potential for use in bioimaging [15].

Nanodiamonds have various medical applications, such as imaging using MRI, dental, skin and bone tissue engineering, cytotoxic, anti- tubercular and anti-diabetic drugs for drug delivery. In addition, siRNA, lysozyme proteins, and poly-phenolic compounds are used as binding biomolecules. Modified nanodiamonds can serve as gene delivery vectors and nucleic acid complexion platforms, although their compatibility with different solvents and polymers needs further investigation. Nanodiamonds can serve as carriers for targeted drug delivery, cancer treatment, and protein separation and purification by binding with various fluorescent molecules such as proteins, antibodies, growth hormones, and DNA molecules. However, the biocompatibility and toxicity of nanodiamonds must be studied on various cell types to ensure safe nanomedical applications [16]. Potential applications of nanodiamonds also include biosensor components, biocompatible composites and implants, and solid supports for peptide synthesis. Important applications of nanodiamonds reported include fluorescent markers for cell imaging, protein immobilization, and medication administration.

1.6 Conclusions and future prospects

This chapter has offered an overview of various 2D carbon materials and nanodiamonds, emphasizing their distinctive attributes and possible applications. The exploration covered different types of 2D carbon materials, such as graphene, GO, and carbon nanotubes, each exhibiting unique structures and properties that render them suitable for diverse purposes. Graphene, renowned for its remarkable mechanical strength, high electrical and thermal conductivity, as well as its extensive surface area, has attracted considerable interest in fields such as electronics, energy storage, sensors, and composite materials. GO, a derivative of graphene, offers additional properties, such as tunable surface chemistry, solubility, and biocompatibility, making it promising for applications in drug delivery, biosensing, and water purification. Carbon nanotubes, cylindrical structures composed of rolled graphene sheets, exhibit remarkable electrical and thermal conductivity, along with exceptional mechanical strength. These properties make carbon nanotubes ideal candidates for applications in nanoelectronics, composite materials, and energy storage. Nanodiamonds are tiny diamond particles with sizes ranging from a few nanometers to a few hundred nanometers. These particles possess extraordinary hardness, chemical stability, and biocompatibility. Nanodiamonds find applications in diverse fields, including electronics, biomedicine, catalysis, and environmental remediation. The chapter has also discussed the synthesis methods for these materials, highlighting their advantages, limitations, and potential for scalability. Additionally, we have explored the characterization techniques used to analyze the structural, electrical, and mechanical properties of 2D carbon materials and nanodiamonds. It is evident that 2D carbon materials and nanodiamonds hold immense promise for a wide range of applications, spanning from electronics and energy to medicine and environmental sustainability. However, there are still several challenges to overcome, such as large-scale synthesis, precise manipulation, and integration into functional devices. Future research efforts should focus on addressing these challenges to unlock the full potential of these materials and drive their commercialization. By harnessing the remarkable properties of these materials, researchers and engineers can pave the way for transformative advancements in various fields, leading to a more sustainable and technologically advanced future.

References

[1] Abbas Q, Shinde P A, Abdelkareem M A, Alami A H, Mirzaeian M, Yadav A and Olabi A G 2022 Graphene synthesis techniques and environmental applications *Materials* **15** 7804

[2] Kumar N, Salehiyan R, Chauke V, Botlhoko O J, Setshedi K, Scriba M, Masukume M and Ray S S 2021 Top-down synthesis of graphene: a comprehensive review *FlatChem.* **27** 100224

[3] Selvaraj V and Krishnan H 2020 Synthesis of graphene encased alumina and its application as nanofluid for cooling of heat-generating electronic devices *Powder Technol.* **363** 665–75

[4] Tao Y, Sui Z-Y and Han B-H 2020 Advanced porous graphene materials: from in-plane pore generation to energy storage applications *J. Mater. Chem.* A **8** 6125–43

[5] Huang H *et al* 2020 The chemistry and promising applications of graphene and porous graphene materials *Adv. Funct. Mater.* **30** 1909035

[6] Novoselov K S, Geim A K, Morozov S V, Jiang D, Zhang Y, Dubonos S V, Grigorieva I V and Firsov A A 2004 Electric field effect in atomically thin carbon films *Science* **306** 666–9

[7] Sun L, Yuan G, Gao L, Yang J, Chhowalla M, Gharahcheshmeh M H, Gleason K K, Choi Y S, Hong B H and Liu Z 2021 Chemical vapour deposition *Nat. Rev. Methods Primers* **1** 5

[8] Gao Z *et al* 2020 Large-area epitaxial growth of curvature-stabilized ABC trilayer graphene *Nat. Commun.* **11** 546

[9] Dideikin T and Vul. A Y 2018 Graphene oxide and derivatives: the place in graphene family *Front. Phys.* **6** 149

[10] Dai H 2002 Carbon nanotubes: opportunities and challenges *Surf. Sci.* **500** 218–41

[11] Hsin Y L, Hwang K C, Chen F R and Kai J J 2001 Production and *in situ* metal filling of carbon nanotubes in water *Adv. Mater.* **13** 830–3

[12] Ishigami M, Cumings J, Zettl A and Chen S 2000 A simple method for the continuous production of carbon nanotubes *Chem. Phys. Lett.* **319** 457–9

[13] Kang J L, Li J J, Du X W, Shi C S, Zhao N Q, Cui L and Nash P 2008 Synthesis and growth mechanism of metal filled carbon nanostructures by CVD using Ni/Y catalyst supported on copper *J. Alloys Compd.* **456** 290–6

[14] Kang J, Li J, Du X, Shi C, Zhao N and Nash P 2008 Synthesis of carbon nanotubes and carbon onions by CVD using a Ni/Y catalyst supported on copper *Mater. Sci. Eng.* A **475** 136–40

[15] Basso L, Cazzanelli M, Orlandi M and Miotello A 2020 Nanodiamonds: synthesis and application in sensing, catalysis, and the possible connection with some processes occurring in space *Appl. Sci.* **10** 4094

[16] Pichot V, Risse B, Schnell F, Mory J and Spitzer D 2013 Understanding ultrafine Nanodiamond formation using nanostructured explosives *Sci. Rep.* **3** 2159

[17] Allen M J, Tung V C and Kaner R B 2010 Honeycomb carbon: a review of graphene *Chem. Rev.* **110** 132–45

[18] Castro Neto A H, Guinea F and Peres N M R 2006 Drawing conclusions from graphene *Phys. World* **19** 33

[19] Abbas Q, Shinde P A, Abdelkareem M A, Alami A H, Mirzaeian M, Yadav A and Olabi A G 2022 Graphene synthesis techniques and environmental applications *Materials* **15** 7804

[20] Novoselov K S *et al* 2004 Electric field effect in atomically thin carbon films *Science (80–)* **306** 666–9

[21] Sinclair R C, Suter J L and Coveney P V 2019 Micromechanical exfoliation of graphene on the atomistic scale *Phys. Chem. Chem. Phys.* **21** 5716–22

[22] Gao C, Zhan B, Chen L and Li X 2017 A micromechanical model of graphenereinforced metal matrix nanocomposites with consideration of graphene orientations *Compos. Sci. Technol.* **152** 120–8

[23] Zhang Y, Zhang L and Zhou C 2013 Review of chemical vapor deposition of graphene and related applications *ACC. Chem. Res.* **46** 2329–39

[24] Bhuyan M S A, Uddin M N, Islam M M, Bipasha F A and Hossain S S 2016 Synthesis of graphene *Int. Nano Lett.* **6** 65–83

[25] Katariya S M 2022 Review on: synthesis approaches and applications of graphene and its derivatives *Int. j. res. trends innov.* **7** 1–6 ISSN: 2456-3315 http://doi.one/10.1729/Journal.29746

[26] Chen D, Feng H and Li J 2012 Graphene oxide: preparation, functionalization, and electrochemical applications *Chem. Rev.* **112** 6027–53

[27] Gerani K, Mortaheb H R and Mokhtarani B 2017 Enhancement in performance of sulfonated PES cation-exchange membrane by introducing pristine and sulfonated graphene oxide nanosheets synthesized through hummers and staudenmaier methods *Polym.—Plast. Technol. Eng.* **56** 543–55

[28] Chen X, Qu Z, Liu Z and Ren G 2022 Mechanism of oxidization of graphite to graphene oxide by the hummers method *ACS Omega* **7** 23503–10

[29] Botas C, Álvarez P, Blanco P, Granda M, Blanco C, Santamaría R, Romasanta L J, Verdejo R, López-Manchado M A and Menéndez R 2013 Graphene materials with different structures prepared from the same graphite by the hummers and brodie methods *Carbon* **65** 156–64

[30] Brodie B C 1860 Sur le poids atomique du graphite *Ann. Chim. Phys.* **59** 466–72

[31] Yoo B M, Shin J E, Lee H D and Park H B 2017 Graphene and graphene oxide membranes for gas separation applications *Curr. Opin. Chem. Eng.* **16** 39–47

[32] Bellier N, Baipaywad P, Ryu N, Lee J Y and Park H 2022 Recent biomedical advancements in graphene oxide- and reduced graphene oxide-based nanocomposite nanocarriers *Biomater. Res.* **26** 1–23 65

[33] Chen C-Y, Guo X-Y, Lu G-Q, Pedersen C M, Qiao Y, Hou X-L and Wang Y-X 2017 Graphene oxide: a novel acid catalyst for the synthesis of 2,5-dimethyl-*N*-phenyl pyrrole by the Paal–Knorr condensation *New Carbon Mater.* **32** 160–7

[34] Savi P, Giorcelli M and Quaranta S 2019 Multi-walled carbon nanotubes composites for microwave absorbing applications *Appl. Sci.* **9** 851

[35] Aqel A, El-Nour K M M A, Ammar R A A and Al-Warthan A 2012 Carbon nanotubes, science and technology, part I: structure, synthesis and characterization *Arab. J. Chem.* **5** 1–23

[36] Jagadeesan A K, Thangavelu K and Dhananjeyan V Carbon nanotubes: synthesis, properties and applications *21st Century Surface Science—A Handbook* (Rijeka, IntechOpen) DOI: 10.5772/intechopen.92995

[37] Arora N and Sharma N N 2014 Arc discharge synthesis of carbon nanotubes: comprehensive review *Diam. Relat. Mater.* **50** 135–50

[38] Wang X-D, Vinodgopal K and Dai G-P 2019 Synthesis of carbon nanotubes by catalytic chemical vapor deposition Perspective of Carbon Nanotubes (Rijeka: IntechOpen) DOI: 10.5772/intechopen.86995

[39] Shariatinia Z 2021 Applications of carbon nanotubes *Micro and Nano Technologies* ch 7 321–64 (Amsterdam: Elsevier) https://doi.org/10.1016/B978-0-12-821996-6.00016-6

[40] Moniruzzaman M and Winey K I 2006 Polymer nanocomposites containing carbon nanotubes *Macromolecules* **39** 5194–205

[41] Sun L, Wang X, Wang Y and Zhang Q 2017 Roles of carbon nanotubes in novel energy storage devices *Carbon* **122** 462–74

[42] Mochalin V N, Shenderova O, Ho D and Gogotsi Y 2012 The properties and applications of nanodiamonds *Nat. Nanotechnol.* **7** 11

[43] Aharonovich I, Greentree A D and Prawer S 2011 Diamond photonics *Nat. Photonics* **5** 397

[44] Georgakilas V, Perman J A, Tucek J and Zboril R 2015 Broad family of carbon nanoallotropes: classification, chemistry, and applications of fullerenes, carbon dots, nanotubes, graphene, nanodiamonds, and combined superstructures *Chem. Rev.* **115** 4744–822

[45] Baidakova M and Vul' A 2007 New prospects and frontiers of nanodiamond clusters *J. Phys. D: Appl. Phys.* **40** 6300

[46] Schrand A M, Hens S A C and Shenderova O 2009 Nanodiamond Particles: Properties and Perspectives for Bioapplications *Crit. Rev. Solid State Mater. Sci.* **34** 18

[47] Wen B, Zhao J and Li T 2007 Relative stability of hydrogenated nanodiamond and nanographite from density function theory *Chem. Phys. Lett.* **441** 318

[48] Mishra R, Chhalodia A K and Tiwari S K 2018 Recent progress in nanodiamonds: synthesis, properties and their potential applications *Veruscript Funct. Nanomater.* **2** #8W2EG0

[49] Kaur R and Badea I 2013 Nanodiamonds as novel nanomaterials for biomedical applications: drug delivery and imaging systems *Int J Nanomedicine* **8** 203–20

[50] Rehor I, Slegerova J, Kucka J, Proks V, Petrakova V *et al* 2014 Fluorescent nanodiamonds embedded in biocompatible translucent shells *Small* **10** 1106–15

[51] Srivastava V, Gusain D and Sharma Y C 2015 Critical review on the toxicity of some widely used engineered nanoparticles *Ind. Eng. Chem. Res.* **54** 6209–33
[52] Rondin L, Tetienne J P, Hingant T, Roch J F, Maletinsky P and Jacques V 2014 Magnetometry with nitrogen-vacancy defects in diamond *Rep. Prog. Phys.* **77** 056503
[53] Turcheniuk K and Mochalin V N 2017 Biomedical applications of nanodiamond *Nanotechnology* **28** 252001
[54] Jablon B M and Navon O 2016 Most diamonds were created equal *Earth Planet. Sci. Lett.* **443** 41–7
[55] Kaur R, Chitanda J M, Michel D, Maley J, Borondics F *et al* 2012 Lysine-functionalized nanodiamonds: synthesis, physiochemical characterization, and nucleic acid binding studies *Int. J. Nanomed.* **7** 3851
[56] Tang C J, Abe I, Fernandes A J, Neto M A, Gu L P *et al* 2011 A new regime for high rate growth of nanocrystalline diamond films using high power and $CH_4/H_2/N_2/O_2$ plasma *Diam. Relat. Mater.* **20** 304–9
[57] Pichot V, Comet M, Risse B and Spitzer D 2015 Detonation of nanosized explosive: new mechanistic model for nanodiamond formation *Diam. Relat. Mater.* **54** 59–63
[58] Park J-W, Kim K-S and Hwang N-M 2016 Gas phase generation of diamond nanoparticles in the hot filament chemical vapor deposition reactor *Carbon* **106** 289–94
[59] Narayan J and Bhaumik A 2015 Novel phase of carbon, ferromagnetism, and conversion into diamond *J. Appl. Phys.* **118** 215303
[60] Amans D, Chenus A C, Ledoux G, Dujardin C, Reynaud C, Subtlemontier O *et al* 2009 Nanodiamond synthesis by pulsed laser ablation inn liquids *Diam. Relat. Mater.* **18** 177–80
[61] Wang C X, Liu P, Cui H and Yang G W 2005 Nucleation and growth kinetics of nanocrystals formed upon pulsed laser ablation in liquid *Appl. Phys. Lett.* **87** 201913
[62] Wang J B, Zhang C Y, Zhong X L and Yang G W 2002 Cubic and hexagonal structures of diamond nanocrystals formed upon pulsed laser induced liquid–solid interfacial reaction *Chem. Phys. Lett.* **361** 86
[63] Pearce S R J, Henley S J, Claeyssens F, May P W, Hallam K R, Smith J A and Rosser K N 2004 Production of nanocrystalline diamond by laser ablation at the solid liquid interface *Diam. Relat. Mater.* **13** 661
[64] Yang L, May P W, Yin L, Smith J A and Rosser K N 2007 Growth of diamond nanocrystals by pulsed laser ablation of graphite in liquid *Diam. Relat. Mater.* **16** 725–9
[65] Yang G W and Wang J B 2001 Pulsed-laser-induced transformation path of graphite to diamond via an intermediate rhombohedral graphite *Appl. Phys.* A **72** 475
[66] Wang C X, Yang Y H and Yang G W 2005 Thermodynamical predictions of nano-diamonds synthesized by pulsed-laser ablation in liquid *J. Appl. Phys.* **97** 066104
[67] Amans D, Chenus A, Ledoux G, Dujardin C, Reynaud C, Sublemontier O, Masenelli-Varlot K and Guillois O 2009 Nanodiamond synthesis by pulsed laser ablation in liquids *Diam. Relat. Mater.* **18** 177–80
[68] Sun J, Hu S-L, Du X-W, Lei Y-W and Jiang L 2006 Ultrafine diamond synthesized by long-pulse-width laser *Appl. Phys. Lett.* **87** 183115
[69] Taylor A C, Edgington R and Jackman R B 2015 Patterning of nanodiamond tracks and nanocrystalline diamond films using a micropipette for additive direct-write processing *ACS Appl. Mater. Interfaces* **7** 6490–5
[70] Huang Q, Yu D, Xu B, Hu W, Ma Y *et al* 2014 Nanotwinned diamond with unprecedented hardness and stability *Nature* **510** 250

[71] Khachatryan A K, Aloyan S G, May P W, Sargsyan R, Khachatryan V A and Baghdasaryan V S 2008 Graphite-to-diamond transformation induced by ultrasound cavitation *Diam. Relat. Mater.* **17** 931–6

[72] Iijima S, Aikawa Y and Baba K 1990 Early formation of chemical vapor deposition diamond films *Appl. Phys. Lett.* **57** 2646–8

[73] Mitev D P, Townsend A T, Paull B and Nesterenko P N 2014 Screening of elemental impurities in commercial detonation nanodiamond using sector field inductively coupled plasma-mass spectrometry *J. Mater. Sci.* **49** 3573–91

[74] Mitev D P, Townsend A T, Paull B and Nesterenko P N 2014 Microwave-assisted purification of detonation nanodiamond *Diam. Relat. Mater.* **48** 37–46

[75] Park T J, Banerjee S, Hemraj-Benny T and Wong S S 2006 Purification strategies and purity visualization techniques for singlewalled carbon nanotubes *J. Mater. Chem.* **16** 141–54

[76] Kim S, Shibata E, Sergiienko R and Nakamura T 2008 Purification and separation of carbon nanocapsules as a magnetic carrier for drug delivery systems *Carbon* **46** 1523–9

[77] Choudhary D and Bellare J 2000 Manufacture of gem quality diamonds: a review *Ceram. Int.* **26** 73–85

[78] Shenderova O A and McGuire G E Science and engineering of nanodiamond particle surfaces for biological applications (Review) *Biointerphases* 2015**10**(3) 030802

[79] Zhou K, Ke P, Li X, Zou Y and Wang A 2015 Microstructure and electrochemical properties of nitrogen-doped DLC films deposited by PECVD technique *Appl. Surf. Sci.* **329** 281–6

[80] Tsubone D, Hasebe T, Kamijo A and Hotta A 2007 Fracture mechanics of diamond-like carbon (DLC) films coated on flexible polymer substrates *Surf. Coat. Technol.* **201** 6423–30

[81] Roy R K, Choi H W, Park S J and Lee K R 2007 Surface energy of the plasma treated Si incorporated diamond-like carbon films *Diam. Relat. Mater.* **16** 1732–8

[82] Bello I, Chong Y M, Ye Q, Yang Y, He B *et al* 2012 Materials with extreme properties: their structuring and applications *Vacuum* **86** 575–85

[83] Qureshi A, Kang W P, Davidson J L and Gurbuz Y 2009 Review on carbon-derived, solid-state, micro and nano sensors for electrochemical sensing applications *Diam. Relat. Mater.* **18** 1401–20

[84] Alkahtani M H, Alghannam F, Jiang L, Rampersaud A A, Brick R *et al* 2018 Fluorescent nanodiamonds for luminescent thermometry in the biological transparency window *Opt. Lett.* **43** 3317–20

[85] Deák P, Aradi B, Kaviani M, Frauenheim T and Gali A 2014 Formation of NV centers in diamond: a theoretical study based on calculated transitions and migration of nitrogen and vacancy related defects *Phys. Rev.* B **89** 075203

[86] Sipahigil A, Jahnke K D, Rogers L J, Teraji T, Isoya J *et al* 2014 Indistinguishable photons from separated silicon-vacancy centers in diamond *Phys. Rev. Lett.* **113** 113602

[87] Jensen K, Kehayias P and Budker D 2017 Magnetometry with nitrogen-vacancy centers in diamond *High Sensitivity Magnetometers* ed A Mason (Cham: Springer) pp 553–76

[88] Zhu Y, Li J, Li W, Zhang Y, Yang X *et al* 2012 The biocompatibility of nanodiamonds and their application in drug delivery systems *Theranostics* **2** 302

[89] Xing Y and Dai L 2009 Nanodiamonds for nanomedicine *Nanomedicine* **4** 207–18

[90] Wehling J, Dringen R, Zare R N, Maas M and Rezwan K 2014 Bactericidal activity of partially oxidized nanodiamonds *ACS Nano* **8** 6475–83

[91] Ho D 2009 Beyond the sparkle: the impact of nanodiamonds as biolabeling and therapeutic agents *ACS Nano* **3** 3825–9

[92] Ho D N 2010 *Applications in Biology and Nanoscale Medicine* (New York: Springer)

[93] Zhong Y L, Loh K P, Midya A and Chen Z K 2008 Suzuki, coupling of aryl organics on diamond *Chem. Mater.* **20** 3137–44

[94] Krysoca H, Kavan L, Zivcova Z V, Yeap W S, Verstappen P, Maes W *et al* 2015 Dye sensitization of boron-doped diamond foam: champion photo electrochemical performance of diamond electrodes under solar light illumination *RSC Adv.* **5** 81069–77

[95] Montes-Gutierrez J A, Alcantar-Pena J J, de Obaldia E, Zuniga-Rivera N J, Chernov V, Melendrez-Amavizca R *et al* 2018 Afterglow, tehrmoluminescence and optically stimulated luminescence characterization of micro-nano- and ultrananocrystalline diamond films grown on silicon by HFCVD *Diam. Relat. Mater.* **85** 117–24

[96] Janssen W, Turner S, Sakr G, Jomard F, Barjon J, Degutis G *et al* 2014 Substitutional phosphorous incorporation in nanocrystalline CVD diamond thin films *Phys. Status Solidi RRL: Rapid Res. Lett.* **8** 705–9

[97] Tsubouchi N, Ogura M, Mizouchi N and Watanabe H 2009 Electrical properties of a B doped layer in diamond formed by hot B implantation and high temperature annealing *Diam. Relat. Mater.* **18** 128–31

[98] Miele E, Malerba M, Dipalo M, Rondanina E, Toma A and Angelis F D 2014 Controlling wetting and self-assembly dynamics by tailored hydrophobic surfaces *Adv. Mater.* **26** 4179–83

[99] Baptisa F R, Belhout S A, Giordani S and Quinn S J 2015 Recent developments in carbon nanomaterial sensors *Chem. Soc. Rev.* **44** 4433–53

[100] Simoni N B, Silva T A, Oliveira G G and Fatbello-Filho O 2017 A nanodiamondbased electrochemical sensor for the determination of pyrazinamide antibiotic *Senssors Actuators* B **250** 315–23

[101] Jiang L, Nelson G W, Abda J and Foord J S 2016 Novel modifications to carbonbased electrodes to improve the electrochemical detection of dopamine *ACS Appl. Mater. Interfaces* **8** 26338–8348

[102] Vlasov I I *et al* 2010 Nitrogen and luminescent nitrogen-vacancy defects in detonation nanodiamond *Small* **6** 687–94

[103] Schrand A M, Lin J B, Hens S C and Hussain S M 2011 Temporal and mechanistic tracking of cellular uptake dynamics with novel surface fluorophorebound nanodiamonds *Nanoscale* **3** 435–45

[104] Zhu Y, Li J, Wenxin Li Y, Zhang X Y, Chen N, Sun Y *et al* 2012 The biocompatibility of nanodiamonds and their application in drug delivery systems *Theranostics* **2** 302–12

Chapter 2

Structural and physical properties of diamanes

Muqaddas Farooq, Umer Mehmood and Rabia Nazar

Diamanes, or two-dimensional (2D) diamonds, are sp^3 hybridized thinnest diamond nanofilms with unique physical and structural properties. Bulk properties inherited from diamond structure are combined with the nanoscale features of graphene. At the present research stage, diamane properties are mainly studied theoretically. Diamanes can have different structures depending on the hybridization states, which directly influence their properties and stability. Bilayer diamanes have the same exceptional mechanical stiffness as diamond, while their indentation modulus (1079 $\pm$ 69) surpasses diamond. Bending stiffness appears because of the nanometer thickness of diamanes. Critical tension and ultimate tensile strength are surface functionalization dependent structural properties. Physical properties include thermal and electronic properties. Diamanes have functional group-dependent thermal conductivity. H-diamanes have the highest thermal conductivity (2×10^3 W m^{-1} K^{-1}) compared to F-diamanes and Cl-diamanes. Specific thermal properties are achievable by the tuning of bigraphene layers stacking and layers twisting. Bigraphene layer alteration affects the thermal conductivity, while layer twisting affects the thermal expansion coefficient. Zero thermal expansion in a wide range of temperatures is possible for partially diamondized twisted bigraphene. Efforts are developing in the direction of finding a twist angle for partially diamondized bigraphene where sp^3–sp^2 ratio would provide zero thermal expansion. Based on theoretical studies, it is predicted that the bandgap of diamanes is less than that of bulk diamond. Diamanes electronic properties are strongly dependent on the type of surface termination. The unique combination of properties provokes an extensive range of applications for diamanes, including nano-optics, nanoelectronics, resonator-based sensors, etc. This chapter mainly covers the structural and physical properties of diamanes and the factors influencing them.

2.1 Introduction

Graphene is a 2D carbon lattice-layered material with semi-metallic properties [1]. Graphene synthesis in the early-2000s led to continuous advances in the family of

doi:10.1088/978-0-7503-5939-9ch2

nanomaterials. Graphene is a hexagonal honeycomb structure comprising a thick network of carbon atoms with zero bandgap [1–3]. Due to its potential properties, graphene has been extensively researched and used since 2004 [4–7]. Its electronic behavior with zero bandgap limits its applications where semiconducting behavior is a prime requirement, such as in nanoelectronics, nano-optics, and nano-sensors [8]. Meanwhile, 2D derivatives of graphene have semiconducting behavior. In recent years, there has been a significant focus on investigating graphene-based 2D materials including bilayer graphene (BLG) [9] and heterostructures of graphene [10–12]. In these structures, carbon atoms maintain their sp^2 hybridization state. Since 2009, cutting-edge research has focused on the new class of graphene derivatives obtained by compressing graphene layers together to form a diamond-like structure that contains interlayer bonds and sp^3 hybridized carbon atoms [13–16].

Functionalized single-layer graphene (SLG) studies aided in the implication and formation of diamanes with the adsorption of hydrogen on BLG in 2009 [13]. First principle investigations based on density functional theories predicted the opening of the bandgap in graphene derivatives by converting a BLG or multilayer graphene into thinnest diamond nanomembranes, called diamanes [13, 17, 18]. Diamane is a novel 2D material that mirrors thin diamond film properties, which is also termed diamondene [19]. Later, diamanes based on the multilayer graphene [14] and Lonsdaleite-like diamanes structures [20, 21] have been studied in detail.

Diamane's physical properties are significantly affected by functional groups because their adsorption induces changes in chemical bonds, forms a single C–C bond between the layers, and opens the bandgap [22]. Extensive investigations have been performed to reveal the interesting physical and structural properties. Briefly, diamanes are the counterparts of diamonds having a unique amalgam of bulk diamond and graphene nanoscale properties [23–25].

Diamanes have acquired electrical properties synchronized with sp^2–sp^3 hybridization. Diamanes are 2D ferromagnetic semiconductors possessing spin-polarized bands [17]. Compared to diamonds, they have better electrical properties attributed to the low effective mass of electrons and bandgap [26]. The diamanes bandgap is smaller than the diamond with a difference of 0.8 eV. Diamonds have the highest thermal conductivity at room temperature and the lowest thermal expansion. Likewise, diamanes possess exceptional thermal properties. Diamanes theoretically predicted thermal conductivity is 2240 W m^{-1} K^{-1} [27].

Parallel to the physical properties, particular attention is focused on the mechanical properties of diamanes. Diamanes display copious structural properties similar to diamond crystal [8, 13]. Exploration has revealed that thetransverse stiffness and hardness of diamanes are relative to diamond (a well-renowned material for its exceptional stiffness) [28]. Theoretical simulations demonstrate that diamane bending stiffness and elastic constants are much higher than graphene owing to the diamond-like structure [13]. In terms of buckling behavior, diamanes nanotubes have also been studied [29]. Their indentation modulus is analogous to diamond and much higher than graphene [16, 30].

How have the properties of diamanes captivated attention in electronic devices? The answer to this question lies in their functionalized structure. The saturation of

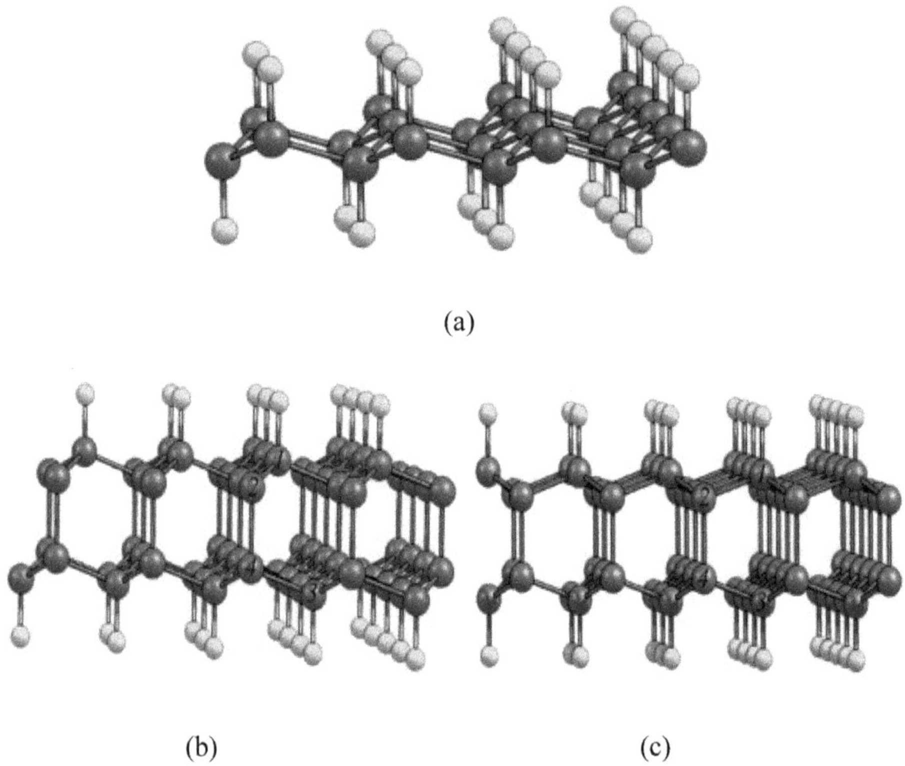

Figure 2.1. (a) Fully hydrogenated SLG (Graphane). (b) AB-Bernel stacking of BLG-Diamanes., (c) AA-Lonsdaleite stacking of BLG-Diamanes. Reprinted from [13], Copyright (2009), with permission from Springer Nature.

their dangling bonds via functional groups opens the bandgap. Diamanes have a different structure than their nano counterparts. Graphene has a flat honeycomb-like structure, whereas functionalized single-layer graphene (SLG; graphane) has slightly puckered carbon sheets and diamanes have diamond-like tetrahedral geometry (figure 2.1). The presence of interlayer sp^3 bonding endows them with tunable and exceptional electronic, mechanical, and thermal properties. This makes them a potential candidate in technological and industrial applications.

Accordingly, diamanes have found many applications in nano-photonics [31] and resonator-based nano-sensors [32]. Functionalized diamanes are suitable for use in field emission displays field-emission displays (FEDs). The small polaron charge carrier, tunable bandgap, high thermal conductivity [27, 33–36] and optical characteristics [37–39] of diamanes make them suitable for potential applications in electromechanical devices [32].

2.2 Critical influencing parameters

Diamanes have emerged as a fascinating material due to their unique combination of properties. Unlike graphene, diamanes have a diamond-like structure, which has led to research and practical applications. Diamanes have opened up new avenues of research because their properties are altered by various factors. In this section, the factors and their effect on physical and structural properties are summarized.

2.2.1 Functional groups

The functional groups involve the adherence of chemical groups or molecules on the surface of diamanes, which alters the properties of diamanes and enables distinctive properties for potential applications. The choice of functional group is crucial to tune the desired physical and structural properties because it marks a significant impact on the atomic structure. Localized states within the Brillouin zone are controllable by changing the nature of functionalized moiety and the level of functionalization [40]. The uniqueness arises in Dirac points and bandgaps due to the choice of functional groups and structural transitions from graphene to graphane, and from graphane to graphene (figure 2.2) [41]. This affects the charge transportation properties, thus controlling the bandgap energy and conductivity of diamanes. For exemple, hydrogen functionalization introduces profound changes in the structure of diamanes. Graphane-a fully H-SLG and Diamane-a fully H-BLG display a prominent change in their physical and structural properties compared to graphene (table 2.1).

The Fermi level is revised with the addition of electron withdrawing or electron donating groups. This affects the charge carrier's concentration and mobility. Chemical adsorption of functional groups such as H, F, Cl, or O on the carbon rings in a single-layer or BLG strongly changes the physical properties by opening the bandgap. Functional group concentration is a key regulator for property alteration. Pristine diamane with a suitable C–H ratio is an insulator. In line with this statement, consider AB-stacked H-Diamanes with 111 conformations that appeared to be a direct bandgap semiconductor with a localized bandgap of

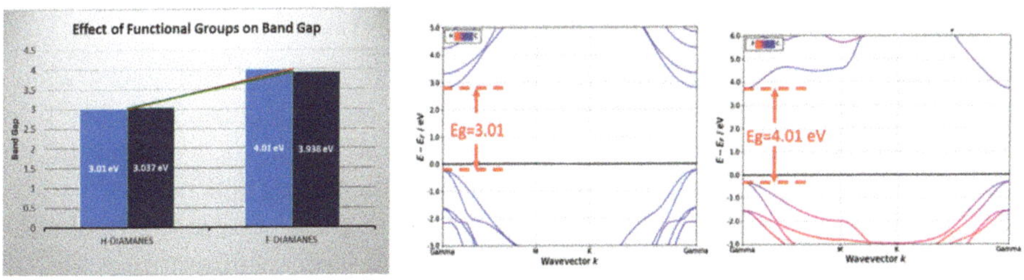

Figure 2.2. (A) Effect of functional group nature on bandgap (data extracted and plotted from [41]), (B) calculated bandgap for chair-1 HD [41], and (C) calculated bandgap for chair-1 FD. Reproduced from [41] with permission from the Royal Society of Chemistry.

Table 2.1. A comparison of the properties of graphene, graphane, and diamanes.

Property	Graphene	Graphane	Diamanes
Bandgap (eV)	0 [42]	3.12 [43]	3.12–4.5 [43]
Elastic constant (N m^{-1})	238 [13]	449 [14]	715 [8]

2.54–3.55 eV in case of full hydrogenation [44]. Meanwhile, semi-hydrogenated diamane film exhibited a ferromagnetic semiconductor behavior within the infrared region. Moreover, the bandgap is dependent on the type of functional group absorbed. Various reported manipulations of bandgap have involved the study of the effects of different functional groups (figure 2.2). Fluorine's ability to covalently bond with carbon is higher than hydrogen, and therefore fluorinated diamanes (FDs) have a higher bandgap as compared to diamanes functionalized with hydrogen because the type of functionalization affects the electron and hole mobility [45].

In 2009, hydrogenated BLG was modeled for the first time with a smaller bandgap than diamond with a difference of 0.8 eV [13]. Chemisorption of chromophore groups introduces the changes in absorption and emission spectra. This provides the diamanes with a controlled response towards different light wavelengths. Energy storage systems and thermo-electric devices necessitate materials with higher thermal conductivity and thermal resistance. Diamanes already have higher thermal conductivity owing to the sp^3 hybridized carbon atoms. Introducing various functional groups regulates their thermal properties. Moreover, the chemisorption of hydrogen or fluorine on the surface of carbon rings in graphene makes their structure similar to that of diamond. The presence of functional groups reinforces the lattice structure of diamanes, intensifying their resilience to mechanical stress. Functional groups instigate the chemical interactions driving toward interlayer coupling, while developing multilayer diamanes with improved structural properties. This property is useful for potential applications in protective coatings, flexible electronics, etc. Compared to graphene, diamanes have a higher stiffness coefficient with an elastic constant of 715 N m^{-1}. Thereby, the presence of hydrogen in bilayer diamane provided a higher Young's modulus. Another researcher found that Young's modulus of fluorinated diamonds is 493.19 N m_{-1} [46], which is relatively high.

2.2.2 Number of layers

In general, the bandgap decreases with the increase in the number of layers, which is ascribed to the decrease in quantum confinement effect (figure 2.3). Multilayer diamanes possess higher electronic properties than single-layer diamanes, thus opening their wide acceptance in optoelectronics. Increasing the number of layers tends to the transition from direct to indirect bandgap. Increased thickness is responsible for variations in electron carrier mobility and quantum tunneling [43]. Diamane film switches from semimetal to wide bandgap semiconductor with the increase in thickness or the number of layers.

Overall thermal transportation declines with the increase in the number of layers. The decreasing trend has been rationalized with elevated phonon scattering due to interlayers. Diamanes have inherited high structural strength due to diamond-like lattice structure. Multilayer diamanes surpass single-layer diamanes in terms of improved mechanical stability due to coupling between layers. The movement of adjacent carbon atoms in BLG is restricted due to the coupling between the top and bottom layers. Interlayer coupling notably enhanced resistance to structural deformation. Coupling makes the BLG flatter with a more rigid structure compared to SLG.

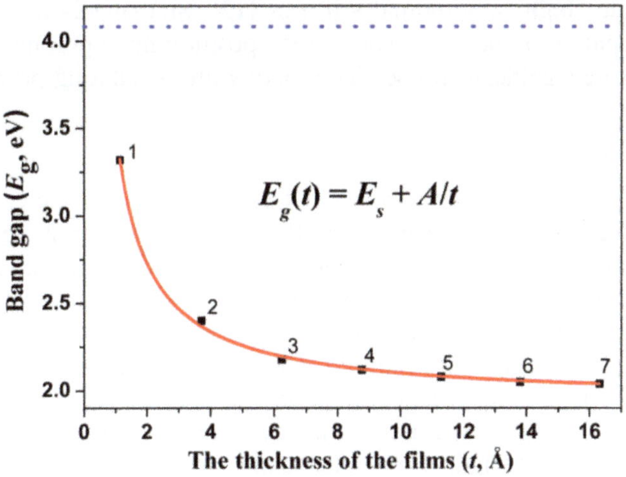

Figure 2.3. Variation in bandgap with thickness. Reprinted from [47], Copyright (2018), with permission from Elsevier.

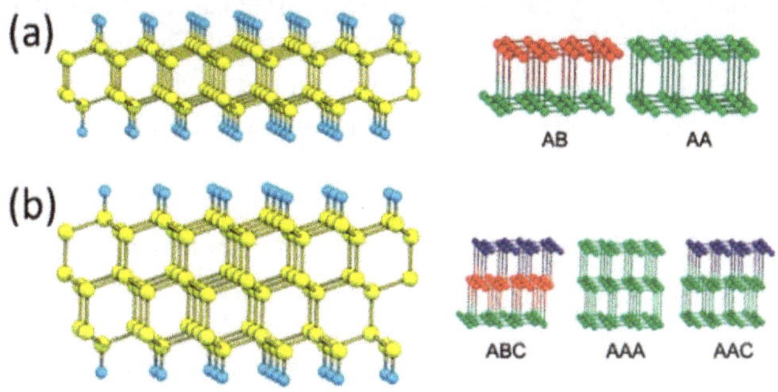

Figure 2.4. Types of stacking sequences. Reprinted with permission from [14]. Copyright (2011) American Chemical Society.

2.2.3 Stacking sequence

The stacking sequence refers to the arrangement of carbon atoms that involves connectivity changes. Three primary stacking sequences AA, AB, and ABC have been studied concerning diamanes (figure 2.4).

The most extensively studied stacking modes are AA and AB. Lonsdaleite diamanes are produced using AA stacking, while Bernel diamanes have AB stacking. Compellingly, stacking sequence influences physical and chemical properties. AA stacking sequence brings forth a stable and flat structure owing to the alignment of carbon atoms directly one above another. While AB stacking has a slightly buckled structure. AB and ABC stackings have certain levels of distortions, so they are less table than AA stacking [22, 31]. Ultrathin diamane films with AA stacking are semimetals, while AB-stacked diamanes are wide bandgap semiconductors [48–50]. Diamanes are much stiffer and harder than diamond but AA stacking provides high longitudinal stiffness [13]. Stacking is a more important parameter compared to thickness. The relative positioning of graphene layers makes AB stacking more significant for semiconducting and insulating properties [51].

2.2.4 Conformation

Conformation is the spatial arrangement of carbon atoms in the crystal lattice. It does not involve bond forming or a breaking process, rather it arises due to the rotation of single bonds. Conformation's role is paramount in regulating the physical and structural properties. Manipulating the arrangement of carbon atoms in terms of bond length and bond angle unlocks the ability to tailor their properties. Cubic or hexagonal lattices are induced when AB and AA stacked diamanes are synthesized [52]. Diamanes with cubic lattice have three plane variations, named 100, 110, and 111 [53]. Hexagonal diamanes have three plane variations, named 0001, 1010, and 2110. Chair 1 is called chair and chair 2 is known as zigzag conformation, which leads to the formation of a cubic lattice. Meanwhile, boat 1 and boat 2 are known as boat and armchair, respectively, and produce diamanes with a hexagonal lattice (figure 2.5) [54, 55].

Stacking, the nature of crystal lattice, planar structure, and conformation play a significant role in manipulating electrical properties. FDs with AB-Bernel staking owing to the higher bandgap relative to hydrogenated diamanes (HDs) have been considered. A cubic lattice with two planes 110 and 111 has been compared with AA stacked diamanes with hexagonal crystal lattice (1010). Bernel stacked F-diamanes

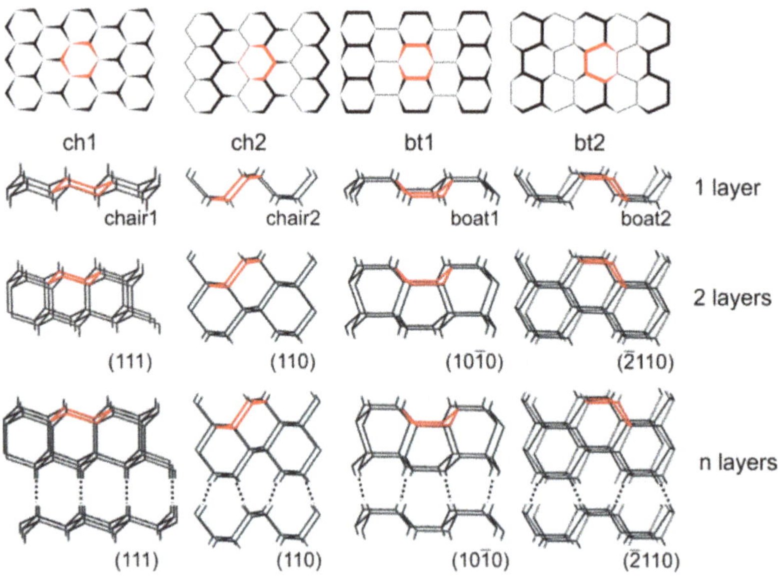

Figure 2.5. (a) Top perspective drawing view and (b) side views of graphene functionalized with different layers. Conformational unit is marked in red. Single-layer functionalized graphene is named by conformation, whereas larger-layer films are named by the corresponding orientation of diamond. Reprinted with permission from [56]. Copyright (2015) American Chemical Society.

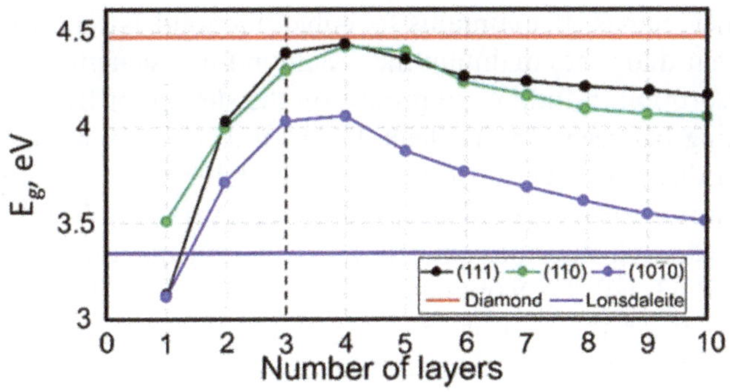

Figure 2.6. Bandgap variation with number of layers. Reprinted with permission from [57]. Copyright (2017) American Chemical Society.

showed better semiconducting properties. It has been reported that for cubic diamanes (111), the effective mass of electrons increases with the increase in a number of layers up to a certain thickness. Thereby, its bandgap tends to the bulk diamond value. In contrast, AA-Lonsdaleite diamanes occupy a low bandgap region [57] (figure 2.6).

Variations in diamane conformations are induced using external stimuli such as mechanical strain, electric field, and temperature. A 60° twist angle in normal diamane provides a superlattice with Moire's pattern. This conformation modifies the hybridization of orbitals while providing a significant impact on the Brillouin zone, thus varying the electronic properties. A 60° rotation when combined with AB stacking produces energetically favorable diamanes on account of the spatial overlap and alignment of carbon atom positions. It maximizes the interlayer interactions and curtails the total energy compared to other stackings with the same conformational rotation, affecting stability and structural properties. Moire's diamanes with partial functionalization on the outer surface maintain insulating properties along with a wider bandgap. A bandgap of 1.21 eV for Moire's diamanes have been theoretically reported since 2012 [31, 58–61] The impact of conformation on properties is combined with other factors such as stacking sequence and the extent of external stimuli involved.

2.2.5 Stability

Pristine diamanes are unstable due to the unavailability of a thermally stable structure. Functionalized diamanes are more stable than graphene. In accordance with computed formation energies, bilayer diamane's existence and stability are more favorable than graphene's [14]. Bernel stacked diamanes have the lowest formation energy. Henceforth, AB-stacked diamanes are thermodynamically more stable. In the race to achieve thermodynamically stable diamanes, various functional moieties have been attached to diamanes forming hybrid structures of diamanes. Early research on their synthesis and stability began with the theoretical prediction of HDs because their formation energy is −0.03 eV/atom lower than graphene [43]. Per conformations, diamanes with (110) cubic planes are more stable than other planes. It maintains its cubic diamond lattice with three layers, while (111) planar diamanes maintain their diamond-like structure for more than 12 layers [47]. Adsorption of functional groups on the surface of BLG develops strong chemical bonding between the graphene layers and removes weak van der Waals forces. Sp^2 hybridized BLG is transformed into sp^3 hybridized diamanes through the covalent bonding between carbons of neighboring layers [62]. In the last 5–6 years, the effects of hydroxyl, fluorine, oxygen, and nitrogen functionalization have also been studied [49, 63, 64]. F-Diamanes are more stable than H-Diamanes owing to the major difference in their formation energies. F-diamanes and H-diamanes have −0.91 and −0.11 eV/atom formation energies, respectively. Additionally, stable diamanes are synthesized using dopants such as N, B, S, and Li. Their energy stability in turn impacts the electronic, phononic, and structural properties [65, 66].

2.3 Physical properties

2.3.1 Electrical properties

Diamanes have prominently different conduction properties from those of bulk diamonds. Electrical conductivity depends upon the presence of free electrons and the ease of their movement, which is assessed by the effective mass of electrons. The effective mass of conduction electrons for diamanes is comparatively less than diamond, which means that electrons can move easily under an applied electric field [67]. The effective mass of electrons refers to the interaction of electrons with a crystal lattice. Diamane's lesser effective mass of electrons (weaker interaction with crystal lattice) is attributed to its honeycomb-like structure with periodic vacancies. Periodic vacancies refer to the missing carbon atoms in the crystal lattice. These vacancies act as electronic traps, creating localized electronic states in the bandgap. These localized electronic states reduce the effective mass of electrons in diamanes, rendering them more mobile under applied electric fields. Moreover, to date, it has been shown that diamanes are semiconductors with a narrower bandgap relative to diamond, and diamond is an insulator with a bandgap of 5.47 eV [68].

The first principle studies projected an idea regarding the effective mass of electrons and bandgap values of diamanes, yet there is still no clarity about an exact bandgap value. Pristine diamanes have a narrow bandgap that is modifiable. Diamane's electronic properties are ideally modulable, ranging from conductivity to insulator or ferromagnetic nature. Modifications highly influence the location of valence and conduction bands at the top and bottom of the graphene layer, respectively. The desired electronic properties could be incorporated into diamanes by controlling critical influencing parameters. Diamane's bandgap modulation is facile relative to diamond and graphene [13]. The bandgap is tuneable by controlling the thickness, functional group nature, and conformation (as discussed above). The presence of dopants or defects also manipulates the electronic states to reduce the bandgap of diamanes [65].

Bernel stacked BLG could be used as a semiconductor by its functionalization which breaks its inversion symmetry and regulated bandgap. Bernel stacking has the widest bandgap compared to Lonsdaleite stacked diamanes [22].

The dopant or functional group nature used to modify graphene layers affects electron mobility. Ultrathin HDs have been reported with the doping of lithium and phosphorous. The results showed use of silicon dopant widens the bandgap in diamanes. Moreover, the bandgap of FDs is comparatively higher than HDs. So far, HDs have higher electrical conductivity and their electronic properties are regulated by the concentration of hydrogen atoms adsorbed along with the conformation. Chair configuration has dominat semiconducting properties (table 2.2).

Unique physical characteristics are obtained in diamanes by the incorporation of functional groups. Functionalization with different strategies plays a different role in controlling electronic properties. Fluorination from both sides of graphene layers makes it an insulator, while from one side it makes it conductive. In the case of semi-FDs, one side is conductive and the other side is an insulator called Janus F-diamanes. High electronegativity of fluorine provides high electron density, developing the good interaction of diamanes with other media. Janus and non-Janus structures have different physical properties in comparison to normal diamanes. The electronic and thermal properties of Janus diamanes lie inbetween non-Janus and twisted diamanes. Janus diamanes are conductive, twisted diamanes are highly conductive, and non-Janus are insulators [69, 70].

Thickness plays a dominant role in controlling bandgap. Bandgap and quantum confinement decrease with an increase in the number of layers and switch from direct bandgap to indirect bandgap. This section will next explain what the correlation between bandgap and quantum confinement effect is. Quantum confinement is the restricted or confined movement of electrons due to the small size of the material. Diamanes with small thicknesses have restricted movement of electrons and quantized energy levels, producing a high bandgap. Increasing the size or thickness of diamanes leads towards the unquantized energy levels, thus providing a sound decrease in bandgap. Bandgap relation with thickness variation is given as,

$$E_{\text{gap}}(n) = E_{\text{gap}}^{\infty} + \frac{A}{n},$$

$$E_g(t) = E_s + A/t$$

Here, E_{gap} and A are the curve fitting parameters and n stands for the number of layers [44, 47, 71].

Fermi velocity, a speed at which electrons move within the highest occupied energy level (Fermi level), either increases or decreases depending on the stacking (AA or AB) during functionalization. Fermi velocity is relative to band structure. Functionalization

Table 2.2. Effect of functional group concentration w.r.t conformations on bandgap (data extracted from [20]).

Full H-diamanes		Semi H-diamanes	
Chair	Boat	Chair	Boat
3.24 eV	2.92 eV	0.54 eV	2.35 eV

provides the precise tuning of sp^3 hybridized carbon in diamanes while introducing some alterations in symmetry, and hence electronic structure. In addition to the aforementioned factors, the use of pressure and interlayer hybridization has also been studied to alter the electronic properties of the pristine graphene layer. In addition, band structure is also modulable by using external electric fields.

Diamane's bandgap is further tunable by the manipulation of the Brillouin zone within the band structure through functionalization. Band structure refers to the energy levels of electrons. The band structure consists of multiple energy bands that are separated by a bandgap. The band structure is demonstrated in graphical form with the energy of electrons on the y-axis and the momentum of electrons on the x-axis. Band projection can be used to study electronic properties in the Brillouin zone of band structure [67].

Predominantly, theoretical analysis for electronic properties has been persistently carried out in terms of band structure, bandgap, or work function. In the case of a wide bandgap or work function, diamanes need a large amount of energy to excite the electrons from the valence band to the conduction band, and vice versa. Moreover, studies have also been rationalized in terms of projected density distributions and the charge carrier's mobility in conduction band maximum (CBM) and valence band minimum (VBM). The conduction band is the minimum energy band, which is either empty or partially filled with electrons. Here the electrons are free to move within the crystal lattice. The valence band maximum is the highest energy band where the electrons are responsible for chemical bonding. Both of these parameters are critical in understanding the electronic properties of diamanes as they are formed due to the specific number of electrons in the crystal lattice.

As mentioned earlier, the properties of diamanes have been studied theoretically. To understand how the electronic properties are evaluated, various current research investigations will be discussed here. Various quantum mechanical models, simulations, and approximations are used to analyze electrical behavior regulating factors. Theoretical first principle calculations are used to understand the possibility of using diamanes in electronic applications. For bandgap evaluation, density functional theory (DFT) and its modified co-relations are used. DFT is the computational method, with the Kohn–Sham electronic (KS) band structure as the key concept involved. DFT-PBE and DFT-LDA are the two computational approaches that have been principally used to predict the electronic properties of diamanes. According to these studies, the bandgap value does not converge to the bulk diamond value with the variation in thickness, while the disadvantage of this approach is that it underestimates the bandgap. HSE06 correction factor is considered to incorporate the effect of bandgap, but it yields approximately the same results as by DFT-PBE and DFT-LDA (table 2.3).

Table 2.3. Comparison of evaluation methods for HD.

Functionalized diamanes	Bandgap-PBE (eV)	Bandgap-HSE06 (eV)	References
HD	3.037	3.835	[7]
	3.017	3.907	[40]

The tight binding method is a computational technique to measure the wavefunctions of the electrons by combining the wavefunctions of individual atoms. The GW correction factor yields the monotonically decreasing trend of bandgap for some conformations of FD films converging to the bulk diamond value. The GW (G represents the Green's function of the system and W represents the screened Coulomb interaction) results obtained are in good agreement with the tight binding method (table 2.4).

Electronic properties for five basic crystallographic conformations have been enlightened to elaborate on the effect of functionalization type and thickness on work function and electron density [73]. Two cubic diamond structures (110 and 111) and three Lonsdaleite diamond structures (0001, 1010, and 2110) with varying thicknesses have been analyzed. Five of these conformations for hydrogenated and FDs with thicknesses from one to six layers have been compared. AA stacking, called Lonsdaleite leads to the formation of 0001 and 2110 diamanes, while AA stacking leads to the 1010 diamanes formation. Analysis carried out using DFT with the aforementioned approximations has provided useful information about electronic band structures and bandgap of diamanes. All five conformations for H-diamanes possessed a direct bandgap according to TB09-KS band structure analysis. Here, the bandgap decreased from mono- to tri-layered diamanes, followed by the increase in bandgap for three to six layers. Thickness prominently altered the bandgap behavior. In addition, in KS analysis for F-diamanes the direct bandgap behavior shifted to indirect bandgap with the increase in thickness, except for 2110 Lonsdaleite diamond. FD possessing 2110 conformation behaved as a direct bandgap semiconductor for one to six layers. GW approximation revealed different results where the bandgap increased and then decreased with the increase from monolayer to three layers for FD. Meanwhile, the direct bandgap for HD monotonically decreased with the increase in the number of layers. Furthermore, 2110 HD possessed direct bandgap behavior. GW-TB09 corrected results and provided the reason for the difference between opposite results. The difference in bandgap behavior is attributed to the different electronic states. It is a surface electronic state for FD and it is a bulk electronic state for HD.

Likewise, the work function has been calculated by using DFT-PBE and then GW approximation on PBE. For HD the work function decreased rapidly with the increase in the number of layers. Meanwhile, for FD work the function decreased linearly slowly with the increase in the number of layers except for cubic 110 and hexagonal 2110. Cubic 110 and Lonsdaleite 2110 displayed behavior similar to the bandgap. From monolayer to three layers, it increased and then decreased for FDs. Foremost, FDs have a higher work function compared to HDs (figure 2.7).

The effect of functional group type and doping on diamanes electrical properties has also been theoretically explored [41]. Diamanes anchored with different anion groups

Table 2.4. Comparison of evaluation methods for FD.

Functionalized diamanes	Bandgap-PBE (eV)	Bandgap-GW (eV)	References
FD	3.1	7.2	[57]
	3.1	7.3	[72]
	3.2	7.42	[54]

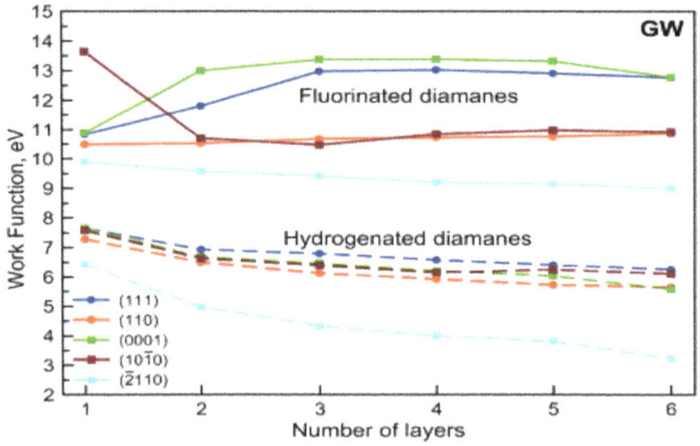

Figure 2.7. Work function for HD and FD. Reproduced from [73]. CC BY 4.0.

have been systematically studied with DFT. The authors studied a total of 30 diamane films with six different crystallographic conformations anchored with five different functional groups i.e., H, F, Cl, OH, and NH_2. Electronic properties investigation has been carried out using band structure calculations via GGA-PBE and HSE06. The results indicate that bandgap values calculated using hybrid functional are higher than bandgaps obtained by GGA-PBE approximation with an approximate difference of 1.3 eV. Among them, 12 conformers are confirmed to be stable and present direct semiconductor features with bandgaps ranging from 2.527 to 4.153 eV. The bandgap obtained for one of the hexagonal geometries of BLH (boron-doped diamond-like carbon (DLC) films, often abbreviated as BLH diamonds) diamanes is the least (2.527 eV) and the highest bandgap is shown by boundary luminescence free (BLF) diamanes (4.153 eV). Even the highest bandgap reported (i.e., 5.47 eV) is less than the bandgap of a diamond, thus confirming the semiconducting behavior of diamanes anchored with functional groups. Due to the broad range of workable bandgaps, diamanes have the potential to be used in optoelectronic applications. DFT with GGA-PBE approximation for exchange–correlation functions has been used. DFT with GGA-PBE underestimates the bandgap and overestimates electronic repulsions.

For a semiconductor, the mobility of the carriers is an important criterion for potential in electronics and optoelectronics devices with high frequency. The carriers' mobility refers to the ease of the charge carriers' movement through the structure in the presence of an electric field. The authors evaluated carrier mobility with respect to deformation potential theory:

$$\mu = \frac{e\hbar^3 C}{K_b T m^* m_d {E_d}^2}$$

$$C = \left(\frac{\partial^2 E}{\partial \varepsilon^2}\right) \Big/ S_0$$

$$m^* = \hbar^2 \left(\frac{d^2 E}{dk^2}\right)^{-1}$$

$$m_{\mathrm{d}} = \sqrt{m_x^* m_y^*}$$

$$E_{\mathrm{d}} = \frac{\mathrm{d}E_{\mathrm{edge}}}{\mathrm{d}\varepsilon}$$

where C is the in-plane stiffness, S_0 is the area of optimized supercell, E is the total energy, M^* is the effective mass, M_d is the average effective mass, k is the wave vector, E is the electronic energy of wave vector, E_d is the deformation potential constant, and $\mathcal{E}$, h, K_b are physical constants with their usual meanings, and the temperature (T) is 300 K.

Both electron and hole carrier mobility have been studied, with a result of remarkably high electron carrier mobility in both directions. Moreover, the electron carrier mobility of cubic BLF diamanes is exceptionally high, 16 546.713 $cm^2\ V^{-1}\ s^{-1}$ along the y-direction, which is remarkably larger than that of diamond. This makes it suitable for applications in electronics. Doping is crucial for fabricating electronic devices. Boat2-H diamane doped with B and N to have p- and n-type conductivity has shown the potential to be used in electronic and optoelectronic devices with a moderate activation energy of 0.34 and 0.37 eV, respectively. Activation energy at the T-point in the Brillouin zone is representative of electronic structure. By having insight into the activation energy of doped diamanes, researchers suggest that diamanes are propitious for electronic devices and engineering materials.

The bandgap of the diamond using DFT-HSE06 and PBEsol functionals is predicted to be 5.33 eV. This value is very close to the 5.5 eV, experimentally determined bandgap value. This proves the efficiency of this method to theoretically determine the electronic band structures for diamanes. The electronic properties of one-four-layered diamanes have been explored in terms of electronic band structures using HSE06 and PBEsol functionals [8]. The results obtained justified boron-doped low-hydrogenated diamond (BLHD as a wide bandgap semiconductor with a direct gap at T-point. The bandgaps determined by HSE06 and PBEsol are 3.835 and 3.037 eV, respectively. These results are in good accordance with PBE data 2.985 eV. Similarly, the bandgap for boundary luminescence free diamanes (BLFDs) is reported to be 3.938 eV, this reported value is also in good accordance with PBE data 4.040 eV. Following the HSE06 functional, the bandgap decreased with the increasing number of layers from monolayer to four-layered hydrogenated and chlorinated diamanes. In the case of FDs, the bandgap increased while moving from two layers to three layers and then decreased for four-layered FDs. In other words, the bandgap of FD switched from direct to quasi-direct and then to an indirect bandgap while moving from two layers two four layers. Partial density of states and partial density distributions at CBM and VBM are also critical quantities to analyze conductive behavior. Projected charge density distribution at VBM is a representative of electron density that provides information about chemical bonding between the atoms. Projected charge density distribution is primarily representative of electron density that is available for conduction at CBM. Simply, it provides an insight into the mobility of charge carriers. The authors also calculated the projected charge density distributions of VBM and CBM for 3L- and 4L diamonds. The results show that VBM and CBM of each diamane show the same character as in their corresponding 2L-diamanes, except for CBMs of 3L- and 4L-F-diamane, which are different than that of 2L-F-diamane due to the direct to indirect bandgap transition of F-diamanes with thickness. For both handed Cl-diamanes, VBM represents an almost non-surface state, therefore for these lattices the quantum confinement effect dominates the bandgap when thickness increases from 2L to 4L. However, for F-diamanes, VBM and CBM present hybridization between surface states of F atom and non-surface states that are uniformly distributed over different carbon layers. Therefore, F-diamanes show a more complex electronic response due to the remarkable coexistence of quantum confinement and surface state effects as the thickness changes (figure 2.8).

One-sided hydrogenated graphene, named graphone [74], is used as a precursor for a semi-hydrogenated graphene bilayer [75]. Unlike fully HDs, semi-hydrogenated graphene has been named as diamanes [17]. Diamane's physical properties have been studied and compared with diamanes for both Bernel stacked-chair and Lonsdaleite stacked-boat conformations. With respect to the impact of conformations, Lonsdaleite stacked-boat diamanes have higher conductivity with a bandgap of 2.92 eV relative to Bernel stacked-chair diamanes with a bandgap of 3.24 eV. Diamane's bandgap is higher than diamones, thus making semi-hydrogenated structures electrically more useful. In addition, the impact of interlayer bonding has been signified. Boat-I structure referred to as diamones with interlayer bonding and boat-II without interlayer bonding. Diamones with chair and boat-I conformation have higher conductivity with a ban gap of ~0.50 eV. In contrast, boat-II has lower conductivity with a bandgap of 2.35 eV. Hence, interlayer bonding is responsible for the increased bandgap with a difference of 129% [20, 43]. The decrease in conductivity due to interlayer bonding is supported by the DFT tight binding calculations. Interlayer C–C bonds appear due to atomic defects leading to a significant decrease in electronic transportation between the layers [76].

Semi-hydrogenated BLG exhibited a 2D ferromagnetic semiconducting behavior. This nature is attributed to the recurring dangling bonds on the bottom layer of

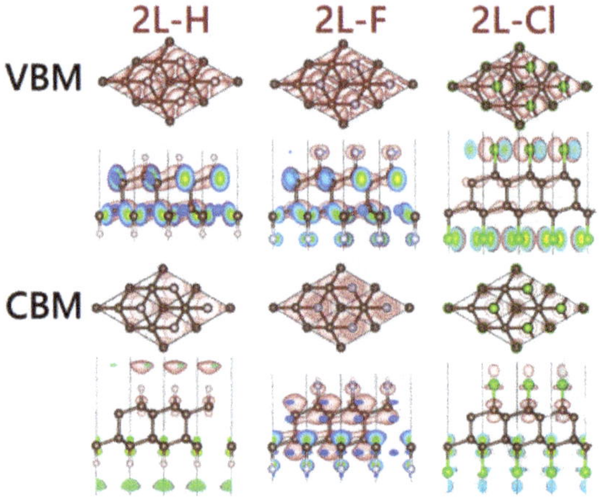

Figure 2.8. Top and side views for charge density distributions of VBM and CBM for bilayer H, F, and Cl-diamanes. Reprinted from [8], Copyright (2020), with permission from Elsevier.

diamones [18]. Its ferromagnetic semiconducting nature makes it an ideal candidate for spin electronics. Ferromagnetic diamones have adjustable bandgaps and magnetic moments by tunable electric fields or magnetic fields. By the increase in electric field, diamanes bandgap declined monotonically from 3 to 0 eV. The critical value of an electric field is 1.05 V $Å^{-1}$ for the transition of fully HDs from semiconducting to metallic. Meanwhile, diamones have non-monotonic dependence of bandgap on the electric field [20].

Diamondol is a multilayer graphene semi-covered by the OH group and has similar properties to those of diamones. Moreover, it has also been characterized as a ferromagnetic semiconductor with the periodicity of dangling bonds. Spin-polarized bands attributed to the different bandgaps for varying spins make diamondol an ideal candidate for spin electronics. The bandgap of HDs becomes a direct bandgap with the increase in thickness and decrease in hydrogen content. The direct bandgap is higher than the indirect and its properties resemble that of the bulk diamond. A direct bandgap is favorable for optoelectronics. However, only with the increase in the number of layers does a bandgap become indirect from a direct bandgap.

Electronic properties are mainly affected by the nature of the graphene that has been used as a starting material. Normal BLG and twisted BLG provide a difference in their electronic properties. Bernel stacked normal BLG bandgap remains zero along with energy-momentum dispersion. Meanwhile, graphene layers within Moire's pattern exhibit arranged superlattices composed of diamond-like crystals, with the arrangement variation being dependent on twist angle. Enhancement of bandgap at a definite degree of twist is possible, which resulted in the superposition of layers. The changes in electronic structure influence the electronic spectra, frequency shifts, presence of active phonon modes, and quantum peculiarities of HDs. The twist angle interlayer interaction of BLG is also significant. Interlayer interaction of BLG provides superconductivity at a twist angle of 1.1. Superconductivity is also achievable by the coupling of electronic phonons. Diamanes owing to these characteristics are useful for the next generation of nanoelectronics.

The twisted structure interlayer connection is significantly different from Bernel and Lonsdaleite stacked structures. The simplest unit cell is considered with a twist angle of 21.8°. Comparatively twisted diamanes show higher bandgaps than normal diamanes. Twisted diamanes show a bandgap of more than 4 eV. Band structure and electronic densities have been studied to compare the electronic properties of twisted and normal diamanes: 4.19 and 3.12 eV are the bandgaps for twisted and normal fully HDs, respectively; and 4.30 and 4 eV are the bandgaps for twisted and normal fully FDs, respectively. The authors opted for DFT to analyze the electronic structure of Moire's diamanes. Unit cells used for their synthesis are twisted BLG with a twist angle of 21.8°. The presence of a large number of narrow peaks of densities of states connected with flattened minizones provides their applicability in optical and optoelectronic devices.

Moire's diamanes with 29.4° twist angle, named diamanes quasicrystals, have an increased number of strained C–C bonds relative to AB-stacked diamanes. Bernel stacked diamanes have equivalent interlayer bond lengths and bonds are oriented normal to the surface. Interlayer bond lengths in diamanes quasicrystals range from 1.55 to 1.66 Å and have various spatial orientations. Thus, the bandgap of diamanes quasicrystals (3.6 eV) is higher than Bernel stacked diamanes (3.1 eV), yet their bandgaps are smaller than diamond (4.2 eV).

2.3.2 Thermal conductivity

Diamonds have excellent thermal properties, making them attractive for research. The thermal properties of diamanes in terms of thermal conductivity, thermal expansion, and stability have been explored and are briefly described here. By exploiting their unique thermal properties, they could be deployed in various electronic devices, high-power lasers, etc. The governing mechanisms for the thermal behavior of diamanes have been explored for use in practical applications.

Thermal conductivity is an anisotropic property of diamane that depends on the direction in which heat is transported. Briefly, thermal conductivity is regulated by effective heat transportation and phonon group velocity. In turn, the controlling mechanisms for effective heat transport are acoustic mode and optical mode. Dominance and overlapping of acoustic mode are necessary for high thermal conductivity because acoustic phonons are core heat carriers. Elevation in group velocity enhances thermal velocity as the width of dedicated frequency for acoustic bands is also increased with the increase in group velocity. Group velocity is a basic parameter to determine the rate at which thermal energy would be transported. Acoustic mode and optical mode play a crucial role in effective heat transport. Acoustic modes are the atomic vibrations generated in diamanes when heated. A large number of well-defined acoustic vibrations smoothly transport the heat energy throughout the structure at lower frequencies. Optical modes are also the kind of vibrations that play a frequency-dependent role in heat transportation. At high frequency, interaction between the vibrating atoms or phonons is insufficient for effective heat transfer and there is a hindrance in the movement of phonons, while at low frequency phonons have enough time for effective heat energy transfer. Here, hybrid phonon modes emerge due to the ample interaction of phonons. Hybrid phonon modes elevate thermal conductivity. These modes facilitate the rapid propagation of vibrations throughout the structure, making them diamanes with high group velocity.

As earlier said, diamanes when heated atoms vibrate, their vibrational energy in the form of small packets is called phonons. The direction of phonon movement is of prime importance. When phonons move in a direction perpendicular to the surface of material it is called out-of-plane phonon movement, while when the phonon movement is parallel to the surface of material it is called in-plane phonon movement. Precisely, high thermal conductivity at room temperature is achievable with acoustic mode and out-of-plane phonon transportation, along with a suitable stacking order. In terms of optical modes, high thermal conductivity is achievable at low frequency when it produces hybrid phonon modes with high group velocity.

Additionally, how the atoms are stacked also confers thermal conductivity as it affects the phonon movement. Lonsdaleite stacking (AA) has higher thermal conductivity compared to Bernel stacking (AB) because the horizontal reflection symmetry in AA stacking restricts the phonon scattering channels.

Interlayer bonds incorporation into diamanes significantly affects the thermal conductivity depending upon the type of bond and its orientation. Interlayer bonds could behave like scattering centers for phonons, where they would scatter in random directions, thus decreasing the thermal conductivity of a material. In contrast, if interlayer bond orientation supports the out-of-plane phonon movement, then it could increase the density of phonons. Thus, high-density phonons provide high thermal conductivity.

Thermal conductivity varies in inverse proportion to the temperature. Phonon scattering increases with the temperature rise, decreasing its overall ability to transport heat through the diamane structure. Umklapp (U) scattering process and temperature power factor are used to analyze the behavior of thermal conductivity with temperature. The U process is the scattering of phonons with high energy and momentum which reduces the mean free path contributing to high thermal resistance. At low temperatures, the U-scattering process is minor and increases with the increase in temperature thus lowering thermal conductivity. Figure 2.9 shows the room temperature thermal conductivity of HD to be 1145 W m^{-1} K^{-1} at 300 K, reduced to 160 W m^{-1} K^{-1} at 1200 K. There is a decrease in thermal conductivity for N-substituted diamanes from 1540 W m^{-1} K^{-1} at 300 K to approximately 500 W m^{-1} K^{-1} at 600 K [77] (figure 2.9).

Like other properties of diamanes, stacking thermal conductivity is also influenced by the number of layers and type of functionalization. Various comparative studies have been carried out for different kinds of functionalized diamanes. The thermal properties of hydrogenated, fluorinated, and chlorinated monolayer diamanes have been demonstrated and compared in terms of phonon dispersion. Variations in acoustic and optical modes for C_2H, C_2F, and C_2Cl diamanes showed differences in their thermal conductivity behavior. Acoustic branches are much wider in C_2H diamane compared to other functionalized diamanes, capitulating

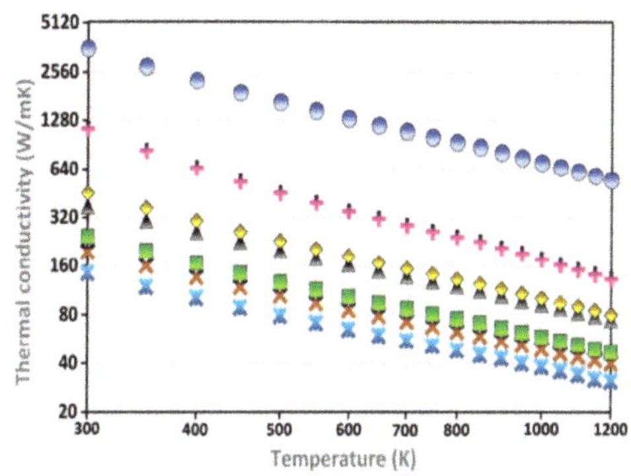

Figure 2.9. Thermal conductivities as a function of temperature for C_2H, C_4HF, C_2F, C_4HCl, C_4FCl, C_2Cl. Reprinted from [35], Copyright (2020), with permission from Elsevier.

higher group velocity and thus higher thermal conductivity. In contrast, C_2F and C_2Cl diamanes bear narrow dispersion of acoustic branches along with lower optical modes crossing each other. Narrow dispersion and crossing of branches enhanced scattering probability, thus suppressing thermal conductivity. The reported thermal conductivities of hydrogenated, fluorinated, and chlorinated monolayer diamanes are 1145 W m^{-1} K^{-1}, 377 W m^{-1} K^{-1}, and 146 W m^{-1} K^{-1}, respectively. Yet, HDs exhibited higher thermal conductivity with extravagant thermal properties for high-temperature applications (figure 2.10).

Phonon-dispersion relations are also studied in terms of acoustic and optical modes branching. The nature of branching regulates the phonon scattering mechanism. The presence of N atoms in diamanes created defects in their crystal structure that led to elevated phonon transportation. The presence of nitrogen led to the quadratic curved branch splitting of transverse acoustic mode, which is called out-of-plane(flexural) acoustic branch (ZA). Here, the presence of ZA branches provided the out-of-plane phonon vibration and reduced phonon–phonon scattering with the overall increased mean free path for phonon transportation. Thus, N-substituted diamanes have higher thermal conductivity due to the quadratic curved branch splitting mechanism. Their reported thermal conductivity is 1540 W m^{-1} K^{-1} at 300 K, comparatively 33% lower than bulk diamond but 15%–35% higher than HDs.

Unpassivated diamanes have an 8.3×10^{-6} K^{-1} coefficient of thermal expansion along with the isotropic in-plane thermal behavior. The coefficient of thermal expansion for diamanes along one direction is calculated as,

$$\alpha_i = \frac{L_i(T + \Delta T) - L_i(T)}{\Delta T \times L_i(T)}$$

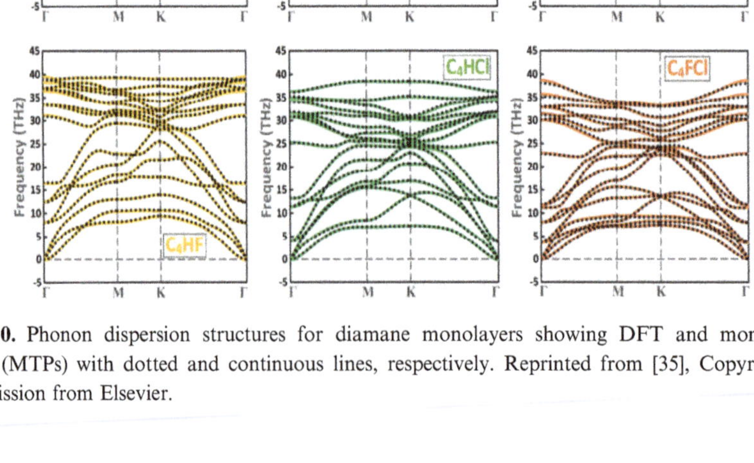

Figure 2.10. Phonon dispersion structures for diamane monolayers showing DFT and moment tensor potentials (MTPs) with dotted and continuous lines, respectively. Reprinted from [35], Copyright (2020), with permission from Elsevier.

Moreover, thermal expansion behavior switches from isotropic to slightly orthotropic when the temperature is higher than 250 K [28].

2.3.3 Optical properties

According to the literature, diamanes and their derivatives show unprecedented optical properties but are still not clear and further research investigations are needed. The wide bandgap range and direct bandgap behavior are the key contributors to its optical properties. Like electronic properties, the optical properties of diamanes are also tunable. Using functionalized diamanes flexible and transparent material for the use in nano-photonics have been prepared. The prerequisites for the smooth applications in optical devices are fulfilled by diamanes in terms of wide bandgap spectra, and multiple resonance peaks in valence and conduction bands. Moreover, fluorinated, hydrogenated, and chlorinated BLG display improved optical properties due to attached functional groups. Absorption of light is critical in controlling the optical properties of light in terms of the amount of light to be transmitted or reflected from a material. Chlorinated diamanes absorb light over a wide energy range in the visible regions compared to fluorinated and HDs. FD and HD are incapable of absorbing light because of the drastic changes in their electronic structure. FD being an insulating diamond (F atoms are tightly bonded to carbon atoms via a covalent bond, providing a decrease in the density of states at the Fermi level) has its span of performance in nanopatterning techniques. In diamanes the concept of silicon vacancy exists, where the carbon atom is replaced by the silicon atom and the neighboring carbon atom is missing. Silicon vacancy provides single photon emission by controlling the energy level silicon-vacancy defect. BLG, or twisted BLG, has additional energy bands called narrow mini bands inside ultrawide 2D bandgap. These narrow mini bands contribute towards optical properties. Substitution of atoms or functionalization brings excellent optical properties.

2.4 Structural properties

The mechanical performance and durability of a material are critical selection parameters for salient applications in nanotechnology. Diamane's structural properties are stupendous. Their sp^3 hybridized tightly bonded carbon atoms coupled with 2D nature converse them with outstanding mechanical robustness [14, 51, 62, 78, 79]. Their structure lattice gifts them the ability to withstand colossal mechanical stress and to resist deformation. Exciting avenues have been opened up for ultra-strong materials with the diamanes synthesis. Nanodevice fabrications using diamanes require the investigation of their tensile strength, elasticity, and vibration. Diamanes are composed of ultrathin multiple layers combined in a honeycomb lattice. Here, the involvement of interlayer size and spacing needs to be realized. Interactions are dominant when the interlayer distance is small. This makes diamanes brittle and easily prone to fracture. In the case of large interlayer distance, interactions between adjacent carbon layers are weakened endowing them with flexibility, relative sliding of sp^3 hybridized carbon atoms, and shearing between the layers. Flexibility imparts diamanes with ductile behavior. Moreover, residual stress accumulation tends to weaken mechanical properties. The presence of interlayer bonds imparts diamanes with a high coefficient of stiffness and makes them harder than graphene [13, 22, 65, 80]. The presence of covalent bonds between adjacent carbon atoms in different layers locks the multiple layers together, restricting the motion of carbon atoms along a vertical direction and resisting deformation under the load applied perpendicular to the plane of material. This provides diamanes with a high coefficient of stiffness. High hardness is also attributed to the existence of interlayer bonds that reinforce the overall structure of diamanes [14, 18, 21, 81]. Along with remarkable structural strength, they also have good elastic properties. Their flexibility and bending characteristics replace graphene in nanotechnology.

Functional groups mark a colossal impact on the hybridization state of bonded carbon atoms in diamanes. Hybridization is the arrangement of electron orbitals around carbon atoms. This dominates the bonding characteristics, and hence overall mechanical properties [65, 82, 83]. The key factor to be investigated with regard to the structural properties of diamanes is the coefficient of finite interlayer connected domain of carbon [84, 85]. This coefficient is a representative of a degree of bonding and connectivity between adjacent layers of carbon atoms. High sp^2 hybridization generates a disturbance in symmetry and puts together a trigonal dominant geometry with flexibility as a dominant trait. This kind of diamanes is called mechanically isotropic [86, 87]. This behavior is similar to graphene with high thickness and no functional group [8]. Meanwhile, high sp^3 hybridization models a structure dominantly occupied with tetrahedral geometry. Sp^3 dominant hybridization state has a structure with much higher stiffness relative to diamond [40, 88]. In this state, multilayers of diamanes that are tightly connected have high elasticity and Young's modulus. Hybridization controls a unique blend of mechanical stiffness and flexibility, which gives diamanes exceptional resilience and makes them applicable for vast range of applications.

AA stacked and Lonsdaleite stacked diamanes have higher buckling resistance owing to the high shear strength but low Young's modulus and tensile strength attributed to their atomic structure. In this mode of stacking, atoms of multiple graphene layers fully overlap each other. Meanwhile, AB-Bernel stacked diamanes have low shear strength. They are flexible due to the easy sliding of carbon layers precisely aligned one above another and one of the carbon layers entirely displaced in an armchair direction by a distance equal to the C–C bond length [13, 14, 62, 89, 90].

Unpassivated diamanes have orthotropic mechanical behavior. In the armchair direction, diamanes are less stiff than graphene while in the zigzag direction diamanes are stiffer than graphene. Studies have proved that diamanes are more useful in a zigzag direction because they endure high strain before deformation, making them useful in flexible devices. Observing mechanical behavior with temperature shows that elastic properties decrease as temperature increases [28].

Functionalization with a suitable moiety imparts additional mechanical properties. Hydrogenation and fluorination enhance the mechanical properties. Meanwhile, chlorination reduces the mechanical stability of diamanes. A researcher reported that FDs are more flexible than unfunctionalized diamanes [91]. The nature of the functionalization group affects the effective mass of the charge carriers' mobility, elastic modulus, and deformation potential constant. Interestingly, for H-diamanes, a very small effect on the charge carriers' mobility has been observed for hydrogenated and Lonsdaleite diamanes.

When diamanes are deployed in energy storage devices, the relationship between mechanical properties and vibrational characteristics is to be considered. Lonsdaleite diamanes due to their strongly bonded structure have higher resonant frequencies under mechanical stress or electric field. Efficient energy storage devices are required to efficiently store and release energy with a low dissipation factor. Diamanes with robust mechanical characteristics fulfill low energy dissipation requirements and retain higher resonant frequencies or vibrations for a longer period of time. The duration of these energy storage devices could be ensured with the use of diamanes. Lonsdaleite diamane's exceptional mechanical stiffness, great flexibility, and low damping ensure its mechanical integrity under repeated cycles. Reportedly, diamanes nano-ribbons with a width of less than 100 nm provided high vibrational characteristics with minimum energy dissipation. An exponential incline in vibrational characteristics is achievable by increasing the lattice spacing both for diamanes and diamane nano-ribbons, but it decreases the stiffness of diamanes. Meanwhile, the effect of stacking sequence on diamanes vibrational characteristics is negligible. Due to high resonant frequencies, diamanes are expected to be effectively used in ultrasensitive resonator-based sensors.

The manipulation of mechanical behavior for twisted BLG is possible by the modification of symmetry. Symmetry alteration is possible by introducing the elective functional groups yet preserving their periodicity and Moire' pattern. Twisted BLG (AB-stacked)-Moire's diamanes with three different angles 21.8°, 27.8°, and 29.8° have been investigated and compared. Reportedly, they possess elastic characteristics with higher stiffness. Moire's diamanes are turned into isotropic from anisotropic with an increase in twist angle. With the increase in twist angle, the Young's modulus decreased from 0.8 to 0.6 [92]. Diamane quasicrystals with Moire' pattern were reported for the first time in 2021. Diamane quasicrystals comprising of BLG twisted at an angle of 30° have stronger mechanical properties than Bernel stacked diamanes. Moreover, an increase in interlayer bond density boosts shear modulus, hence offering a prominent increase in tensile strength and Young's modulus.

2.5 Conclusion

This chapter presented an overview of the physical and structural properties of diamanes, while focusing on the influencing parameters. Diamanes are 2D diamonds with extravagant mechanical properties and tunable electrical properties. Diamanes possess all mandatory properties for effective use in nanotechnology applications. Their distinctive properties have opened up tremendous applications in electronics, and electromechanical microsystems. Fluorination is preferred for insulating applications because a 30% increase in bandgap is attained with the transition from monolayer graphane to bilayer diamanes. Based on their properties, F-diamanes are preferable because they restrict the carrier's mobility and provide high mechanical stiffness specifically in the longitudinal direction. F-diamanes could be anticipated as insulating layers due to their wide bandgap range. AB staked diamanes provide semiconducting behavior with the widest bandgap, while AA stacked offers semimetallic properties. Their potential applications include nanoelectronics, optoelectronics, and electromechanical systems.

References

[1] Liu L, Zhou H, Cheng R, Yu W J, Liu Y, Chen Y, Shaw J, Zhong X, Huang Y and Duan X 2012 High-yield chemical vapor deposition growth of high-quality large-area AB-stacked bilayer graphene *ACS Nano* **6** 8241–9

[2] Lin Y M *et al* 2010 100-GHz transistors from wafer-scale epitaxial graphene *Science* **327** 662 https://www.science.org/doi/10.1126/science.1184289

[3] Sun Z, Pint C L, Marcano D C, Zhang C, Yao J, Ruan G, Yan Z, Zhu Y, Hauge R H and Tour J M 2011 Towards hybrid superlattices in graphene *Nat. Commun.* **2** 559

[4] Castro Neto A H, Guinea F, Peres N M R, Novoselov K S and Geim A K 2009 The electronic properties of graphene *Rev. Mod. Phys.* **81** 109

[5] Lee C, Wei X, Kysar J W and Hone J 2008 Measurement of the elastic properties and intrinsic strength of monolayer graphene *Science* **321** 385–8

[6] Balandin A A, Ghosh S, Bao W, Calizo I, Teweldebrhan D, Miao F and Lau C N 2008 Superior thermal conductivity of single-layer graphene *Nano Lett.* **8** 902–7

[7] Ersan F, Üzengi Aktürk O, Aktürk E and Ciraci S 2018 Metal–insulator transition and heterostructure formation by glycines self-assembled on defect-patterned graphene *J. Phys. Chem.* C **122** 14598–605

[8] Mortazavi B *et al* 2020 First-principles investigation of mechanical, electronic and optical properties of H-, F- and Cl-diamane *Appl. Surf. Sci.* **528** 147035 https://www.sciencedirect.com/science/article/pii/S016943322031792X

[9] Zhang Y, Tang T-T, Girit C, Hao Z, Martin M C, Zettl A, Crommie M F, Shen Y R and Wang F 2009 Direct observation of a widely tunable bandgap in bilayer graphene *Nature* **459** 820–3

[10] Zollner K, Gmitra M and Fabian J 2019 Heterostructures of graphene and hBN: electronic, spin–orbit, and spin relaxation properties from first principles *Phys. Rev.* B **99** 125151

[11] Davletshin A R, Ustiuzhanina S V, Kistanov A A, Saadatmand D, Dmitriev S V, Zhou K and Korznikova E A 2018 Electronic structure of graphene- and BN-supported phosphorene *J. Phys.: Condens. Matter* **534** 63–7

[12] Kistanov A A, Cai Y, Zhang Y-W, Dmitriev S V and Zhou K 2017 Strain and water effects on the electronic structure and chemical activity of in-plane graphene/silicene heterostructure *J. Phys.: Condens. Matter* **29** 095302

[13] Chernozatonskii L A, Sorokin P B, Kvashnin A G and Kvashnin D G 2009 Diamond-like C2H nanolayer, diamane: simulation of the structure and properties *JETP Lett.* **90** 134–8

[14] Chernozatonskii L A, Sorokin P B, Kuzubov A A, Sorokin B P, Kvashnin A G, Kvashnin D G, Avramov P V and Yakobson B I 2011 Influence of size effect on the electronic and elastic properties of diamond films with nanometer thickness *J. Phys. Chem.* C **115** 132–6

[15] Rysaeva L K, Lisovenko D S, Gorodtsov V A and Baimova J A 2020 Stability, elastic properties and deformation behavior of graphene-based diamond-like phases *Comput. Mater. Sci.* **172** 109355

[16] Cellini F, Lavini F, Cao T, De Heer W, Berger C, Bongiorno A and Riedo E 2018 Epitaxial two-layer graphene under pressure: diamene stiffer than diamond *FlatChem.* **10** 8–13

[17] Barboza A P M, Guimaraes M H D, Massote D V P, Campos L C, Barbosa Neto N M, Cancado L G, Lacerda R G, Chacham H, Mazzoni M S C and Neves B R A 2011 Room-temperature compression-induced diamondization of few-layer graphene *Adv. Mater.* **23** 3014–7

[18] Kvashnin A G, Chernozatonskii L A, Yakobson B I and Sorokin P B 2014 Phase diagram of quasi-two-dimensional carbon, from graphene to diamond *Nano Lett.* **14** 676–81

[19] Martins L G P *et al* 2017 Raman evidence for pressure-induced formation of diamondene *Nat. Commun.* **8** 1–9

[20] Samarakoon D K and Wang X-Q 2010 Tunable band gap in hydrogenated bilayer graphene *ACS Nano* **4** 4126–30

[21] Kvashnin A G and Sorokin P B 2014 Lonsdaleite films with nanometer thickness *J. Phys. Chem. Lett.* **5** 541–8

[22] Qin G, Wu L and Gou H 2021 Diamane: design, synthesis, properties, and challenges *Functional Diamond* **1** 83–92

[23] Sun Z, Pint C L, Marcano D C, Zhang C, Yao J, Ruan G, Yan Z, Zhu Y, Hauge R H and Tour J M 2011 Towards hybrid superlattices in graphene *Nat. Commun.* **2** 559

[24] Khaliullin R Z, Eshet H, Kühne T D, Behler J and Parrinello M 2011 Nucleation mechanism for the direct graphite-to-diamond phase transition *Nat. Mater.* **10** 693–7

[25] Lavini F, Rejhon M and Riedo E 2022 Two-dimensional diamonds from sp^2-to-sp^3 phase transitions *Nat. Rev. Mater.* **7** 814–32

[26] Wort C J H and Balmer R S 2008 Diamond as an electronic material *Mater. Today* **11** 22–8

[27] Zhu L, Li W and Ding F 2019 Giant thermal conductivity in diamane and the influence of horizontal reflection symmetry on phonon scattering *Nanoscale* **11** 4248–57

[28] Shi J, Cai K and Xie Y M 2018 Thermal and tensile properties of diamondene at finite temperature: a molecular dynamics study *Mater. Des.* **156** 125–34

[29] Cai K, Wang L and Xie Y M 2018 Buckling behavior of nanotubes from diamondene *Mater. Des.* **149** 34–42

[30] Wu Y C, Shao J L, Zheng Z and Zhan H 2021 Mechanical Properties of a Single-Layer Diamane under Tension and Bending *J. Phys. Chem.* C **125** 915–22

[31] Muniz A R and Maroudas D 2012 Opening and tuning of band gap by the formation of diamond superlattices in twisted bilayer graphene *Phys. Rev.* B **86** 075404

[32] Zheng Z, Zhan H, Nie Y, Xu X, Qi D and Gu Y 2020 Single layer diamond—a new ultrathin 2D carbon nanostructure for mechanical resonator *Carbon* **161** 809–15

[33] Zhu L and Zhang T 2019 Suppressed thermal conductivity in fluorinated diamane: optical phonon dominant thermal transport *Appl. Phys. Lett.* **115** 151904

[34] Zhang T and Zhu L 2020 Sensitively tuning the thermal conductivity of diamane via engineering the mass of functional groups *Nanotechnology* **31** 435409

[35] Raeisi M, Mortazavi B, Podryabinkin E V, Shojaei F, Zhuang X and Shapeev A V 2020 High thermal conductivity in semiconducting Janus and non-Janus diamanes *Carbon N. Y.* **167** 51–61

[36] Hu Y, Li D, Yin Y, Li S, Ding G, Zhou H and Zhang G 2020 The important role of strain on phonon hydrodynamics in diamond-like bi-layer graphene *Nanotechnology* **31** 335711

[37] Gupta S, Yang J-H and Yakobson B I 2019 Two-level quantum systems in two-dimensional materials for single photon emission *Nano Lett.* **19** 408–14

[38] Qiu D, Wang Q, Cheng S, Gao N and Li H 2019 Electronic structures of two-dimensional hydrogenated bilayer diamond films with Si dopant and Si-V center *Res. Phys.* **13** 102240

[39] Li J, Yin H, Gao N, Zhang M, Mu J, Gao L and Li H 2019 First-principles calculations for Li, P dopants and vacancy defect in ultra-thin hydrogenated diamond nanofilms: structural, electronic and optical properties *Diam. Relat. Mater.* **99** 107526

[40] Sorokin P B and Yakobson B I 2021 Two-dimensional diamond—diamane: current state and further prospects *Nano Lett.* **21** 5475–84

[41] Ge L, Liu H, Wang J, Huang H, Cui Z, Huang Q, Fu Z and Lu Y 2021 Properties of diamane anchored with different groups *Phys. Chem. Chem. Phys.* **23** 14195–204

[42] Gui G, Li J and Zhong J 2008 Band structure engineering of graphene by strain: First-principles calculations *Phys. Rev.* B **78** 075435

[43] Chernozatonskii L A, Demin V A and Kvashnin D G 2021 Fully Hydrogenated and Fluorinated Bigraphenes–Diamanes: Theoretical and Experimental Studies *C* **7** 17

[44] Li J, Li H, Wang Z and Zou G 2013 Structure, magnetic, and electronic properties of hydrogenated two-dimensional diamond films *Appl. Phys. Lett.* **102** 7

[45] Chernozatonskii L A, Sorokin P B and Artukh A A 2014 Novel graphene-based nano-structures: physicochemical properties and applications *Russ. Chem. Rev.* **83** 251

[46] Tuan V Vu, Huynh V Phuc, Sohail Ahmad, Vo Quang Nha, Chu Van Lanh, Rai D P, Kartamyshev A I, Khang D Pham, Le Cong Nhan and Nguyen N Hieu 2021 Outstanding elastic, electronic, transport and optical properties of a novel layered material C4F2: first-principles study *RSC Adv.* **11** 23280–7

[47] Qiu D, Zhang Z, Li J, Cheng S and Li H 2018 Structural and electronic properties of two-dimensional (110) diamond nanofilms by first-principles calculations *Diam. Relat. Mater.* **84** 55–61

[48] Piazza F, Gough K, Monthioux M, Puech P, Gerber I, Wiens R, Paredes G and Ozoria C 2019 Low temperature, pressureless sp^2 to sp^3 transformation of ultrathin, crystalline carbon films *Carbon* **145** 10–22

[49] Bakharev B V *et al* 2019 Chemically induced transformation of chemical vapour deposition grown bilayer graphene into fluorinated single-layer diamond *Nat. Nanotechnol.* **15** 59–66

[50] Piazza F, Cruz K, Monthioux M, Puech P and Gerber I 2020 Raman evidence for the successful synthesis of diamane *Carbon* **169** 129–33

[51] Chernozatonskii L A, Mavrin B N and Sorokin P B 2012 Determination of ultrathin diamond films by Raman spectroscopy *Phys. Status Solidi* **249** 1550–4

[52] Odkhuu D, Shin D, Ruoff R S and Park N 2013 Conversion of multilayer graphene into continuous ultrathin sp3-bonded carbon films on metal surfaces *Sci. Rep.* **3** 1–7

[53] Lurie P G and Wilson J M 1977 The diamond surface: I. The structure of the clean surface and the interaction with gases and metals *Surf. Sci.* **65** 453–75

[54] Leenaerts O, Peelaers H, Hernández-Nieves A D, Partoens B and Peeters F M 2010 First-principles investigation of graphene fluoride and graphane *Phys. Rev.* B **82** 195436

[55] Artyukhov V I and Chernozatonskii L A 2010 Structure and layer interaction in carbon monofluoride and graphane: A comparative computational study *J. Phys. Chem.* A **114** 5389–96

[56] Antipina L Y and Sorokin P B 2015 Converting chemically functionalized few-layer graphene to diamond films: A computational study *J. Phys. Chem.* C **119** 2828–36

[57] Kvashnin A G, Avramov P V, Kvashnin D G, Chernozatonskii L A and Sorokin P B 2017 Features of electronic, mechanical, and electromechanical properties of fluorinated diamond films of nanometer thickness *J. Phys. Chem.* C **121** 28484–9

[58] Chernozatonskii L A, Demin V A and Kvashnin D G 2020 Ultrawide-bandgap Moiré diamanes based on bigraphenes with the twist angles Θ ~30° *Appl. Phys. Lett.* **117** 25
[59] Chen M, Muniz A R and Maroudas D 2018 Formation and mechanical behavior of nanocomposite superstructures from interlayer bonding in twisted bilayer graphene *ACS Appl. Mater. Interfaces* **10** 28898–908
[60] Liu S, Chen Y and Liu Y 2022 In-plane and interlayer mechanical behaviors of diamane superlattice generated in twisted bilayer graphene *J. Appl. Phys.* **132** 23
[61] Chen M, Weerasinghe A, Muniz A R, Ramasubramaniam A and Maroudas D 2021 Molecular-dynamics simulations on nanoindentation of graphene-diamond composite superstructures in interlayer-bonded twisted bilayer graphene: implications for mechanical metamaterials *ACS Appl. Nano Mater.* **4** 8611–25
[62] Leenaerts O, Partoens B and Peeters F M 2009 Hydrogenation of bilayer graphene and the formation of bilayer graphane from first principles *Phys. Rev.* B **80** 245422
[63] Ke F *et al* 2019 Synthesis and electronic structure characterization of diamane
[64] Piazza F and Monthioux M 2021 Ultra-thin carbon films: the rise of sp^3-C-based 2D materials? *C* **7** 30
[65] Nitrogen-vacancy centers and dopants in ultrathin diamond films: electronic structure *J. Phys. Chem.* C **117** 21376–81
[66] Modifying electronic and elastic properties of 2-dimensional [110] diamond by nitrogen substitution *C* **7** 8 https://doi.org/10.3390/c7010008
[67] Cheng T, Liu Z and Liu Z 2020 High elastic moduli, controllable bandgap and extraordinary carrier mobility in single-layer diamond *J. Mater. Chem.* C **8** 13819–26
[68] Clark C D, Dean P J and Harris P V 1964 Intrinsic edge absorption in diamond *Proc. R. Soc.* A **277** 312–29
[69] Son J *et al* 2021 Tailoring Single- And Double-Sided Fluorination of Bilayer Graphene via Substrate Interactions *Nano Lett.* **21** 891–8
[70] Bianco A, Chen Y, Chen Y, Ghoshal D, Hurt R H, Kim Y A, Koratkar N, Meunier V and Terrones M 2018 A carbon science perspective in 2018: current achievements and future challenges *Carbon* **132** 785–801
[71] Sun Z, Zhang C, Qiu D, Gao N and Li H 2020 Modulated band structures of two-dimensional atomically thick (100) diamond nanofilms with surface functionalization *Diam. Relat. Mater.* **101** 107641
[72] Karlický F and Otyepka M 2013 Band gaps and optical spectra of chlorographene, fluorographene and graphane from G0W0, GW0 and GW calculations on top of PBE and HSE06 orbitals *J. Chem. Theory Comput.* **9** 4155–64
[73] Tantardini C *et al* 2023 Electronic Properties of Functionalized Diamanes for Field-Emission Displays *ACS Appl. Mater. Interfaces* **15** 16317–26
[74] Kharche N and Nayak S K 2011 Quasiparticle band gap engineering of graphene and graphone on hexagonal boron nitride substrate *Nano Lett.* **11** 5274–8
[75] Zhou J, Wang Q, Sun Q, Chen X S, Kawazoe Y and Jena P 2009 Ferromagnetism in semihydrogenated graphene sheet *Nano Lett.* **9** 3867–70
[76] Cruz-Silva E, Botello-Méndez A R, Barnett Z M, Jia X, Dresselhaus M S, Terrones H, Terrones M, Sumpter B G and Meunier V 2010 Controlling edge morphology in graphene layers using electron irradiation: from sharp atomic edges to coalesced layers forming loops *Phys. Rev. Lett.* **105** 045501

[77] Khamkaeo S, Pakornchote T, Ektarawong A, Bovornratanaruks T and Hongo K 2023 First principles study on thermal conductivity of nitrogen substituted diamane *J. Phys. Conf. Ser.* **2431** 012057

[78] Muniz A R, Machado A S and Maroudas D 2015 Mechanical behavior of interlayer-bonded nanostructures obtained from bilayer graphene *Carbon N. Y.* **81** 663–77

[79] Li H, Li J, Wang Z and Zou G 2012 Layer number-dependent structural evolution of two-dimensional diamond films *Chem. Phys. Lett.* **550** 130–3

[80] Nascimento J M and Nunes R W 2013 Lubrication of Stone–Wales transformations in graphene by hydrogen and hydroxyl functional groups *Nanotechnology* **24** 435707

[81] Gao Y *et al* 2017 Ultrahard carbon film from epitaxial two-layer graphene *Nat. Nanotechnol.* **13** 133–8

[82] Sorokin P B and Chernozatonskii L A 2013 Graphene-based semiconductor nanostructures *Phys.-Uspekhi* **56** 105

[83] Ke F *et al* 2020 Synthesis of Atomically Thin Hexagonal Diamond with Compression *Nano Lett.* **20** 5916–21

[84] Piazza F, Monthioux M, Puech P, Gerber I C and Gough K 2021 Progress on diamane and diamanoid thin film pressureless synthesis *C* **7** 9

[85] Tiwari S K, Pandey R, Wang N, Kumar V, Sunday O J, Bystrzejewski M, Zhu Y and Mishra Y K 2022 Progress in diamanes and diamanoids nanosystems for emerging technologies *Adv. Sci.* **9** 2105770

[86] Raeisi M, Mortazavi B, Podryabinkin E V, Shojaei F, Zhuang X and Shapeev A V 2020 High thermal conductivity in semiconducting Janus and non-Janus diamanes *Carbon N. Y.* **167** 51–61

[87] Naima, Tyagi P K and Singh V 2022 N-type diamane: an effective emitter layer in crystalline silicon heterojunction solar cell *Carbon Trends* **9** 100209

[88] Piazza F, Monthioux M and Puech P 2022 2D (<10 nm) sp^3-C-rich carbon materials, possibly hydrogenated: a review *Carbon Trends* **9** 100219

[89] Zhu L, Hu H, Chen Q, Wang S, Wang J and Ding F 2011 Formation and electronic properties of hydrogenated few layer graphene *Nanotechnology* **22** 185202

[90] Muniz A R, Singh T, Aydil E S and Maroudas D 2009 Analysis of diamond nanocrystal formation from multiwalled carbon nanotubes *Phys. Rev.* B **80** 144105

[91] Tiwari S K *et al* 2022 Progress in Diamanes and Diamanoids Nanosystems for Emerging Technologies *Adv. Sci.* **9** 2105770

[92] Artyukh A A and Chernozatonskii L A 2022 Mechanical characteristics of diamond-like Moiré films *JETP Lett.* **116** 737–44

Chapter 3

Synthetic approaches for diamane

Aditya L Shinde and Tapas Ghatak

3.1 Introduction

The contemporary era has witnessed the advent of a pioneering nanomaterial, diamane, featuring a binary crystalline carbon configuration that is covalently interconnected through sp^3 bonding. The remarkable attributes of this substance have elicited noteworthy attention from the scientific fraternity. Numerous experimental methodologies have been devised for the synthesis of diamane, which have been substantiated by computational and theoretical analyzes. The synthesis of this novel sp^3 hybridized carbon nano-allotrope can be achieved through the selective functionalization [1–4], substitution [5], and Birch reduction [6] of a graphene system under meticulously controlled pressure conditions. In 2009, Chernozatonskii *et al* [7] theoretically studied the hydrogenated bilayer graphene (BLG), i.e., diamane. Their investigation employed AB and AA stacking configurations of graphene as precursors for the hydrogenation process [3], resulting in the formation of diamanes. Unfortunately, the actual synthesis of diamane was not carried out until 2018 [8]. During the early-2000s, the principal objective of the comprehensive theoretical inquiry was to assess diamane. The preliminary advancements in diamanes have piqued the curiosity of scholars, prompting investigations into their chemical, mechanical, thermal, and physical characteristics. The year 2020 marked a significant milestone in the scientific community as Bakhrav and colleagues conducted a pioneering investigation into the prospective applications of fluorinated diamonds [1]. Their research culminated in the successful synthesis of a novel structure characterized by interlayer bonding of C–C, which was achieved through the meticulous process of fluorinating BLG [1].

Diamane, also known as BLG or double-layered diamond, is a material with unprecedented thinness and stiffness for its mechanical characteristics. The term ‘diamane’ was derived from the substance ‘graphene’ due to its remarkable resemblance to a diamond-like structure, which is formed by combining a BLG with hydrogen. The synthesis of diamane can be achieved through a multitude of materials, it is, however, noteworthy that graphite as well as graphene are the most commonly

doi:10.1088/978-0-7503-5939-9ch3

utilized precursors. The synthesis of pristine BLG can be achieved through the implementation of both top-down and bottom-up methodologies. Subsequent to this, an array of chemical procedures may be utilized to accomplish the functionalization of the synthesized substance. Throughout the synthesis process, a multitude of challenges must be confronted in order to produce excellent diamanes. These difficulties often involve the careful selection of precursors, substrate effects, the development of wrinkles, the growth kinetics of the materials, etc. The inherent instability of atomically thin diamonds in their pristine and fully free state is often attributed to the non-existence of thermally stable structures in two-dimensional (2D) systems. The attainment of thermal stability is realized through the utilization of a varied spectrum of hydrocarbons, hybridizations, and functional moieties in the fabrication of the diamane framework. In recent years, functional groups such as hydroxyl, fluorine, oxygen, nitrogen, and others have been incorporated to produce diverse hybrid structures of diamanes [1, 9, 10]. In order to refine the customization of diamanes, the incorporation of dopants into graphene is a viable strategy to enhance the stabilization of the hybridization process. In recent years, there has been a predominant focus in scholarly research on the consequences of incorporating N, B, S, and Li atoms into diamond doping [11–13]. The exploration of alternative dopants has emerged as a promising avenue of investigation.

In diamane, the C_2H layered unit is formed by the bonding of two carbon atoms from separate sub-atomic lattices with hydrogen atoms, while other carbon layer forms a covalent bond with the sub-lattices of adjacent carbon atoms in the graphene structure [7]. However, there is no alteration in the two isolated pairs of electrons on the carbon surface. This results in a predominance of vertical (111) stacking of the CCCC atoms until a total of four carbon atoms are packed into a unit cell, which forms diamane [5, 14]. The hexagonal or cubic lattice of the hybridized layer of diamond or lonsdaleite is observed when half of the carbon atoms are hydrogenated and the other half are bonded to adjacent carbon of other carbon layers. AB stacking and AA stacking sequences are required for the formation of the diamanes. Diamane and its isomer, diamane-II, differ significantly in their physical and chemical properties due to the AB or AA stacking sequence of the C_2H unit [7, 15–17].

The nano-allotrope known as diamane displays a multitude of advantageous characteristics, thus positioning it as a highly promising material. The remarkable electrical characteristics, expansive bandgap, adjustable thermal conductivity, and exceptional thermal insulating properties render them highly suitable for a broad spectrum of purposes, encompassing but not restricted to quantum apparatus, photonics, aerospace technologies, nano-electronics, optical waveguides, and thermal regulation mechanisms.

3.2 Synthesis of diamane

3.2.1 Fluorination

The desired product, i.e., diamane, is obtained through the process of fluorination, which involves the reaction of materials with compounds containing fluorine [17].

Bakharev *et al* presented the first experimental proof that fluorination of BLG results in the development of the F-diamane by forming C–C interlayer bonds. A monolayer of fluorinated diamond, also known as 'F-diamane', is synthesized using sheets of (AB)-stacked BLG which are fluorinated by the chemical vapor deposition (CVD) technique on a single-crystal Cu–Ni (111) surface that induces a formation of interlayer C–C bonds. The present study employs a home-made fluorination system to achieve controlled and reproducible fluorination of graphene. This setup enables facile regulation of the partial pressure of XeF_2 vapor and the reaction chamber temperature. The structures of fluorinated graphene with varying C/F ratios were synthesized by changing factors such as fluorination temperature, XeF_2 partial pressure, and exposure time.

Fluorination of graphene films was carried out at a 65°C temperature and a 50–60 Torr of vapor pressure using XeF_2. AR-XPS was utilized to conduct a qualitative study of the arrangement of chemical bonds across various layers, which involved a meticulous analysis of the peak intensities acquired at grazing emission angles of 0° and 50° in relation to both the surface normal and the detector path. The results indicate that measurements obtained at an emission angle of 50° exhibit greater surface sensitivity compared to those obtained at 0°, owing to the variation in photoelectron depths [18]. The three distinct structures observed in fluorinated BLG can be categorized as sample A, sample B, and sample C. Sample A corresponds to a fluorination process lasting for more than 12 h, while sample B is obtained after approximately 6 h of fluorination. Finally, sample C is achieved through a fluorination duration of 2–3 h. Figure 3.1 depicts the detailed C1s and F1s spectra for samples A, B, and C, respectively.

The observed C–F peak at 288.1 eV can be attributed to the presence of C atoms bonded to F atoms. Meanwhile, the C–C bond peaks were detected within the range

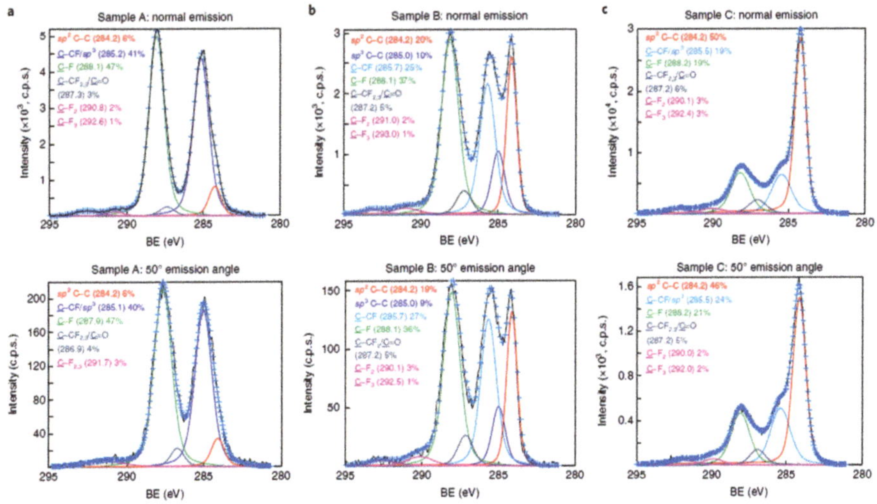

Figure 3.1. AR-XPS spectra of (a) sample A (fluorination time; > 12 h), (b) sample B (fluorination time; ~6 h), and (c) sample C (fluorination time; 2–3 h). Reprinted from [1], Copyright (2020), with permission from Springer Nature.

of 284–286 eV of binding energy [19–25]. The peaks observed at 291 and 293 eV in graphene sheets are ascribed to the C–F_2 and C–F_3 groups, respectively, which are formed at boundary lines and structural imperfections within the material [21]. The carbon atoms in proximity to C–F groups in samples B and C are identified by the peaks observed at 285.5–285.6 eV. Meanwhile, carbon atoms adjacent to C–F_2 or C–F_3 groups are indicated by the peaks observed at 287.2–287.4 eV [20–24]. The high electronegativity of fluorine suggested that the electronic composition of the bare carbon was perturbed in the close proximity of covalent carbon-fluorine as well as C–$F_{2,3}$ bonds, leading to strong shifts of peaks (labeled C–CF and C–$CF_{2,3}$) to a higher binding energy in comparison to carbon–carbon double bonds. Sample A having perfect C_2F stoichiometry was formed when fluorination time of BLG extended to >12 h, which was carried over a single-crystal Cu–Ni (111) surface applying aforementioned conditions.

Figure 3.2 presents the Raman spectra of as-grown BLG on the Cu–Ni (111) along with samples A, B, and C. As-grown BLG exhibits properties similar to the typical AB-stacked BLG, including position, FWHM, the form of the 2D band, and the ratio of peak intensities of G/2D [26–28]. Sample C, with a fluorine concentration of 15%–16%, showed a prominent D band at 1360 cm^{-1} in its Raman spectra, indicating that CVD-grown BLG on the Cu–Ni (111) surface was significantly modified by exposure to XeF_2. When the fluorination period was increased to 6 h (figure 3.2(c)), the D peak became more prominent and shifted to a

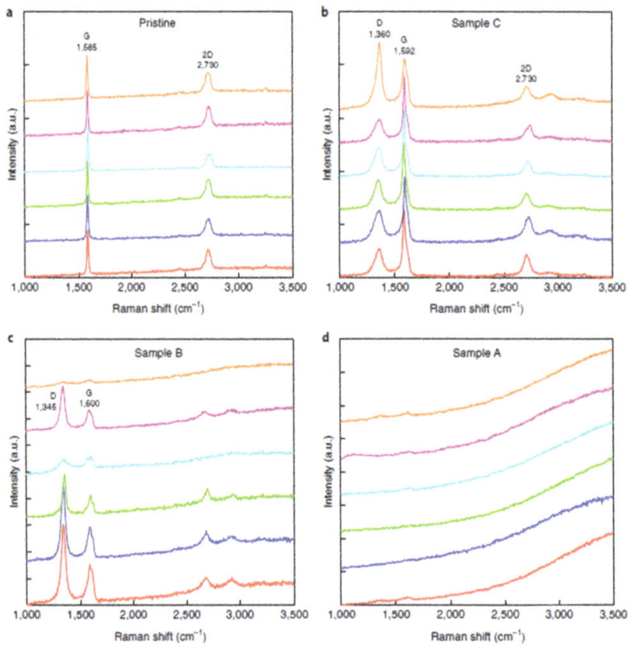

Figure 3.2. Raman spectra at 6 different positions on Cu–Ni (111) (a) as-grown BLG, (b) sample C (fluorination time 2–3 h), (c) sample B (fluorination time ~6 h) and (d) sample A (fluorination time >12 h). Reprinted from [1], Copyright (2020), with permission from Springer Nature.

position around 1350 cm^{-1}, while the 2D/G′ Raman peak was weakened. For sample B, the Raman signature peaks exhibited significant suppression in certain regions, suggesting a high degree of transparency in the fluorinated BLG structure. The Raman spectroscopy of the sample A with a fluorine content between 32% and 33% showed a uniform structure across a broad region with reduced Raman signals, similar to fluorographene [29, 30]. Sample A, which is partially fluorinated BLG, shows a significant change in Raman spectra due to the development of F-diamane, which is attributed to the 'zipping' of highly fluorinated BLG due to fluorine chemisorption.

Son *et al* developed a selective functionalization technique to fabricate single- and double-sided F-BLG through substrate interaction modification. A flake of exfoliated BLG with a flake of 10 nm hBN was placed on a flake of pre-exfoliated hBN. The BLG flake was divided in half, one half resting over the hBN while the remaining half was placed on its base SiO_2, for a direct comparison of the effect of the chemical functionalization of the substrates. When exposed to XeF_2 gas, hBN was rapidly etched, while graphene was only functionalized [31]. Utilizing patterned electrodes, the top hBN layer was initially subjected to etching with XeF_2 gas through the graphene etching stop method [31]. Subsequently, the metals were deposited into the etched areas, as depicted in figure 3.3. Upon completion of the

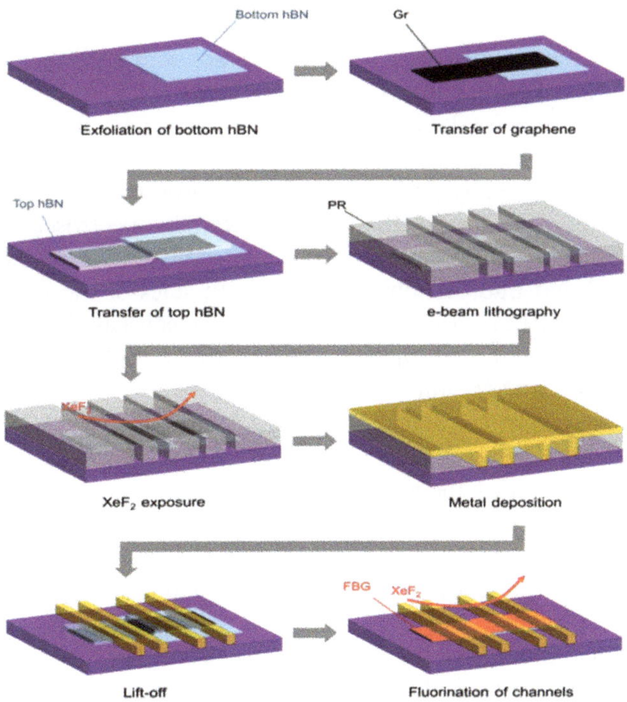

Figure 3.3. A representation of the development of BLG device and optical microscope photographs of the device before and after XeF_2 gas exposure. Reprinted with permission from [4]. Copyright (2020) American Chemical Society.

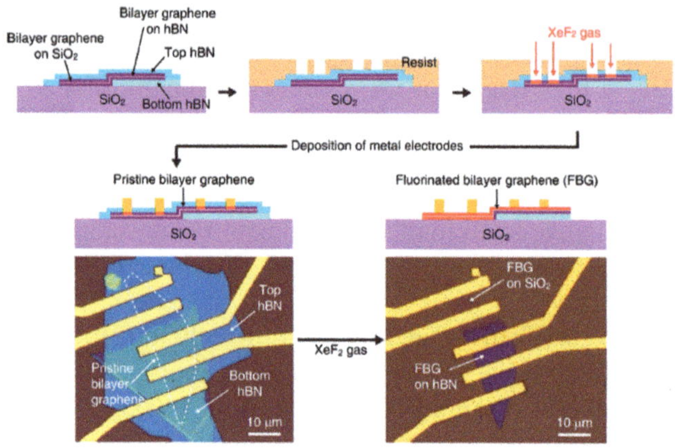

Figure 3.4. A schematic illustration of fabrication process of a BLG device. Reprinted with permission from [4]. Copyright (2020) American Chemical Society.

fabrication process and preliminary measurements, the devices underwent a secondary exposure to XeF_2 gas to eliminate the upper hBN layer, as illustrated in figure 3.4. The fluorination of the BLG was performed after the top hBN layer was entirely removed by etching. The gas impermeability property of graphene allowed the underlying hBN layer beneath the fluorinated graphene to be protected from XeF_2 [31, 32].

The selection of the supporting substrate, which can be either rough SiO_2 or atomically smooth hBN, facilitated the modulation of gas diffusion across the interface between the base layer and the substrate. This, in turn, governs the fluorination of the underlying surface. BLG grown on atomically rough silicon dioxide (SiO_2) and atomically smooth hexagonal boron nitride (hBN) substrates in the presence of XeF_2 gas were evaluated using Raman spectroscopy and electrical transport for the fluorination structure and electronic properties. Raman spectral changes indicated that fluorination occurs on both sides of the BLG on SiO_2, but only one side of the BLG on hBN. Differences in electrical characteristics were a direct result of the structural variations. In contrast to double-sided FBG on SiO_2, single-sided FBG on hBN retains its conductivity.

The substrate-dependent structure of FBG as determined by Raman spectroscopy is presented in figure 3.5. The Raman spectra of BLG over SiO_2 and hBN shows the four stages of change during the process of fluorination and annealing (figures 3.5(a) and (b)). The Raman peak located at 1370 cm^{-1} was observed in hBN across all four stages. The Raman spectra of BLG implanted on hBN acquired immediately after stacking revealed the presence of BLG, with a broad 2D peak consisting of four distinct peaks in the first stage [33]. Stage 2 fluorination took place for time durations of 3 and 9 min, during this process the D peak appeared and became prominent while the 2D peak diminished, which suggests that the top of the hBN was etched, and carbon-fluorine bonds were formed within graphene on both SiO_2 and hBN. In the third step, the fluorination behavior in the BLG was different for each substrate. The 2D peak of BLG on SiO_2 was diminished during 21 min

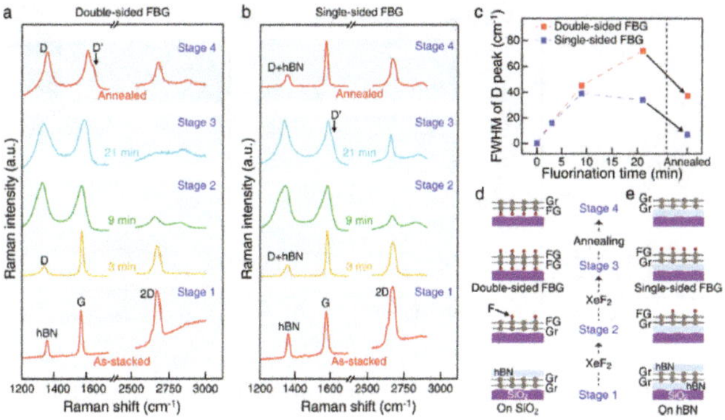

Figure 3.5. Raman spectra of BLG (a) on SiO_2 and (b) on hBN as fluorination time increases. (c) Fluorination-induced D-peak FWHM development for BLG over SiO_2 and hBN. Fluorination evolution of BLG (d) on SiO_2 and (e) on hBN. Reprinted with permission from [4]. Copyright (2020) American Chemical Society.

fluorination. This suppression was evidence for functionalization of both sides of the BLG. Meanwhile, the 2D peak of BLG on hBN remained and its intensity and width were very similar to those of monolayer graphene, which was characterized by a narrow and wide single Lorentzian peak. Following the annealing process, the FWHM of double-sided fluorinated BLG on SiO_2 was found to be 37 cm^{-1}, which was in close proximity to the FWHM of single-sided fluorinated BLG on hBN (34 cm^{-1}) (figure 3.5(c)). The annealing process resulted in the defluorination of the topmost fluorine atoms, leading to the conversion of a double-sided FBG on silicon dioxide to a single-sided FBG. Figures 3.5(c) and (d) depict the development process of the BLG structure via fluorination steps on SiO_2 and hBN substrates, respectively. In the first steps, both bilayers were in their original unaltered states. During stage 2, the process of fluorination resulted in the functionalization of the upper surface of the topmost layer of both bilayers. In stage 3, fluorination occurred on both the lower and upper surfaces of the BLG, which was supported by SiO_2. However, during stage 3, only the top surface of the BLG that was supported by hBN becomes fluorinated. In stage 4, following the annealing process, the F atoms that were on the topmost surface of the BLG on SiO_2 were eliminated, while the F atoms that were confined to the bottom layer were retained.

The comparative analysis of the electrical characteristics of double-sided and single-sided FBG was conducted through the evaluation of the gate-dependant resistivity alteration of BLG field-effect transistors upon fluorination. Figure 3.6(a) indicates that the resistivity of pure BLG over SiO_2 substrate (Step 1) was <1 kΩ μm across the whole gate voltage range. In stage 2, under fluorination, the resistivity increased as the fluorination period advanced, eventually reaching >2 GΩ μm after 21 min of fluorination (Stage 3), with no gate tunability remaining. Figure 3.6(b) revealed that the resistivity and charge neutrality point (CNP) of the pure BLG over hBN substrate (Step 1) were negligible at zero gate voltage because of the absence of charged impurities in the hBN and the pristine state of the heterointerface between the graphene and hBN. Stage 2 involves partially fluorinating the top surface of the BLG,

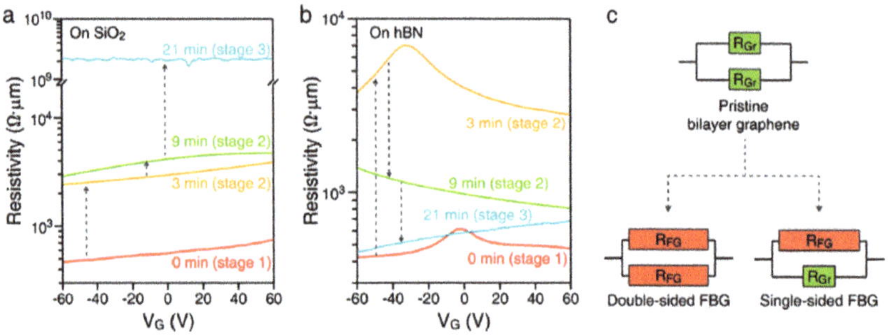

Figure 3.6. The resistivity study of (a) BLG over SiO_2 and (b) BLG over hBN as a function of back-gate bias, with a constant drain-source bias of 10 mV (Vds), as the fluorination time is varied. (c) A diagrammatic illustration of the two-channel model. (RGr = resistance of conducting graphene and RFG = insulating fluorinated graphene.) Reprinted with permission from [4]. Copyright (2020) American Chemical Society.

which results in electrical disturbances and charge localization, with hopping and percolation dominating the conductivity. Step 2 involves partially fluorinating the upper surface of the BLG, which resulted in electrical disorders and charge localization, with hopping and percolation dominating the conductivity. The resistivity of a single-sided FBG over hBN substrate was dropped to level similar to that of the pure BLG after 21 min (Stage 3), which corresponded to the field-effect mobility of 600 cm^2 V^{-1} s^{-1} but with a sigificant positive CNP. Changes in electrical resistivity between the single- and double-sided FBG are depicted in figure 3.6(c). Initially, there was a high vertical resistance connecting two channels in BLG. Due to the simultaneous fluorination, both channels in double-sided FBG developed as insulating. Fluorination of the top layer of single-sided FBG renders it insulating, while the bottom layer, which is left untreated, continues to function as a conductive channel.

Colin *et al* reported that the implementation of pre-exfoliation/fluorination in conjunction with moderate sonication was a promising approach for industrial-scale manufacturing of F-diamane-like nanosheets. In order to synthesize the F-diamane-like nanosheets, either expandable graphite (Exp-able) or thermally expanded graphite (Exp-ed) were used as a precursor. Exp-able was exfoliated using a rapid thermal shock, which was induced by a system under an argon atmosphere to prevent oxidation amid the exfoliation process. The sudden rise in temperature (~10°C s^{-1}) caused the volume to expand at ~650°C [34]. The rapid fluorination was carried out using the static mode, which consist of injecting a predetermined volume of F_2 gas into a NiF_2 passivated closed reactor [35]. By the application of an expanded structure, F atoms are evenly dispersed within the carbon lattice was achieved. This, in turn, facilitated the fabrication of the $(C_2F)_n$ structure through regulated exposure to F_2 gas and temperature of reaction.

Further F-diamane-like nanosheets were prepared by liquid-phase 'top-down' exfoliation where sonication of a suspension of both the F-Exp-able and F-Exp-ed samples in chloroform for 5 min gave stable dispersions of the exfoliated samples [36].

TEM and AFM analysis of exfoliated F-Exp-ed nanosheets revealed their lateral dimensions. (figure 3.7(a)) varied between several hundred nanometers to several

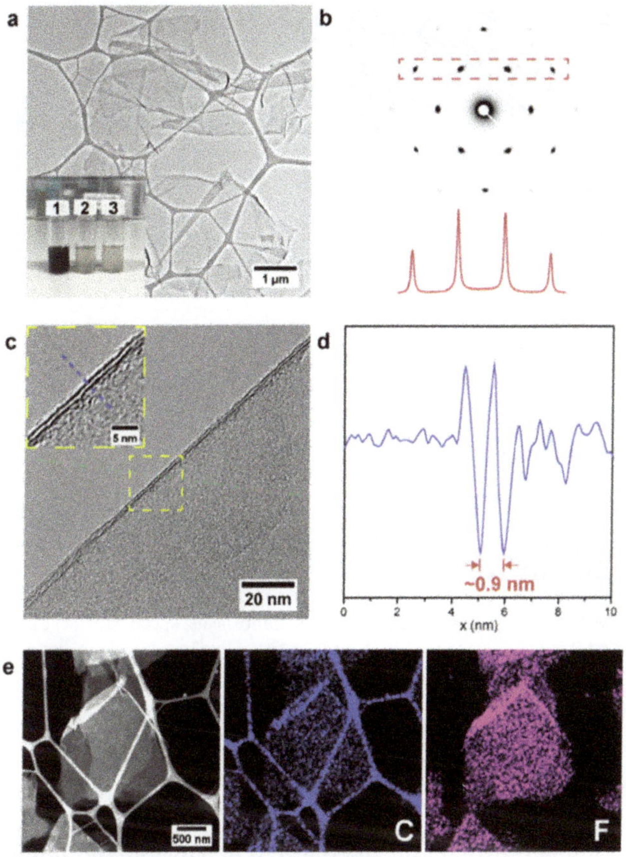

Figure 3.7. (a) TEM photograph of exfoliated F-Exp-ed nanosheets. (b) SAED pattern and intensity profile of exfoliated F-Exp-ed nanosheet. (c) TEM picture of the bent edge of an exfoliated F-Exp-ed nanosheet. (d) Profile plot along line (c) inset, indicating a 0.9 nm interlayer spacing. (e) Dark-field picture and C and F elemental maps of exfoliated F-Exp-ed nanosheets. Reprinted from [2], Copyright (2022), with permission from Elsevier.

micrometers. The nanosheets were characterized using AFM analysis, which indicated that their thickness ranged from 2 nm to predominantly less than 10 nm [36]. Figure 3.7(a) exhibits ultrathin exfoliated F-Exp-ed nanosheets. The sixfold symmetry observed in the SAED pattern (figure 3.7(b)) indicates the hexagonal symmetry. A distinguishing feature of F-diamane was the greater intensity of first-order diffraction spots than second-order spots [1]. The magnified edge of a folded F-Exp-ed nanosheet is shown in figure 3.7(c). A profile plotted along the dashed line shows that the postulated model and observed outcomes for F-diamane are consistent with an interlayer distance of ~9 nm (figure 3.7(d)) [36]. In STEM mode, a HAADF picture and elemental mapping of carbon and fluorine showed a uniform distribution of these elements across the exfoliated nanosheets (figure 3.7(e)). The TEM photographs of exfoliated F-Exp-ed and F-Exp-able nanosheets indicate that their longitudinal nanosheet size was comparable to that

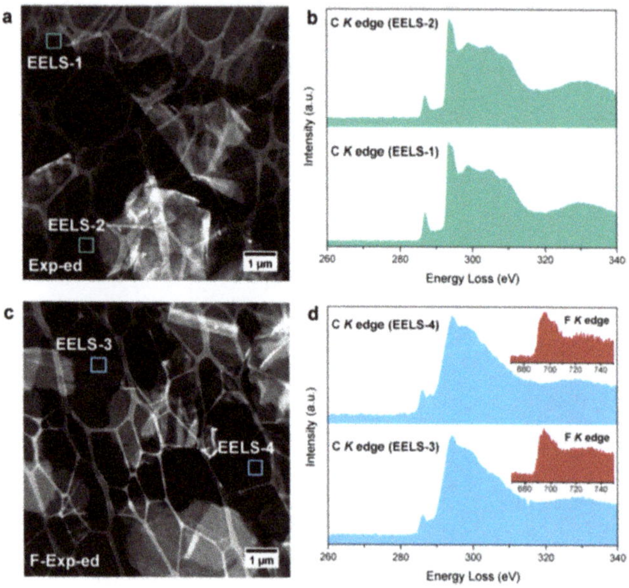

Figure 3.8. The STEM-HAADF pictures of the exfoliated nanosheets, specifically (a) Exp-ed and (c) F-Exp-ed types, respectively. EELS spectra were also obtained from the respective regions indicated in (a) and (c), with (b) and (d) representing the C K edge and F K edge, respectively. Reprinted from [2], Copyright (2022), with permission from Elsevier.

of the exfoliated Exp-ed nanosheet, indicating that fragmentation can be suppressed or reduced with mild sonication. The graphitic structure of the exfoliated Exp-ed nanosheet was retained throughout the thermal expansion and liquid-phase exfoliation, as evidenced by the presence of sixfold symmetric diffraction peaks in the SEAD pattern. The interlayer distance between the exfoliated F-Exp-able nanosheet and the exfoliated F-Exp-ed nanosheet, was ~0.9 nm, both of which correspond to F-diamane-like structures, as shown in the TEM image of the edge of a FLG nanosheet.

Both Exp-ed and F-Exp-ed nanosheets were subjected to EELS, yielding results revealing a significant π^* peak at 286.9 eV and a sharper σ^* peak at 293.6 eV in the energy-loss near-edge structure of the CK edge (260–340 eV) of Exp-ed, with significantly fine edge structures above the σ^* peak. Whereas, EELS of F-Exp-ed, showed the π^* peak (286.1 eV) was weaker and the σ^* peak (294.3 eV) lacks fine structures. As a result of fluorination, the Exp-ed conjugated network undergoes a dramatic rearrangement that was responsible for this shift. As shown in figure 3.8, the fine structure of the FK edge showed that F atoms were there in the structure [37].

3.2.2 Hydrogenation

Hydrogenation often describes a process in which hydrogen reacts with specific components to yield a desired product. Depending on the circumstances, a reaction may be reversible or irreversible, involve adsorption, etc. The phenomenon of hydrogen adsorption induces a modification in the hybridization state of C atoms, sp^2 to sp^3 transition. This alteration then proceeds to propagate across the layers of graphene by establishing interlayer bonds. Piazza *et al* investigated the hydrogenation of few-layer graphene and its transformation into crystalline sp^3 material by monitoring the chemisorption of hydrogen produced by the disintegration of H_2 in a hot-filament reactor operating at lower temperature and pressure. FT-IR microscopy was used to analyze the evolution of carbon-hydrogen bonds, while UV–visible Raman spectroscopy was employed to analyze the structural transformation. The different IR stretching frequencies of C-H bonds are shown in table 3.1.

The FT-IR mapping was carried out for precise identification of the C–H bonding. Defects formed during hydrogenation were identified as C–H bonds and their presence in a broad, circular region with a diameter of 150 mm^2 is shown in figure 3.9(A). This further showed that hydrogenation occurred not just at the domain boundaries of graphene but also in the basal plane. The C(sp^3)–H stretching mode centered at ~2846 cm^{-1} corresponds to the narrow stretching C–H band as shown in figure 3.9(B).

It was found that AB stacking in 2H-on AB stacking is more stable than AA stacking, and that ABC stacking in 2H-on AB stacking is more stable than AAA and ABB stacking, based on the calculation of C–H stretching mode frequencies. Since the IR absorption cross section for the C(sp^3)–H stretching mode is predicted to be much lower than the values for the other modes, as in free molecules, its discovery is a significant milestone. The IR absorption cross section for the C(sp^3)–H

Table 3.1. The stretching frequency of C–H in several different structure. Adapted from [3].

Structure	Frequency (cm^{-1})
Isolated H on graphene	3670
Isolated H on single vacancy in graphene	3210
Graphene	2890
2H-on AB stacking	2880
H-on ABC stacking	2890

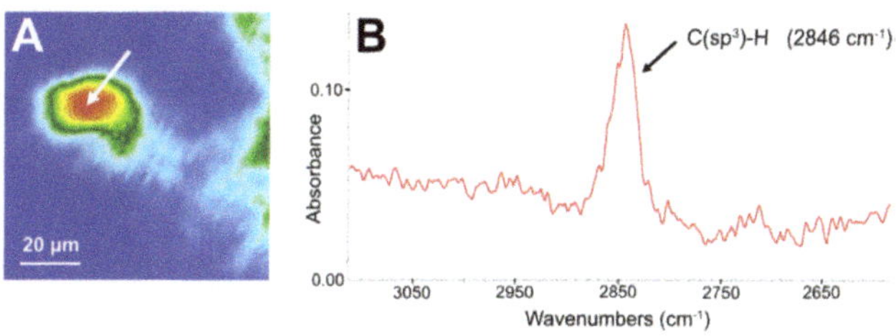

Figure 3.9. (A) FTIR-ATR microscopy pictures of FLG after hot-filament-promoted hydrogenation. (B) Typical absorbance observed in FT-IR spectrum of region (A). Reprinted from [3], Copyright (2019), with permission from Elsevier.

stretching mode is 18.6–31.6 times lower in free molecules than the value for the C(sp^3)–H_3 and C(sp^3)–H_2 modes. The hydrogenation of FLG causes a structural change, which is thought to account for the observed shift of the C(sp^3)–H mode peak, from its position in free molecules to its position by about 54 cm^{-1}.

UV Raman spectra of FLG exposed to H radicals is shown in figures 3.10(B)–(E). Figure 3.10(B) shows a spectrum with two distinct peaks, at ~1063 and ~1323.1 cm^{-1},

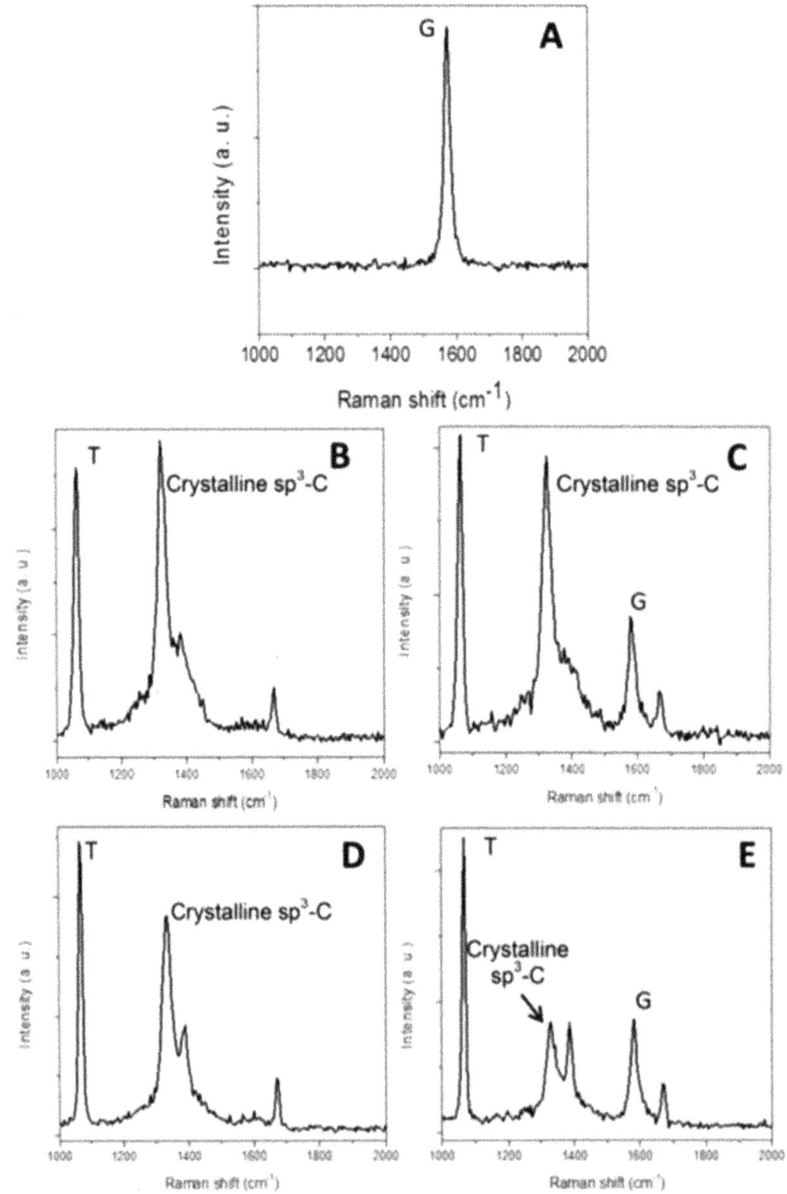

Figure 3.10. Typical UV Raman spectra (A) and (B–E) of FLG at 244 nm before and after hydrogenation, respectively. Reprinted from [3], Copyright (2019), with permission from Elsevier.

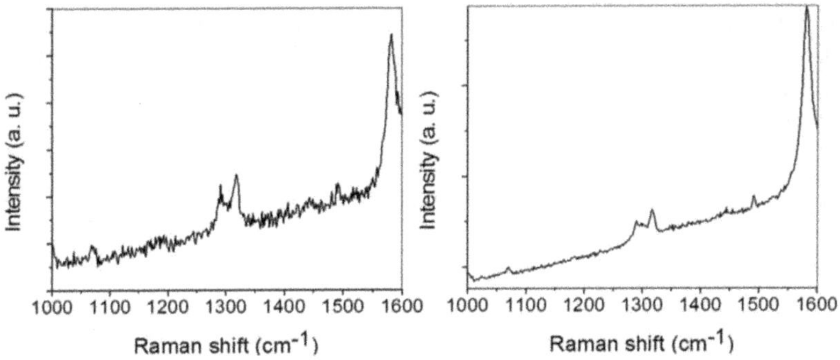

Figure 3.11. Raman spectrum of FLG right after hot filament-promoted hydrogenation at 514.5 nm. Reprinted from [3], Copyright (2019), with permission from Elsevier.

respectively. The FWHM of these two peaks was ~15.6 and ~30.2 cm^{-1}, respectively. The initial peak observed in the given spectra was identified as the T peak [37–39]. This identification was based on the presence of C–C sp^3 vibration, which has been previously reported in ta-C films as a broad band with a low intensity as well as a narrow peak with high intensity in diamond nanoclusters [44].

The second peak located around ~1323.1 cm^{-1} corresponds to the two vibrational modes of lonsdaleite [40–43], which were easy to detect under visible excitation (figure 3.11). These modes are known as the A_{1g} and E_{2g} modes. These modes are associated with the triply-degenerate stretching mode exhibited by the cubic diamond. The vibrational modes in lonsdaleite are categorized in two distinct components, with one component exhibiting in-plane vibrations and the other component exhibiting vibrations perpendicular to the layers [43]. The high relative intensity of the T peak and its small width indicate a C(sp^3) ordered structure in crystalline materials. The bonding strengths of lonsdaleite were similar in both modes and wave number differentiation between A_{1g} and E_{2g} was negligible [43]. The C(sp^3) characteristics observed in the Raman spectra were consistent with the identification of a single component in the FT-IR C–H stretching band, as depicted in figure 3.9.

3.2.3 Birch reduction

Zhang *et al* discovered that graphene hydrogenation can also be done using Birch reduction. The Birch reduction method employed a solution-based approach to generate hydrogenated few-layered graphene, instead of utilizing hydrogen plasma [44]. The SLG and AB-BLG regions were obtained through the process of mechanical exfoliation over SiO_2 (300 nm)/Si, using a graphene flake as the starting material. SiO_2/Si-supported graphenes were subjected to reduction in Li/NH_3 solutions for a duration of 10 min to facilitate Birch-type reactions. During the succeeding phase of protonation, the reduced sheets obtained were quenched by the introduction of water molecule as the source of H^+ ions. As a result of this reaction, the characteristic blue color of the Li/NH_3 mixture disappeared. The application of Birch reduction induces significant structural modifications in both SLG and AB-BLG.

Figure 3.12(a) displays AFM pictures of the material, which showed SLG and BLG regions having darker and lighter contrast. Sharp 2D bands, with a typical $FWHM_{2D}$ of 30 cm^{-1}, appeared in the SLG zones. Figure 3.12(b) shows that across the AB-BLG regions, the 2D bands were uniformly broader ~55 cm^{-1}. Raman analysis was performed on these pristine materials, and the results showed that the positioning of SLG and AB-BLG sections was consistent with the entire spectra at the three points specified in figure 3.12(b) (figure 3.12(c)). Products formed after Birch reduction showed structural alternation in SLG and AB-BLG and exhibit uniform high Raman values of approximately 2 throughout the entire flake. (figure 3.12(d)). The appearance of sp^3-C sites resulted in the breaking of translational symmetry in both the hexagonal lattice and H-SLG and H-BLG. This led to a reduction in the strengths of the 2D bands and a lowering of the A_{2D}/A_G ratio (figure 3.12(e)). Raman spectra were obtained from three distinct locations, as illustrated in figure 3.12(e). The spectra exhibited several similarities, including a prominent and wide D peak at 1322 cm^{-1}, a broad G peak at 1595 cm^{-1}, a diminished 2D peak at 2650 cm^{-1}, and a newly formed weaker D + G band at 2890 cm^{-1} [44]. These results showed that SLG and AB-BLG undergo substantial functionalization under Birch-type reaction conditions in 10 min of reduction time.

Raman spectroscopy can be used to characterize the defect density in graphene for H-SLG and H-BLG by the use of Raman characteristics such as the A_D/A_G and

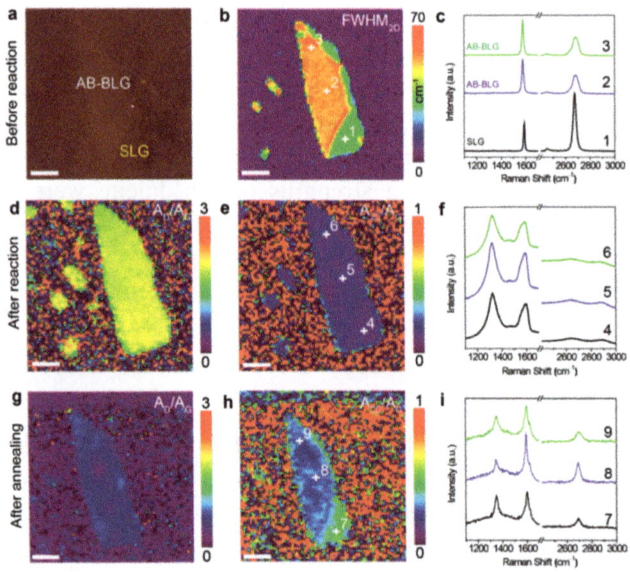

Figure 3.12. (a) An AFM topography image of pristine graphene. (b) Raman spectrum of pristine graphene, showing the 2D peak width $FWHM_{2D}$. (c) Raman spectrum acquired at SLG and AB-BLG locations indicated by figure (b). (d) A Raman spectrum showing (A_D/A_G).and (e) A_{2D}/A_G ratio of Raman map after 10 min of reduction in a Li–NH_3 solution. (f) Raman spectra were obtained at part e locations. (g) A_D/A_G ratio Raman map and (h) Raman spectra showing A_{2D}/A_G ratio after 3 h annealing at 700°C in Ar. (i) Raman spectrum for places highlighted in part (h) (scale bars—3 μm). Reprinted with permission from [6]. Copyright (2016) American Chemical Society.

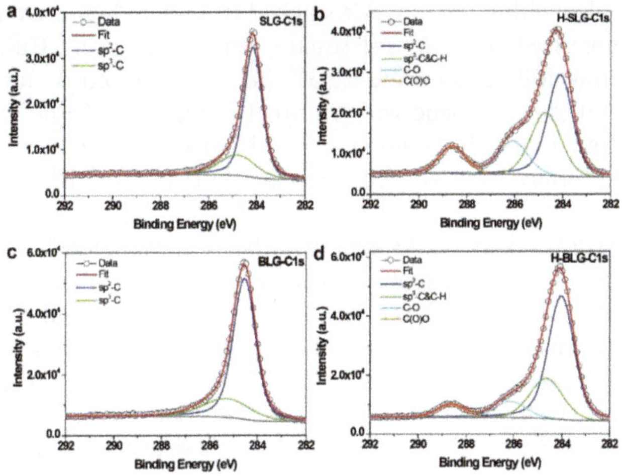

Figure 3.13. XPS C1s spectrum of SLG and BLG on SiO_2/Si grown by CVD technique prior to and following hydrogenation. (a) SLG. (b) H-SLG. (c) BLG. (d) H-BLG. Reprinted with permission from [6]. Copyright (2016) American Chemical Society.

A_{2D}/A_G ratios, the 2D band energies, and the D and 2D peak widths [45–47]. The results indicated that both hydrogenated materials were classified as 'stage II' graphene, and had an mean defect distance (L_D) of 1.5±0.3 nm [45–47]. These 'defects' were identified as sp^3-hybridized carbon atoms bound to hydrogen atoms. Raman analysis showed that H-BLG was more stable at higher temperatures than H-SLG. The A_D/A_G ratios of H-SLG and H-BLG in the Raman map are substantially reduced after annealing at 700°C (figure 3.12(g)), whereas the A_{2D}/A_G ratios exhibit an inverse trend (figure 3.12(h)), which was associated with their dehydrogenation to the respective SLG and BLG starting materials. When H-SLG and H-BLG undergo annealing, the Raman bands (figure 3.12(i)) acquired from different locations (figure 3.12(h)) became narrow, and the 2D peaks shift hypsochromically to 2692 and 2700 cm^{-1}, respectively. Using XPS, the hydrogen coverages of H-SLG and H-BLG were easily estimated to be 30.5% ± 1.5% and 24.1% ± 2.1%, respectively (figure 3.13).

3.2.4 Chemical vapor deposition

CVD is utilized quite frequently in the synthesis process of diamane and its analogs. The present methodology involves the utilization of chemical vapors derived from various sources to synthesize graphene layers, which subsequently lead to the formation of diamane. Diamane was synthesized using a wide variety of templates and catalysts (including carbon monoxide, nickel, iron, and their alloys). Huang *et al* revealed that the CVD of large-area and high-quality AB-BLG and ABA-stacked trilayer graphene sheets requires single-crystal Cu/Ni (111) alloy foils with regulated concentrations of nickel. A series of Cu/Ni (111) alloy foils was fabricated through a thermal treatment of a Cu (111) foil plated with Ni [48]. The quantity of nickel that was electroplated onto the copper was altered to precisely control the percentage of nickel within the foil. AB-stacked BLG with high coverage ⩾95% was successfully synthesized over several square centimetres using Cu/Ni (111) foils. Single-crystal alloy foils comprising 20.3% Ni were utilized to fabricate trilayer ABA-stacked (∼100%) graphene with an areal occupancy of >60% across several square centimetres. ToF-SIMS mapping, HR-TEM and in situ hydrogen etching in an E-SEM, were utilized to examine the frameworks of the BLG and multilayer graphene films.

Figure 3.14(a) displays the SIMS map of carbon following ion beam sputtering for varying periods, with top-down maps making it abundantly evident that the graphene was eliminated almost layer by layer. Even after the first sputtering cycle, the carbon signal remained extremely weak, most likely due to the accumulation of contaminants that settled on the surface sample, while the transfer process and its exposure to air diluted the signal. The topmost layer became visible after 18 s of Cs^+ sputtering and it was eliminated after 36 s of sputtering, while the second layer was eliminated after 54 s of sputtering. Adlayers beneath surface had been revealed as the top layers were peeled away. For the first 45 s of sputtering, a little hexagonal adlayer was not visible. However, after 81 s of sputtering, the adlayer became apparent and was partially etched. This indicated that the larger top layers act as a screen for the smaller adlayer, preventing the adlayer from being etched by the Cs^+ beam unless the topmost layers were eliminated. The adlayers develop underneath the primary monolayer graphene, forming an inverted wedding cake (IWC) structure, as shown in the cross-sectional pictures (figure 3.1(c)) taken from both *x*- and *y*-directions.

Etching a multilayer graphene sheet in a layer-by-layer method gave the impression that the graphene surfaces in a vertical arrangement were not uniformly

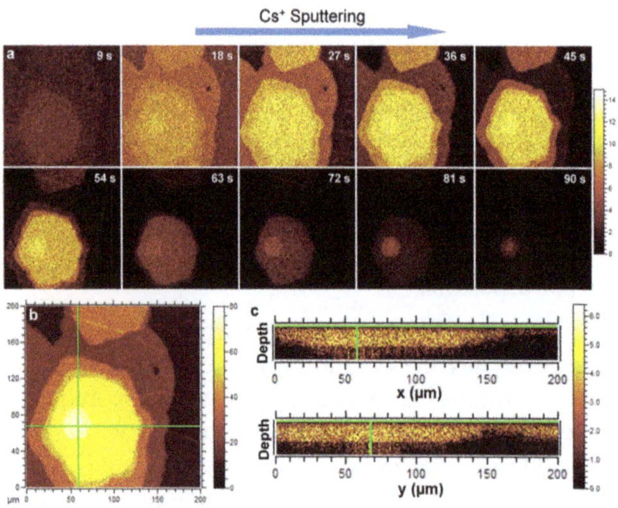

Figure 3.14. Multilayer graphene island mapping by ToF-SIMS on SiO_2/Si substrates. (a) Pictures of graphene's carbon distribution after being sputtered by a Cs+ ion beam at 1 kV for varying periods. (b) A composite image resulting from the overlap of 30 individual images. (c) The *x*–*z* and *y*–*z* cross sections of carbon intensity. Reprinted from [49], Copyright (2020), with permission from Springer Nature.

subjected to reactive H_2 environment. This showed that the IWC stacking arrangement was present (adlayers develop at the graphene-metal connection). Both changes in the areas and the perimeters were consistent with linear and quadratic functions, respectively which was consistent with the first-order kinetics of the detachment-limited etching procedure that was being followed [50]. In multilayer regions 1–3, the etching of the second layers began at approximately 758, 1084, and 1190 s, respectively. The difference in time can be rationalized by pointing out that the distance from the etching front to the edge of each multilayer domain (regions 1, 2, and 3) varies. The average etching rates of 15.5 nm s^{-1}, −4.5 nm s^{-1}, −3.1 nm s^{-1}, and −4.2 nm s^{-1} were calculated for the hole in the monolayer graphene (region 4), 2nd layer in three multilayer domains (regions 1, 2, and 3), respectively, by measuring the slopes of the lines in figure 3.15.

Raman spectra (figure 3.16(a)) without any D peaks confirmed that the graphene produced on Cu/Ni (111) foil (16.6% Ni) was of high quality. The spectra of monolayer graphene display a distinct, sharp, symmetrical, 2D peak that possesses an intensity ratio $I_{2D}/I_G \sim 2$ and a FWHM of ~30 cm^{-1}. The 2D peak for BLG had a broader width (FWHM ~ 51 cm^{-1}), and its I_{2D}/I_G ratio was very close to 1, which was precisely compatible with AB-BLG [51]. Figure 3.16(b) demonstrates that the AB stacking of the newly developed BLG was shown by integrating the asymmetric 2D band comprising four Lorentzian peaks.

The Raman analyzes conducted on both CVD-grown and exfoliated ABA-stacked graphene flakes yielded valuable insights, specifically focused on the 2D band of trilayer graphene, which was identified as ABA through the meticulous

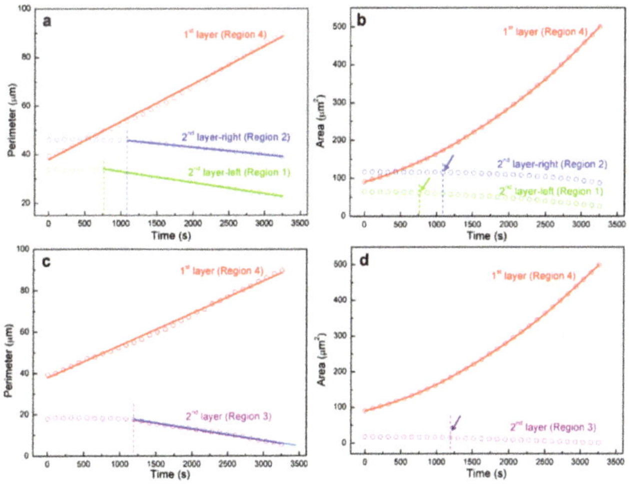

Figure 3.15. H_2-etched variations in graphene layers. (a) Changes of the peripherals of the first layer (monolayer graphene, region 4), the second layers in multilayer regions 1 and 2, and linear fits. (b) Alterations of the first layer in region 4, and second layers in regions 1 and 2. (c) Changes in the peripherals of the first layer (region 4), second layer (region 3), and linear fits. (d) Changes in the region of the first layer (region 4) and second layer (region 3). Reprinted from [49], Copyright (2020), with permission from Springer Nature.

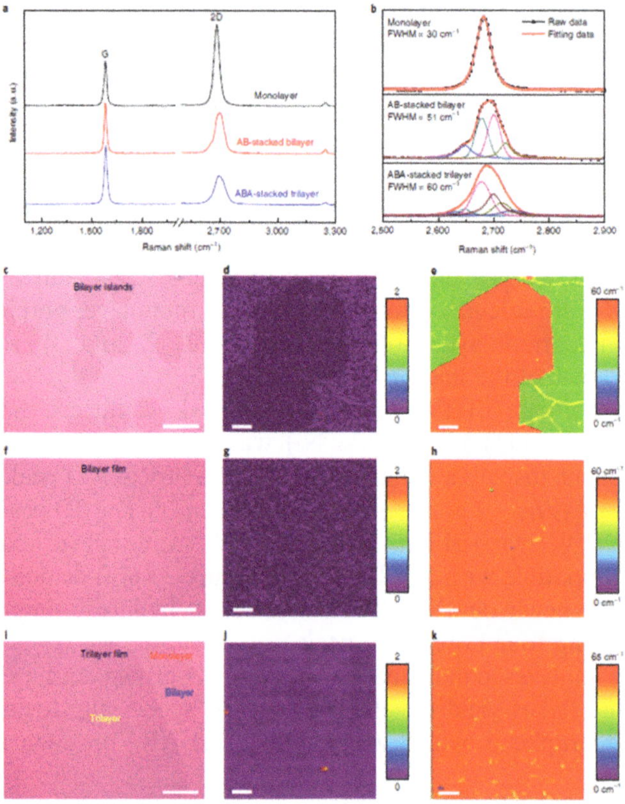

Figure 3.16. (a) The Raman spectrum of monolayer, bilayer and trilayer graphene on SiO_2/Si substrates. (b) Lorentzian function of 2D peaks for monolayer, bilayer, and trilayer graphene films. (c–k) Bilayer islands, (c–e) captured in optical and Raman pictures showing the I_D/I_G ratio and 2D FWHM, respectively. (f–h) A bilayer film and (i–k) a trilayer graphene film, (c, f, i) scale, 50 μm and (d, e, g, h, j, k) scale bars, 10 μm. Reprinted from [49], Copyright (2020), with permission from Springer Nature.

fitting of six Lorentzian peaks (figure 3.16(b)) [33, 52, 53]. Figure 3.16(c) depicts an optical picture of a BLG island that has been transferred. This island was prepared by exposing BLG for 5 min at a temperature of 1075°C (16.6% nickel) to a flow of gas of 200 standard cubic centimeters per minute (sccm) argon, 10 sccm hydrogen, and 1 sccm methane. The 2D FWHM plot clearly demonstrated that the bilayer films were AB-stacked as well as single crystal. A remarkably continuous BLG film developed when the growing duration was increased to 10 min (figure 3.16(f)). Raman plots of the I_D/I_G ratio and the 2D FWHM for the uniform bilayer film (figures 3.16(g) and (h)) exhibited consistent contrast across the entire area. The Raman mapping and optical microscopy results for the trilayer graphene film are displayed in figures 3.16(i)–(k), respectively. An absence of D peaks and consistent contrast indicated the high-quality and uniform ABA-stacking (2D FWHM ~ 60 cm^{-1}) throughout the entire area. A BLG was suspended above a hole in an Au-coated SiN_x support, and its thermal conductivity was measured as shown in table 3.2. For the five

Table 3.2. Thermal conductivity of five different BLG. Adapted from [49].

Sample	Thermal conductivity (W m^{-1} K^{-1})
BLG_1	2298
BLG_2	1933
BLG_3	1954
BLG_4	2222
BLG_5	1975

BLG samples, the calculated thermal conductivities are 2298, 1933, 1954, 2222, and 1975 W m^{-1} K^{-1}. The thermal conductivity of synthesized BLG was significantly higher ~2300 W m^{-1} K^{-1}. In comparison, CVD-grown twisted BLG has thermal conductivities ranging from 657 to 1413 W m^{-1} K^{-1}, while AB-stacked BLG exhibits a thermal conductivity of 1896 W m^{-1} K^{-1}. The thermal conductivity value of the synthesized AB-stacked BLG was almost similar to that observed in an exfoliated BLG flake (~2800 W m^{-1} K^{-1}). The synthesized AB-stacked BLG sheet was found to be of excellent quality and homogeneity.

3.2.5 Use of pressure

As an alternative to the conventional CVD method, diamane has been synthesized using pressure. The fact that this method only needs barely any pressure for stabilization post-synthesis has attracted a lot of attention, but is rarely used. Ke *et al* used the dimondization approach to synthesize the diamane film from a few mechanically exofoliated layers of graphene via compression [54]. They reported the first room-temperature synthesis of pure h-diamane via the pressure-induced sp^2 to sp^3 dimondization transformation of mechanically exfoliated graphene of varying thicknesses from 12 layers to bilayer. The transition pressure was layer-number dependent; in other words, at higher pressures, the transition takes place in thinner samples. The electronic structure of mechanically exfoliated few-layer graphene was monitored to observe its diamondization through the use of electrical transport and optical absorption measurements. The resistance of a sheet of 12-layer graphene exhibits a gradual change until a starting point just above 19.6 GPa, which was followed by a more than three orders of rise in resistance with further compression to 50 GPa. The minimum pressure for a transformation in hexa-, tetra-, and trilayer graphene was ~21.3, ~27.1, and ~34.0 GPa, respectively. However, in the case of BLG, the sheet resistance does not change much, even when subjected to compression up to 60 GPa. Since BLG behaves differently than thicker graphene, this indicated the intrinsic nature of the pressure-induced development of few-layer graphene. The phenomenal increase in resistance of 12-, hexa-, tetra-, and trilayer graphene under uniform pressure can be attributed to the sp^2–sp^3 diamondization transformation that is induced by the pressure. In order to initiate the process of diamondization in BLG, it was necessary to apply higher pressures. Figure 3.17 shows resistance–temperature (*R*–*T*) curves for mechanically exfoliated graphene

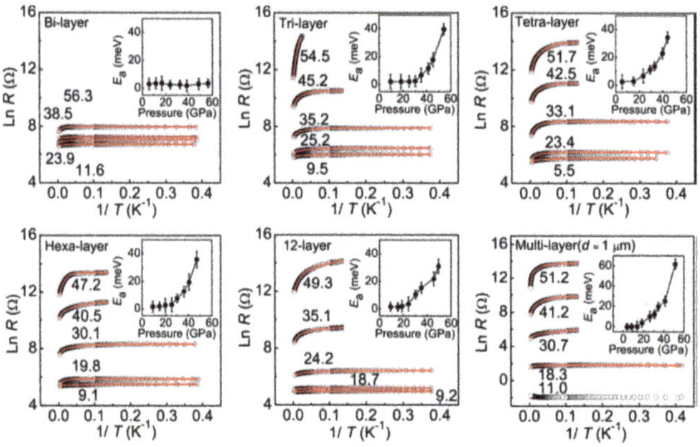

Figure 3.17. Temperature reliance of the sheet resistance of bi-, tri-, tetra-, hexa-, 12-, and multi-layer graphene at specific pressure conditions. The insets represents the extracted activation energies as a function of pressure. Reprinted with permission from [54]. Copyright (2020) American Chemical Society.

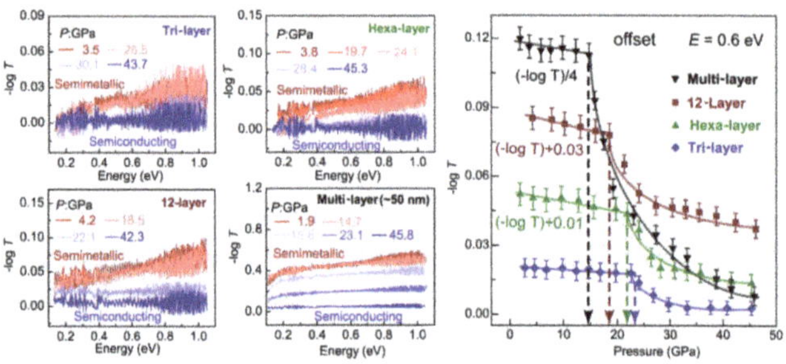

Figure 3.18. IR absorption spectrum of graphene in various layer configurations, including trilayer, hexa-layer, 12-layer, and multilayer graphene under compression (four left-hand panels) and absorbance as a function of pressure at a photon energy of ~0.6 eV (right-hand panel). Reprinted with permission from [54]. Copyright (2020) American Chemical Society.

with layer thicknesses varying from 12 to bilayer, lending more credence to the idea that the sp^2–sp^3 diamondization transition is layer dependent [54].

The IR (0.2–1.0 eV) and the UV–visible (1.4–4.9 eV) absorption spectrums confirmed the pressure-induced sp^2–sp^3 diamondization in mechanically exfoliated trilayer and thicker graphene (figure 3.18). The absorption spectra for every graphene sample exhibited a distinctly strong and asymmetrical peak located at 4.6 eV prior to the transition. (figure 3.19(b)) [55]. In graphene, this peak represents the excitonic resonance of the interband transition. Prior to diamondization, both the peak and absorption spectra exhibited pressure insensitivity. Following the transition, a sudden decrease in absorbance is observed in all samples within the energy ranges of 0.2–1.0 eV (figure 3.18) and 1.4–2.5 eV (figure 3.19(b)), with a gradual approach towards zero as compression continues. Figure 3.19(c) demonstrates that the diamondization pressures for tri-, hexa-, and 12-layer graphene were ~28.1, ~20.6, and ~19.4 GPa, respectively. These pressures were following the

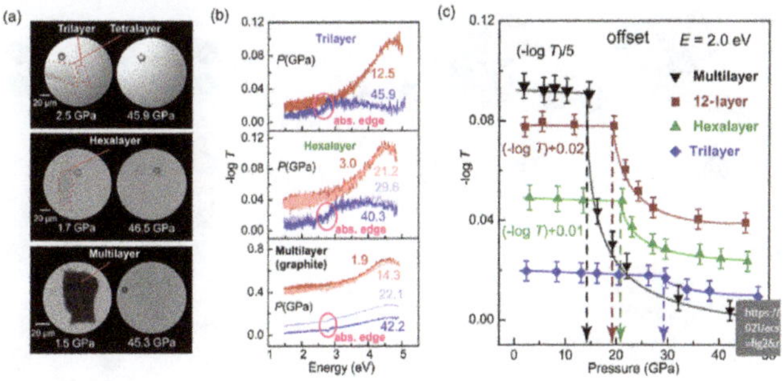

Figure 3.19. (a) Optical microscopy pictures of compressed graphene samples tri-, tetra-, hexa-, and multilayer configurations. (b) Pressure-dependent absorbance of tri-, hexa-, and multilayer graphene. (c) The absorbance of tri-, hexa-, 12-, and multilayer graphene at 2.0 eV as a function of pressure. Reprinted with permission from [54]. Copyright (2020) American Chemical Society.

results obtained from the electrical transport. Electrical transport and optical absorption investigations revealed that mechanically exfoliated graphene of varying layer thickness (three, four, six, 12, or more) undergoes a systematic sp^2 to sp^3 diamondization transformation upon compression.

3.2.6 Substitution by nitrogen

Pakornchote *et al* suggested that passivation by atoms and molecules can be used for the formation of stable diamane and successfully substituted N atoms, which possess one extra valence electron than a carbon atom. This extra electron tended to form a lone pair of electron and lowers the potential energy of the surface. N substitution in diamane, rather than the hydrogenation or fluorination processes, resulted in a 2D carbon nitride that has the potential to adopt the super hard feature that was originally found in bulk carbon nitrides. Substituting an N atom for a surface-level C atom forms a lone electron pair, which had the potential to stabilize the diamane-like structure. Four N-substituted diamane structures ($C_{4-x}N_x$) were derived from the diamane unit cell with $x = 1$ and 2: NCCC, CNCC, NCCN, and CNCN. After structural relaxation, a vertically-stacked pair of diamane as NCCC and CNCC with an extra N atom (C_3N) undergoes conversion and returns to a layer of graphene and CN [56]. Thus, in NCCC, the formation of lone pairs of electrons on one of the sides of the N substitution was inadequate to achieve a local minimum in the energy of the system and to stabilize sp^3 bonding between layers. In order to minimize the potential energy of the surface resulting from unpaired electrons present on the surface of graphene, a rectangular cell (figure 3.20) was used to reconstruct a graphene side into the Pandey chain [57]. After structure relaxation, the rebuilt NCCC ($-C_3N$) did not undergo a transformation back to graphene layers and CN (figure 3.20).

Two different vertically-stacked N-substituted diamane structures, NCCN and CNCN, were investigated for C_2N_2 (figures 3.21(a) and (c)). After structural relaxation, both NCCN and CNCN were energetically stable. But only NCCN was dynamically stable [5]. The collinear spin-polarized technique and the HSE06

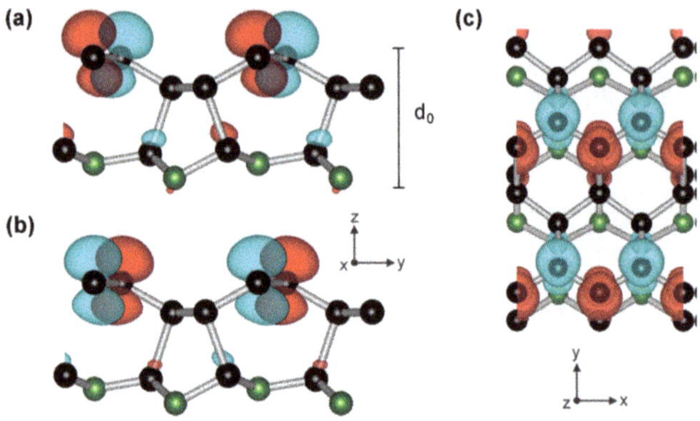

Figure 3.20. The relaxed form of π-C_3N in directions (a) [100] and (d) [001], where the valence-band maximum appears at the K point, and the +ve and −ve magnetization densities are represented by the blue and red envelopes, respectively, and (b) at the J point, the minimum of conduction band arises. C and N atoms are represented as black and green balls, respectively. Reprinted with permission from [5], Copyright (2020) by the American Physical Society.

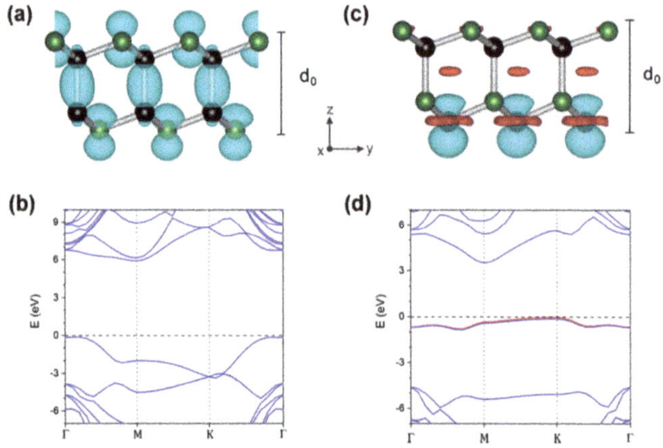

Figure 3.21. (a) Relaxed structure of NCCN shown via projecting in the [100] direction, where blue bands represent the spin-up charge density. (c) The relaxed structure of CNCN shown via projecting in the [100] direction, where both the +ve and −ve magnetization densities of the valence-band maxima at the K point are represented by blue and red bands, respectively. (b) Band structure of NCCN and (d) band structure of CNCN. C and N atoms are represented by black and green balls. Reprinted with permission from [5], Copyright (2020) by the American Physical Society.

functional were implemented to determine the magnetism of NCCN and CNCN. CNCN did not possess magnetism, like π-C_3N, and had a negligible magnetization of 0.007 μ_B. For CNCN, the magnetism was primarily due to the C atoms. The magnetizations of the topmost C atoms had a value of 0.008 μ_B, whereas those of the innermost C atoms were roughly 0.001 μ_B. In the case of NCCN, the absence of magnetism resulted in dispersion in the electronic band that does not break into spin up and spins down (figure 3.21(b)).

Figure 3.22(a) shows that the p_z orbital of both the carbon and nitrogen atoms is the primary source of the valence states. The observed significant density of states close to the Fermi level in CNCN happened because its valence-band maximum was virtually flat (figure 3.22(d)) and the observed phenomenon of spin-up and spin-down splitting was exclusively exhibited by this band. Figure 3.22(b) shows that the valence states of the surface C atom originate mostly from the s and p_z orbitals. A significant contribution is made by the surface p_z orbital of the nitrogen atom to

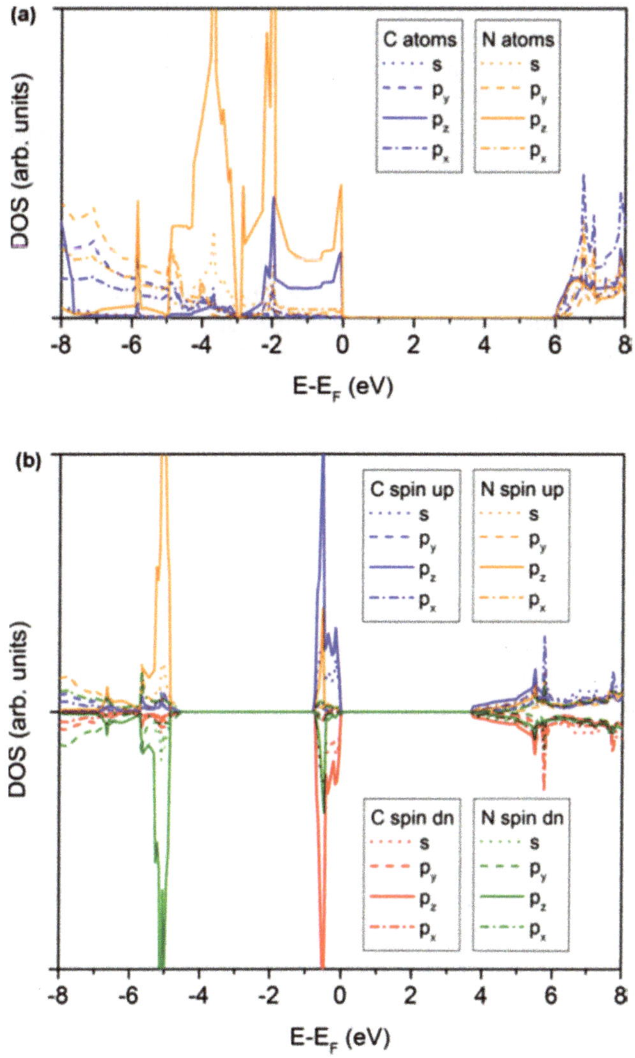

Figure 3.22. Electronic partial DOS of (a) NCCN and (b) CNCN is colored for the spin-up C atom (blue) and N atom (orange) and for the spin-down C and N atoms (red and green). The proposed orbitals for s, p_y, p_z, and p_x orbitals are dotted, dashed, solid, and dashed–dotted, respectively. NCCN has identical spin-up and -down DOS, so only spin-up DOS is shown. Reprinted with permission from [5], Copyright (2020) by the American Physical Society.

the observed feature, which was located in between −4 and −6 eV underneath the Fermi level. The surface potential energy is lower due to the lower energy contribution of the nitrogen atom as compared to a carbon atom. Based on HSE06, the energy gaps of NCCN and CNCN were found to be 6.1 and 4.4 eV, respectively.

References

[1] Bakharev P V *et al* 2020 Chemically induced transformation of chemical vapour deposition grown bilayer graphene into fluorinated single-layer diamond *Nat. Nanotechnol.* **15** 59–66

[2] Colin M *et al* 2022 F-diamane-like nanosheets from expanded fluorinated graphite *Appl. Surf. Sci.* **583** 152534

[3] Piazza F *et al* 2019 Low temperature, pressureless sp^2 to sp^3 transformation of ultrathin, crystalline carbon films *Carbon* **145** 10–22

[4] Son J *et al* 2020 Tailoring single-and double-sided fluorination of bilayer graphene via substrate interactions *Nano Lett.* **21** 891–8

[5] Pakornchote T *et al* 2020 Roles of nitrogen substitution and surface reconstruction in stabilizing nonpassivated single-layer diamond *Phys. Rev.* B **102** 075418

[6] Zhang X *et al* 2016 Birch-type hydrogenation of few-layer graphenes: products and mechanistic implications *J. Am. Chem. Soc.* **138** 14980–6

[7] Chernozatonskii L A *et al* 2009 Diamond-like C 2 H nanolayer, diamane: simulation of the structure and properties *JETP Lett.* **90** 134–8

[8] Piazza F *et al* 2021 Progress on diamane and diamanoid thin film pressureless synthesis *C* **7** 9

[9] Chernozatonskii L A *et al* 2011 Influence of size effect on the electronic and elastic properties of diamond films with nanometer thickness *J. Phys. Chem.* C **115** 132–6

[10] Ke F *et al* 2019 Synthesis and electronic structure characterization of diamane *arXiv preprint* arXiv:1902.01570

[11] Fyta M 2013 Nitrogen-vacancy centers and dopants in ultrathin diamond films: electronic structure *J. Phys. Chem.* C **117** 21376–81

[12] Pakornchote T *et al* 2021 Modifying electronic and elastic properties of 2-dimensional [110] diamond by nitrogen substitution *C* **7** 8

[13] Li J *et al* 2019 First-principles calculations for Li, P dopants and vacancy defect in ultra-thin hydrogenated diamond nanofilms: structural, electronic and optical properties *Diam. Relat. Mater.* **99** 107526

[14] Chernozatonskii L A *et al* 2022 Diamane quasicrystals *Appl. Surf. Sci.* **572** 151362

[15] Muniz A R and Maroudas D 2012 Opening and tuning of band gap by the formation of diamond superlattices in twisted bilayer graphene *Phys. Rev.* B **86** 075404

[16] Chernozatonskii L A, Demin V A and Kvashnin D G 2020 Ultrawide-bandgap Moiré diamanes based on bigraphenes with the twist angles Θ~ 30° *Appl. Phys. Lett.* **117** 253104-1–5

[17] Qin G, Wu L and Gou H 2022 Diamane: design, synthesis, properties, and challenges *Funct. Diam.* **1** 83–92

[18] Hofmann S 1993 Approaching the limits of high resolution depth profiling *Appl. Surf. Sci.* **70** 9–19

[19] Watanabe N 1981 Characteristics and applications of graphite fluoride *Phys. B*+ C **105** 17–21

[20] Asanov I *et al* 1998 X-ray photoelectron study of fluorinated graphite intercalation compounds *J. Struct. Chem.* **39** 928–32

[21] An K H *et al* 2002 X-ray photoemission spectroscopy study of fluorinated single-walled carbon nanotubes *Appl. Phys. Lett.* **80** 4235–7
[22] Sato Y *et al* 2004 On the so-called 'semi-ionic' C–F bond character in fluorine–GIC *Carbon* **42** 3243–9
[23] Okotrub A V *et al* 2013 Anisotropy of chemical bonding in semifluorinated graphite C2F revealed with angle-resolved x-ray absorption spectroscopy *ACS Nano* **7** 65–74
[24] Grayfer E D *et al* 2013 Synthesis, properties, and dispersion of few-layer graphene fluoride *Chem.–Asian* J **8** 2015–22
[25] Claves D 2011 Spectroscopic study of fluorinated carbon nanostructures *New J. Chem.* **35** 2477–82
[26] Wu Y *et al* 2012 Growth mechanism and controlled synthesis of AB-stacked bilayer graphene on Cu–Ni alloy foils *ACS Nano* **6** 7731–8
[27] Liu L *et al* 2012 High-yield chemical vapor deposition growth of high-quality large-area AB-stacked bilayer graphene *ACS Nano* **6** 8241–9
[28] Luo Z *et al* 2011 Large-scale synthesis of bi-layer graphene in strongly coupled stacking order *Adv. Funct. Mater.* **21** 911–7
[29] Nair R R *et al* 2010 Fluorographene: a two-dimensional counterpart of Teflon *Small* **6** 2877–84
[30] Robinson J T *et al* 2010 Properties of fluorinated graphene films *Nano Lett.* **10** 3001–5
[31] Son J *et al* 2018 Atomically precise graphene etch stops for three dimensional integrated systems from two dimensional material heterostructures *Nat. Commun.* **9** 3988
[32] Bunch J S *et al* 2008 Impermeable atomic membranes from graphene sheets *Nano Lett.* **8** 2458–62
[33] Ferrari A C *et al* 2006 Raman spectrum of graphene and graphene layers *Phys. Rev. Lett.* **97** 187401
[34] Herraiz M *et al* 2019 Exfoliated fluorinated carbons with a low and stable friction coefficient *RSC Adv.* **9** 13615–22
[35] Ahmad Y *et al* 2013 Enhanced performances in primary lithium batteries of fluorinated carbon nanofibers through static fluorination *Electrochim. Acta* **114** 142–51
[36] Chen X *et al* 2021 Liquid-phase exfoliation of F-diamane-like nanosheets *Carbon* **175** 124–30
[37] Gilkes K *et al* 2000 Direct quantitative detection of the sp^3 bonding in diamond-like carbon films using ultraviolet and visible Raman spectroscopy *J. Appl. Phys.* **87** 7283–9
[38] Ferrari A C and Robertson J 2001 Resonant Raman spectroscopy of disordered, amorphous, and diamondlike carbon *Phys. Rev.* B **64** 075414
[39] Prawer S *et al* 2000 The Raman spectrum of nanocrystalline diamond *Chem. Phys. Lett.* **332** 93–7
[40] Chernozatonskii L A, Mavrin B N and Sorokin P B 2012 Determination of ultrathin diamond films by Raman spectroscopy *Phys. Status Solidi (b)* **249** 1550–4
[41] Smith D C and Godard G 2009 UV and VIS Raman spectra of natural lonsdaleites: towards a recognised standard *Spectrochim. Acta* A **73** 428–35
[42] Gogotsi Y G, Kailer A and Nickel K G 1998 Pressure-induced phase transformations in diamond *J. Appl. Phys.* **84** 1299–304
[43] Knight D S and White W B 1989 Characterization of diamond films by Raman spectroscopy *J. Mater. Res.* **4** 385–93
[44] Luo Z *et al* 2011 Electronic structures and structural evolution of hydrogenated graphene probed by Raman spectroscopy *J. Phys. Chem.* C **115** 1422–7
[45] Ferreira E M *et al* 2010 Evolution of the Raman spectra from single-, few-, and many-layer graphene with increasing disorder *Phys. Rev.* B **82** 125429

[46] Cançado L G *et al* 2011 Quantifying defects in graphene via Raman spectroscopy at different excitation energies *Nano Lett.* **11** 3190–6
[47] Lucchese M M *et al* 2010 Quantifying ion-induced defects and Raman relaxation length in graphene *Carbon* **48** 1592–7
[48] Huang M *et al* 2018 Highly oriented monolayer graphene grown on a Cu/Ni (111) alloy foil *Acs Nano* **12** 6117–27
[49] Huang M, Bakharev P V, Wang Z J *et al* 2020 Large-area single-crystal AB-bilayer and ABA-trilayer graphene grown on a Cu/Ni(111) foil *Nat. Nanotechnol.* **15** 289–95
[50] McLean J G *et al* 1997 Decay of isolated surface features driven by the Gibbs–Thomson effect in an analytic model and a simulation *Phys. Rev.* B **55** 1811
[51] Lee S, Lee K and Zhong Z 2010 Wafer scale homogeneous bilayer graphene films by chemical vapor deposition *Nano Lett.* **10** 4702–7
[52] Graf D *et al* 2007 Spatially resolved Raman spectroscopy of single-and few-layer graphene *Nano Lett.* **7** 238–42
[53] Gupta A *et al* 2006 Raman scattering from high-frequency phonons in supported n-graphene layer films *Nano Lett.* **6** 2667–73
[54] Ke F *et al* 2020 Synthesis of atomically thin hexagonal diamond with compression *Nano Lett.* **20** 5916–21
[55] Mak K F, Shan J and Heinz T F 2011 Seeing many-body effects in single-and few-layer graphene: observation of two-dimensional saddle-point excitons *Phys. Rev. Lett.* **106** 046401
[56] Pakornchote T *et al* 2019 Phase stabilities and vibrational analysis of hydrogenated diamondized bilayer graphenes: a first principles investigation *Carbon* **146** 468–75
[57] Pandey K 1982 New dimerized-chain model for the reconstruction of the diamond (111)-(2 × 1) surface *Phys. Rev.* B **25** 4338

Chapter 4

Role of temperature and pressure in the fabrication of diamane and its derivative

Radha Gupta, Chandan Kumar Maity, Narendhar Chandrasekar, Ramachandran Balaji and Myung Jong Kim

New materials are the foundation of today's technologically advanced society, and at the moment carbon nanostructures are the top contenders, which have sparked intense research. The latest iterations of sp^3 hybridized carbon known as diamanes and diamanoids can be produced utilizing graphitic system as a precursor and Birch reduction under desired pressure, substitution, and suitable functionalization. The characteristics of diamane are now largely researched theoretically, and its synthesis is the focus of the most experimental work. Utilizing surface chemical adsorption or high-pressure technology, diamane, the two-dimensional (2D) analog of diamond, is produced from few-layer graphene (FLG) or bilayer graphene (BLG). Therefore, the role of temperature and pressure of the synthesis of diamanes is very important. In this chapter we will discuss the different roles of temperatures and pressure for the fabrications of diamanes and its derivatives. Herein, diamanes are also thoroughly discussed in terms of their different derivatives, historical outlook, structural design, and thermodynamics.

4.1 Introduction

The two new nanomaterials that were added to the family of carbon, diamanoids (composed of diamane and graphene) and diamane (sp^3 bonded two crystalline carbon layers), have attracted the interest of scientists. Diamane is a semiconducting material with a straight wide bandgap that is highly desirable for active laser medium in nano-optics, bandgap engineering, and nanoelectronics, according to simulations [1]. Diamane can be employed in thermal management equipment owing to its anticipated superior thermal conductivity [2]. Calculations have revealed that diamane is a superior host system for single photon emission than diamond due to the higher radiative electronic transition rate and lower effective

doi:10.1088/978-0-7503-5939-9ch4

Bohr radius of defects [3]. Diamanes also exhibit excellent mechanical, vibrational, optical, thermal, and electrical properties, based on which quantum technologies can be further developed. Properties such as excellent insulation, manipulatable thermal conductivity, wide bandgap, etc make diamanes perfect contenders for optical waveguides, resonators, and nanoelectrical devices. Diamane has a unique structure, which influences its properties and stability limitations. Diamanes can be synthesized by Birch reduction under programmed conditions, such as temperature and pressure, using a graphitic material as a precursor [4]. The stability is mainly due to graphitization by virtue of energy preference of graphite as compared to diamond. At the same time, in bulk a transformation barrier forces the diamond to be metastable and to exist over centuries. In addition, as the dimensions reduce, the surface effects become more prominent. Without the passivating H or F, phase equilibrium shifts to a higher graphene–diamane phase transition pressure or complete absence of the stability regions. Based on the findings of computational and theoretical investigations, a number of experimental approaches have been found for the synthesis of diamanoids and diamane [5]. In light of this, diamanoids and diamane are those intriguing nanocarbons that may represent a brand-new class of sensational materials comparable to graphene, metal dichalcogenides, MXene, and single-layer hexagonal boron nitride (h-BN).

Recently, MXene and h-BN have been widely used in different applications along with different carbon materials [6–11]. However, looking back, nanocarbons play a significant role as the precursor of diamanoids and diamane. It is difficult to envisage the present commercial facets of the textile industries, automobile, defense, space, pharma-biomedical, optical fiber, mobile phone, and internet without carbon materials [4, 12–14]. Carbon materials such as reduced graphene oxide, graphene, graphene oxide, and CNT are also being widely used for electrochemical energy storage devices [15–18]. By 2040, supercomputers and the next generation of communication devices, among other high-tech products, will largely be made of carbon-based materials instead of non-carbonaceous semiconductors, according to Buckley *et al* [19]. In the transition from CNTs to graphene, Andre Geim *et al* effectively created graphene, a single sheet of carbon atoms organized in a 2D honeycomb lattice [20, 21]. Different functional groups, such as oxygen, fluorine, and hydrogen, can chemically adsorb on the single-layer graphene (SLG) surface, changing the chemical bonding and widening the bandgap [22–27]. Additionally, research has shown that FLG or BLG can create a single C–C connection across layers through adsorption [28]. Graphite and diamond have a 0.03 eV/atom free energy difference in bulk. However, due to the high energy barrier for diamond nucleation, conversion of graphite to diamond typically necessitates high temperatures, high pressures, or plasma chemical vapor deposition (CVD) technology [28]. It has been discovered that flaws or dangling bonds in the precursors can lower the diamond nucleation's energy barrier [28]. According to intuition, functionalizing 2D graphene with heteroatoms (functional groups or foreign atoms) may be an effective method to lower the activation energy of the *para-* and *ortho-*positions of the hexagonal ring, and as a result significantly lower the transition pressure needed to move carbon atoms from the sp^2 bonding state to the sp^3 bonding environment [28].

Currently, numerous potentials use for graphene nanocarbon have been suggested, and more than a thousand review articles have been published in this specialized field of study. However, diamanoids and diamane have only been the subject of a small number of research papers and scholarly commentary to date. In this chapter, we provide a summary of diamane's structural layout and its derivatives. The role of pressure and temperature is very important for the fabrication of diamane and the importance of these two parameters for the fabrication of diamane will be discussed in this chapter. Moreover, this chapter also covers the thermodynamics of pristine diamane.

4.2 Early works on diamane

In 1994, IUPAC made known to the term 'graphene' to describe the terminology, reaction, and nomenclature of graphene and related nanosystems [29]. Although graphene was first studied in 1853, it was experimentally discovered in 2004 by Geim and Novoselov [21, 30]. As 2D structures attracted more attention, researchers began investigating other modifications of graphene. Reversible hydrogenation of SLG in 2009 enhanced its electrical and transport characteristics [22]. Diamanes, such as 'diamane' and 'diamane-II,' were subsequently produced in 2007 by hydrogenating BLG [5]. The study of the different physical, chemical, mechanical, and thermal properties of diamane has become a larger focus of research. Triamane was synthesized in 2012, followed by the exploration of other functional groups, such as F-diamane [4]. Around 2018, diamane was experimentally formed [31]. Diamane's properties were further tuned by substituting atoms with nitrogen, boron, lithium, etc, and twisting BLG to create a Moire's pattern on diamane. 'Diamanoids' were developed in 2019 using FLG [32, 33]. In general, between 2000 and 2010, the research focused primarily on evaluating diamanes and diamanoid compounds.

4.3 Overview of diamane and diamanoids

During the process of discovering new 2D nanomaterials, scientists discovered graphene-based materials known as 'diamanoid' and 'diamane'. These materials have tremendous potential for applications in various fields, including quantum computing technologies, nanoelectronics, electrical, sensory, and biomedical applications. Diamane is essentially a double-layered diamond or graphene and is the thinnest hard material. It is fashioned by converting BLG with sp^2 hybridization to a sp^3 hybridized structure via covalent bonding approaches [4]. Figures 4.1(a)–(c) illustrate the formation of the new carbon system known as diamane, which is composed of the C_2H layered unit, hydrogen atoms, and two carbon atoms from different subatomic lattices. Meanwhile, another non-attached carbon layer interacts by covalent bonding with the sub-lattices of neighboring carbon atoms in graphene to form the new carbon system [5]. The formation of diamane involves a process where on the surface of the carbon layer there are two lone pairs of electrons that are not disturbed. This enables vertical CCCC stacking, notably 111 stacking, which persists up until four carbon atoms are arranged in a unit cell, leading to the

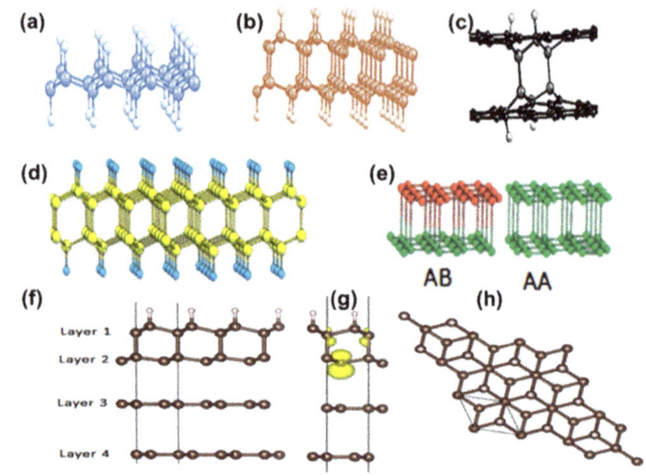

Figure 4.1. (a and b) Diamane structural units in various arrangements. (c) Diagram showing the development of the diamane nucleus as a result of hydrogenation of BLG. The sp^2 carbon atoms are represented by the black circles. Reproduced from [39]. CC BY 4.0. (d and e) Molecular structure of diamane (AB)-based quasi-2D carbon nanofilms with stacked AB layers. Reprinted with permission from [1]. Copyright (2011) American Chemical Society. (f–h) Molecular structure of diamane (AB)-based quasi-2D carbon nanofilms with stacked AB layers. Reprinted with permission from [33]. Copyright (2011) American Chemical Society.

synthesis of diamane. When the other half of the carbon is bonded to neighboring carbon layers and half of the carbon is hydrogenated, the hybridization layer of lonsdaleite or diamond adopts a cubic or hexagonal lattice (figure 4.1 (b)). Although these processes take place at the nanometer scale, their impacts are felt on a much larger scale [5, 31, 32, 34]. The term 'diamane' draws an analogy to 'graphene–graphane,' representing a diamond-like structure created from hydrogen and BLG [4]. Various research groups have used other names, such as interlayer-bonded BLG, hydrogenated FLG, diamondene, diamandol, hydrogenated diamond, and hydrogenated BLG to refer to diamane [35–37]. BLG must be stacked twice to create diamane: AA (Lonsdaleite) stacking and AB (Bernal) stacking (figures 4.1(d) and (e)). Diamane's C_2H unit displays AB stacking, whereas the isomer diamane-II adopts an AA stacking sequence, greatly affecting its physical and chemical properties [4]. The most stable forms of graphene are its AA stacking chain isomers, which are the building blocks of three-dimensional graphene crystals when hydrogenated under pressure [4]. BLG exhibits Moire's patterns when twisted at angles up to 30° [38, 39]. The development of novel structures may result from further functionalization of these twisted layers. Due to the absence of thermally stable structures in this 2D system, atomically thin diamond films are unstable in their pure and completely free states. To fabricate stable diamane structures that are thermodynamically stable, a variety of functional moieties, hybridizations, and hydrocarbon are utilized. Functional groups such as hydroxyl, fluorine, oxygen, and nitrogen are employed to form various hybrid diamanes [40, 41]. However, sustaining stability at low pressure is difficult, making it less favored [40, 41].

Pressure application has also been investigated as an alternative approach for diamane production. Another method to obtain different types of diamane involves substituting atoms as dopants in graphene and stabilizing the hybridization to form modified diamanes. Lithium, sulfur, boron, and nitrogen atoms have already been utilized as dopants over the past 5 years, and further exploration of other dopants is rapidly progressing [4]. These innovative methods offer exciting prospects for discovering novel belongings of diamanes, which hold significant importance for next-generation energy storage technologies and quantum electronics.

A novel 2D carbon structure called a diamanoid is created when there are two to several (3–5) layers of hydrogenated graphene. This new structure has unique features. Figures 4.1(f) and (g) depict the diamanoid structure, also known as partly hydrogenated FLG. Figure 4.2(h) represents the bonding patterns of diamanoid with ABBA stacking using density functional theory (DFT). Hydrogen radicals from a hot-filament technique carried out at low temperature and pressure were initially used in diamanoid synthesis. More modern fabrication procedures have been devised to make diamanoids with different layers [6, 8, 23]. In the hot-filament process, diamanoid is produced by hydrogenating the top carbon layer while the bottom layer forms covalent bonds with the adjacent graphene layer, as shown in figures 4.1(f) and (g). To comprehend the bonding patterns better, figure 4.1(h) presents the diamanoid's DFT computed geometry with an ABBA stacking.

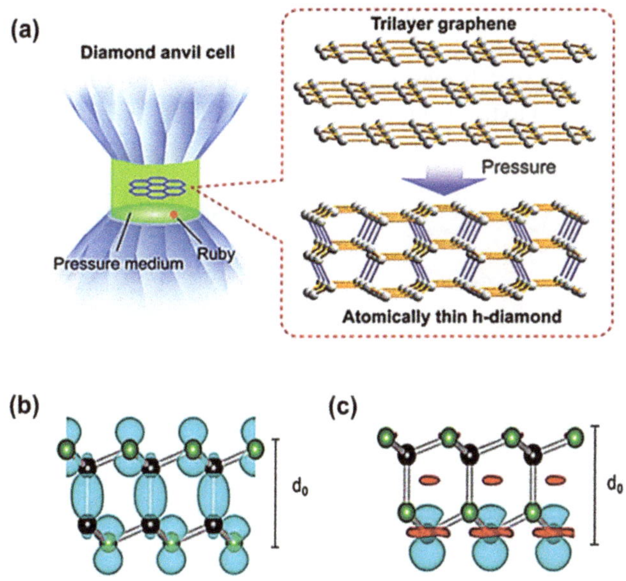

Figure 4.2. (a) The use of pressure during the synthesis of diamane and its impact on molecular organization. Reprinted with permission from [47]. Copyright (2020) American Chemical Society. (b) Projecting in the [100] direction, the relaxed structure of NCCN, whereas the blue envelopes represent the spin up charge density of the valence band maximum at Γ-point. Reprinted with permission from [49], Copyright (2020) by the American Physical Society. (c) Projecting on the [100] direction, the relaxed structure of CNCN, whereas red and blue envelopes represent the negative and positive magnetization density, correspondingly, of the valence band maxima at K-point. The green and black balls signify the N and C atoms, correspondingly.

Initially, the synthesis of diamanoid relied on the hydrogen radicals generated by the low-pressure and temperature and in a hot-filament method, but more recent developments in fabrication techniques have made it possible to create diamanoids with various layers [4].

Diamane often exhibits Bernal stacking, whereas diamanoid lacks this feature because of its structural unsteadiness. However, various stacking arrangements, such as (AABBCC) diamanoids, (ABBA) diamanoids, and (ABB) diamanoids, are advantageous for the generation of diamanoid molecules. The name for (3–5) few-layered graphene lacks a precise definition, leaving the maxima allowed layer in diamanoid synthesis undecided. To ascertain this and carry out more in-depth studies, additional study is required. Although several diamanoids containing fluorine and hydrogen have recently been identified, the scientific community has not given them much attention due to their thermodynamic difficulties. Currently, research on diamanes, diamanoid, and carbon-like thin films mostly depends on simulation and theoretical -based computational studies, emphasizing the need for more experimental studies.

4.4 Structure of diamane

Chemical adsorption of functional groups on graphene's carbon rings, akin to aromatics, typically disrupts properties. Graphane (figure 4.3(a)) and diamane (figures 4.3(b) and (c)) have chair-like C6 rings forming slightly puckered carbon sheets due to hydrogenated (fluorinated) carbon atoms' pyramidalization, as opposed to graphene's flat honeycomb pattern. Introducing heteroatoms increases material thickness with extra atomic layers. Research on hydrogenated (fluorinated)

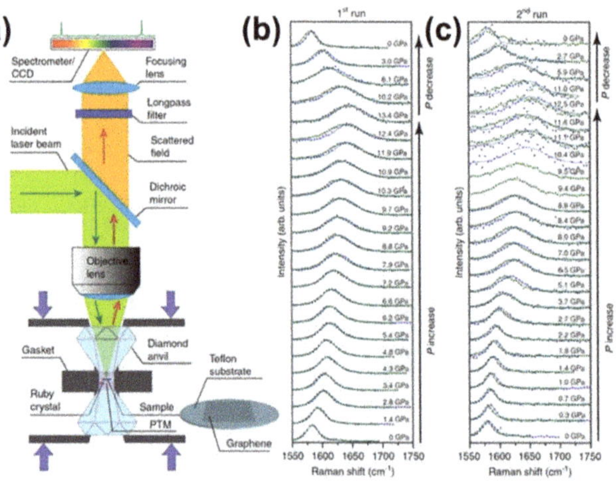

Figure 4.3. (a) The experimental setup is shown schematically. In the diamond anvil cell, the sample is compressed to 15GPa pressure with water as the pressure transmission medium. Raman spectra in (b) and (c) display the progression of the G-band with enhancing pressure. The results are gathered from two separate measurement runs (first and second, correspondingly), each using a different sample. Reproduced from [53].

diamane reveals a structure similar to diamond, having a little crystal lattice deformation brought on by hydrogen on the surface [28]. Hydrogenated BLG exhibits less movement of head-to-head carbon atoms than hydrogenated SLG due to the connection between layers that keeps graphene flatter.

4.5 Derivatives of diamane

4.5.1 From graphene and their derivatives

Most commonly, different processes are used to synthesize graphene, a 2D nanocarbon, from bulk graphite. The top-down method comprises the mechanical and chemical exfoliation of graphite, whereas the bottom-up method uses epitaxial development on insulating surfaces (diamond or SiC) and CVD. Although the stacking sequence and layer count cannot be precisely controlled using the bottom-up strategy, it is nevertheless widely used because of its suitability. Liquid exfoliation has produced single to few-layered platelets from different materials, such as graphite, black phosphorus, h-BN, and transition metal dichalcogenides. Another technique involves exfoliation on a silicon substrate coated with silicon dioxide (SiO_2) over a Si wafer along with h-BN, employing the pickup method to synthesize 2D systems such as graphene [42, 43]. The early technique of using Scotch tape or micromechanical cleavage is challenging to achieve monolayered structures [44]. Additionally, micromechanical cleavage of graphene crystals on an oxidized silicon substrate has been used to create other 2D nanocarbon structures and diamanes [22]. Graphene with three layers and a higher thickness can have anisotropic characteristics depending on how the sheets are stacked. Mechanical milling increases rhombohedral modifications in graphene. The synthesis of tri-layer graphene involves high-temperature annealing of n-type Si-terminated 6H-SiC (0001). Exfoliation of normal graphite using the Scotch-tape method leads to graphene with ABC stacking, but controlling the quality, yield, and number of layers remains challenging. Graphene flakes obtained through the Scotch-tape method can have various stacking orders, including some twisted layers. CVD has been successful in producing BLG and monolayer graphene through epitaxial growth on metal surfaces from hydrocarbon deposition. Graphene islands are obtained by rolling CVD-formed graphene on a polymer after etching away the Cu substrate [44]. Cu–Ni alloy has been utilized in producing graphene through CVD in high pressure [45, 46]. Trilayered graphene can also be produced on Si/SiO_2 surface via the mechanical exfoliation of pristine graphite [47]. Graphene synthesis techniques, mainly CVD, fulfill the demand for diamane and diamanoids synthesis, introducing pressure under suitable conditions (figure 4.2(a)) [4]. The well-established procedures for doping of heteroatoms within graphene can also be executed for diamanes, resulting in various diamane structures, such as NCCC, CNCC, NCCN, and CNCN, as depicted in figures 4.2(b) and (c) [48, 49]. Recent investigations have explored various heteroatoms, with larger atoms such as sulfur causing structural deformation and instability in diamanes, while smaller substitution atoms such as boron and nitrogen contribute to supplementary stability [4].

4.5.2 Hydrogenated and hydroxylated diamane

SLG hydrogenation reveals non-stoichiometric composition with one-sided exposure to hydrogen. Hydrogenation from one or both sides results in rapid hydrogen loss at moderate temperatures [22, 50]. Theoretical findings indicate that hydrogenating BLG is more challenging than SLG [50]. Fully hydrogenated diamane (HD) preparation is hard and needs precise temperature regulation. Pressure application on multilayer graphene and BLG covered by the water molecule reduces electric force microscopy response, but this effect is absent in a dry environment [37]. Simulation indicated that the hydroxyl group from the water molecules triggered the graphene diamondization under pressure, but this structure could not be maintained at ambient pressure. It was projected that hydroxylated diamane would be a 2D ferromagnetic insulator. By employing a diamond anvil cell (DAC), high-pressure tests were conducted to introduce the functionalities on the top of BLG. Water served as both a source of hydrogen atoms or hydroxyl in the test and a pressure transmission medium (PTM). The CVD-prepared BLG was transferred to a Teflon substrate, which prevented functionalization of bottom carbon atoms, leading to chemical adsorption only on top. Raman spectroscopy measurements during pressurization revealed the formation of a mixed sp^2 and sp3 hybridization, with diamane formation observed at 4–5 GPa (figure 4.3). However, since Raman spectroscopy is indirect and cannot precisely determine hydrogen positions, further evidence is needed to confirm the interlayer sp3 bonds in graphene [32].

By using the hot-filament technique under practically ambient circumstances, graphene can be hydrogenated, effectively generating hydrogen atoms [32, 51, 52]. This process avoids the damage owing to ion acceleration at the sample of graphene. Practically, FLG can be deposited on Pt (111) by employing the CVD method. HD can then be produced by adsorbing carbon atoms on graphene surface with deuterium atoms formed through the hot wire procedure. In four-layer graphene, 76% of the carbon atoms showed interlayer sp^3 hybridization, according to spectroscopic methods using a synchrotron. Theoretical calculation suggests that surface adsorption of hydrogens led to the propagation of sp^3 hybridization throughout the carbon layers, stabilizing the bottom carbon atoms through interaction with the substrate. Graphene might potentially be double-sided hydrogenated using the hot-filament method. Nevertheless, because of the alterations in the stacking patterns and amount of layers during graphene production, the hydrogenated products consist of various mixtures of diamane, as confirmed by Fourier transform infrared (FTIR) and Raman analysis [32].

4.5.3 Fluorinated diamane

Fluorinated diamane (FD), which resembles a diamond, is formed by fluorinating BLG or FLG. Graphene can be fluorinated more easily in experiments than it can be hydrogenated and FD can be produced at lower temperatures [24, 40, 54, 55]. FD exhibits superior thermal stability compared to HD, even remaining stable in air at 400°C [28]. Experimentally, diamane, which is fully fluorinated, is easier to achieve than fully HD [41, 42]. Reports suggest different carbon/fluorine ratios in FD, such as 4:1 and 2:1 [28]. The preparation of FD involves both bottom-up and top-down methods, such as the fluorination of graphite by a fluorine radical to produce a molecule that facilitates the intercalation of graphene, which is then separated to produce FD [28]. Graphene layers can be fluoridated using CF_4 plasma, SF_6 plasma, or XeF_2 [28]. Bakharev *et al* employed XeF_2 to successfully fluorinate AB-stacked graphene on CuNi alloy (111) at normal pressure, resulting in stable FD throughout a wide area [40]. The fluorination process led to the generation of carbon–carbon interlayer bond in BLG, resulting in FD. Non-fluorinated and fluorinated BLG diffraction patterns and photomicrographs corresponded to FD- and AB-stacked BLG simulation data. The fluorinated AB-BLG exhibited a C_2F stoichiometry, indicated by low electron energy loss and K edge spectra. By electron beam irradiation, FD, a wide band-gap semiconductor, can be transformed back to AB-BLG [40].

4.5.4 Pristine diamane

In a current study, Gao *et al* used SiO_2 as the substrate for atomic force microscopy to evaluate the impact of pressure on a BLG sample at ambient temperature [56]. As a result of the generation of a diamond-like carbon film, the sample displayed diamond-like hardness and a reversible drop in electrical conductivity upon indentation. However, as the graphene thickness increased (three to five layers), an unfavorable multilayer stacking structure hindered its performance. Previous research on pristine FLG under pressure demonstrated the generation of interlayer sp^3 C–C bonds at 16 GPa with partial sp^2 bonds in seven-layered graphene [24].

Inopportunely, the structure of graphene remained unchanged after the release of pressure. Currently, resistance measurements estimated the pristine graphene's phase transition pressure, as depicted in figure 4.4 [47]. From both theoretical simulation and x-ray diffraction (XRD), it is found that hexagonal pristine diamane (PD) is

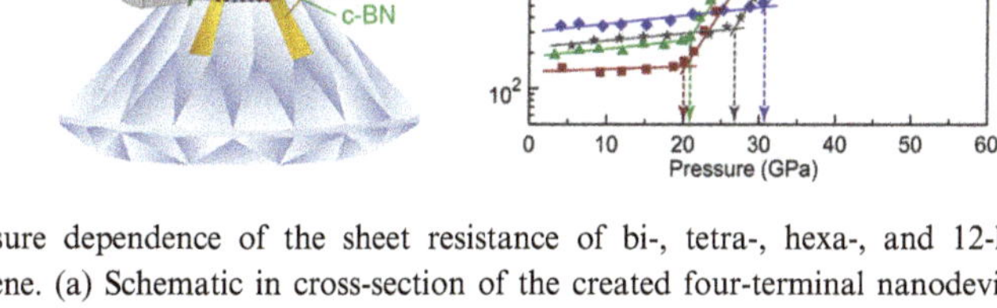

Figure 4.4. Pressure dependence of the sheet resistance of bi-, tetra-, hexa-, and 12-layer mechanically exfoliated graphene. (a) Schematic in cross-section of the created four-terminal nanodevices. Platinum foils were used to extend the Ti/Au film electrodes that were patterned onto the diamond culet. (b) Measured at ambient temperature, the relationship between sheet resistance and pressure for bi-, tetra-, hexa-, and 12-layer graphene. (The pressure dependence of graphite resistance is shown in the inset.) Reprinted with permission from [47]. Copyright (2020) American Chemical Society.

developed with three-layer or thicker graphene during compression. The phase transition pressure decreases with increasing graphene layers, but the hexagonal PD obtained at high pressure can be retained up to 1 GPa, contrary to previous reports. Diamane exhibits diversified conformations and functional groups due to the varied stacking order and layers of graphene, providing versatile properties. PD developed under high pressure cannot be retained under normal pressure because it reversibly converts to graphene upon pressure release, indicating relatively poor stability. PD is the diamane that has proven to be the most stable in studies.

4.6 Thermodynamics of pristine diamane

The conversion of FLG with a clean surface to diamane is interesting. Functionalized diamane is referred to as PD. To transform two to eight layers of graphene with a (111) surface into PD, Kvashnin *et al* examined the thermodynamic phase diagram [57]. Due to the effects of graphitization, diamond nucleation failed to happen in two layers. These effects lead to diamond converting to graphite because of surface influences in the diamond nanostructure [58, 59]. Therefore, to create diamond films, low-layer graphene needs heteroatom stability. Nucleation of diamond is unaffected by surface effects at greater thicknesses. Pristine diamond becomes metastable after five layers, and a semimetal to semiconductor transition takes place. Surprisingly, molecular dynamics simulations imply that the transformation of pristine graphene to diamane is reversible [28]. Theoretical findings show that high pressure conditions are necessary for the formation of PD and that the pressure required for PD to transition into its phase is increased when the number of layers is reduced [28].

4.7 Role of temperature in fabrication of diamane and its derivatives

Temperature plays a vital role in the synthesis of any carbon material. The morphology and physico-chemical characteristics of the carbon materials are highly dependent on the processing temperature. Being a member of carbon family, diamane and its derivatives are no exception [60]. The hydrogenation of graphene sheets is synthesized through various techniques, including the hot-filament process. In the hot-filament process, hydrogenation was mainly carried out in a hot-filament reactor, using two tungsten wires as the heating elements. The filament temperature was kept within the range of 2350°C–2590°C. Meanwhile, the substrate holder exhibits a maximum temperature of 325°C for few-layered graphene and 400°C for two-layered graphene. Overall, the hydrogenation process was continued for 6–20 s [31]. In this aspect, Piazza *et al* reported a low temperature hot-filament process for successive transformation of sp^2 carbon to sp^3 carbon in crystalline carbon film [32]. Specifically, the holder temperature varies with the distance between the filament and substrate in the hot-filament process. In contrast to other functionalization processes, the diamane formation through hot filament-induced hydrogenation process is possible at relatively lower temperature. The production of homogeneous diamane films is possible with high-quality two-layered graphene at low temperature, as well as low pressure. However, once the films are formed, the existence of large stress has been observed for the hydrogenation of multi-domain two-layered graphene. The structural change of the diamane through this temperature-dependent hot-filament-induced hydrogenation process can be visible through UV Raman spectroscopy. As shown in figure 4.5, the crystalline sp^3-C peak can be visible at 1319–1337 cm^{-1}. Therein, the G-band reveals the partial conversion only. Two peaks typical of twisted BLG, labeled TBL1 and TBL2 (twisted bilayer), are also observed at ∼1385 and ∼1669 cm^{-1}. An example of the fitting made by using symmetric Voigt functions (dashed line) to estimate peak positions is provided in figure 4.5(d). Therefore, it is evident that conventional G-band peak at 1582 cm^{-1} is drastically altered after the hydrogenation process. Overall, the UV Raman spectra confirms the presence of sp^3-C peak and formation of genuine diamane. Most importantly, researchers have found unaltered UV Raman spectra for such sp^3-C material after a time duration of five years, indicating the high stability of diamane.

Researchers have further utilized the Raman spectra to identify the formation of high-quality diamane, synthesized from the BLG through similar approach [61]. The theoretical first principle calculations also supported the successive hydrogenation process, and also confirmed the hydrogenated AB configuration as the stable form. The micro-Raman mapping further revealed significant structural charge through the hydrogenation process. As depicted in figure 4.6, the Raman spectrum display absence of characteristic G peak. Interestingly, the characteristic D peak, which generates from the defects in graphene structure, is also missing. Instead of such characteristic peaks, the spectrum reveals a sharp peak at around 1344–1367 cm^{-1}.

Overall, the synthesis of high-quality diamane is highly dependent on the quality of BLG. The single-crystal AB flakes should be in larger portion in the layered

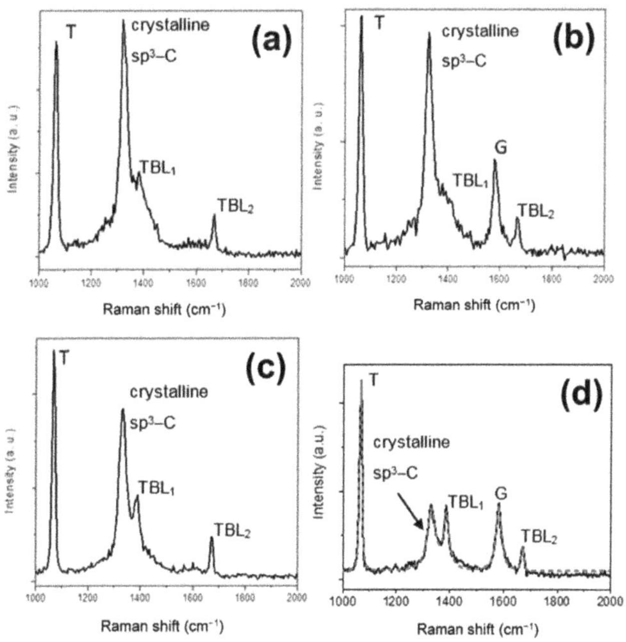

Figure 4.5. (a–d) Characteristic UV Raman plots of FLG after the exposure to the hydrogenation reaction induced by hot filament. Reprinted from [32], Copyright (2019), with permission from Elsevier.

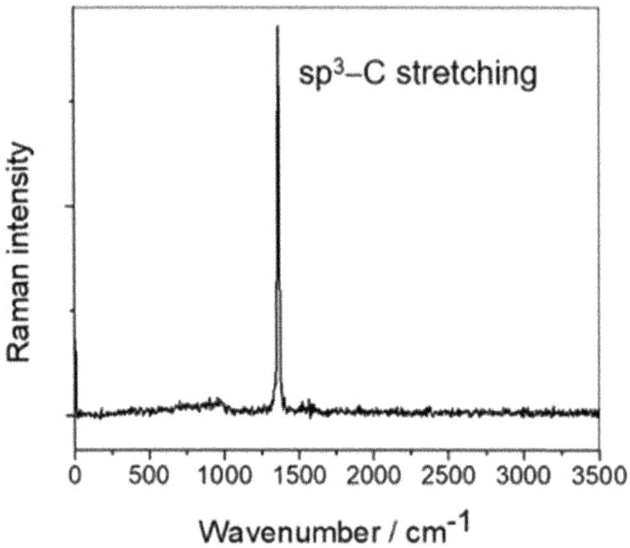

Figure 4.6. Raman spectrum of hydrogenated BLG as proof that real diamane exists. Reprinted from [61], Copyright (2020), with permission from Elsevier.

graphene precursor. Modern advanced characterization techniques such as UV Raman, micro-Raman etc play a significant role in detecting genuine diamane. The DFT calculation and theoretical simulations also played a crucial role in supporting the existence of diamanes. Among the synthetic approaches, the hot-filament process is the most popular for developing diamane from few-layered graphene as well as BLG. In this process, the temperature of the substrate holder is highly significant, which varies with the layers of graphene precursors.

4.8 Role of pressure in fabrication of diamane and its derivatives

Diamane has a unique structure, which influences its stability, properties, and limitations. The stability is mainly due to the graphitization by virtue of energy preference of graphite as compared to diamond. At the same time, in bulk a transformation barrier forces the diamond to be metastable and to exist over centuries—as the dimensions reduces, the surface effects become more prominent. In the absence of the passivating gases, the phase equilibrium shifts to a higher graphene–diamane phase transition pressure or complete lack of stability regions.

Pressure mediated synthesis of diamanes is relatively challenging due to the possibility of formation unstable structure. Exposure of a BLG to H^+ radicles produced from a hot filament was utilized in a low-pressure method. Tungsten filaments were used as heating source and pressure was reduced to 10^{-1} Torr before the introduction of hydrogen gas [31]. Temperature was maintained at 2350°C–2590°C, 50 Torr pressure was maintained and hydrogen gas flow was set at 1 SCCM.

Exposure to higher pressures has resulted in interlayer bonding in carbon-based materials, as reported by Martins *et al*, who used a DAC that produced 15 GPa pressure utilizing water as the pressure transition medium [53]. With higher pressure, the G-band widened and the phenomena was seen to reverse upon release of pressure. This confirms the notion by the earlier studies that graphene undergoes phase transition in the pressure range of 10–20 GPa [62, 63]. The G-band dispersion due to the variation in excitation energy is portrayed in figure 4.7. The phase transition is portrayed by an instantaneous broadening of the G-band caused by the substrate deformation strain, as reported by Lu *et al* [64]. However, the experiment done by Lu *et al* reported graphene nanoplatelets in which the authors used liquid argon as the phase transition medium. The variations in the peak broadenings by virtue of pressure change are portrayed in figure 4.8.

In Tiwari *et al*'s. comprehensive review on diamane and diamanoids, the various synthesis strategies and characterization of diamanes have been detailed alongside the unique properties of diamane and diamanoids [4]. In summary, this manuscript discourages low-pressure synthesis because the stability of the formed structures are severely compromised. As a substitute to pressure, the use of functionalization is encouraged to obtain a stable structure. Varlamova *et al* investigated the effects of pressure alongside the precursor and temperature on the formation of diamane by DFT methods [65]. The chemical

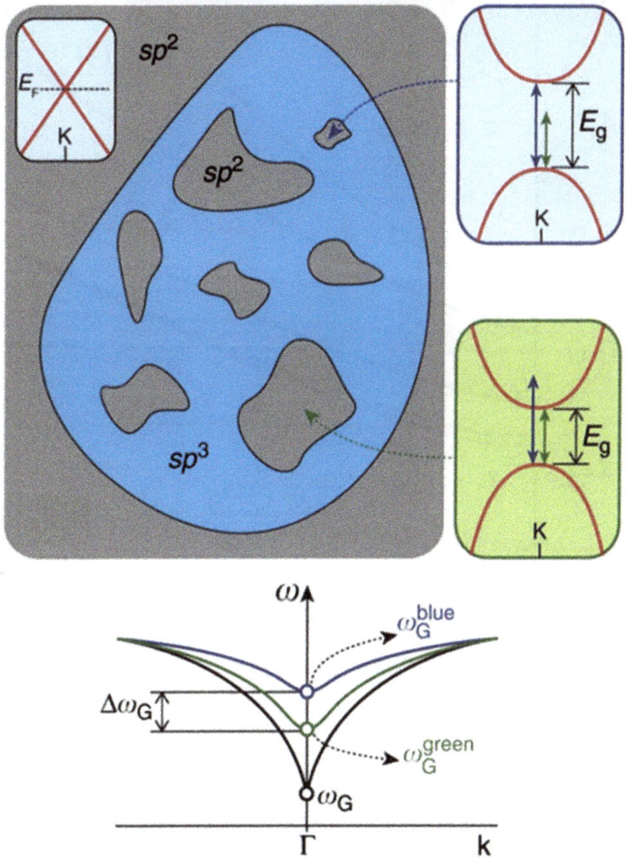

Figure 4.7. Effect of excitation energy on G-band dispersion. Reproduced from [53]. CC BY 4.0.

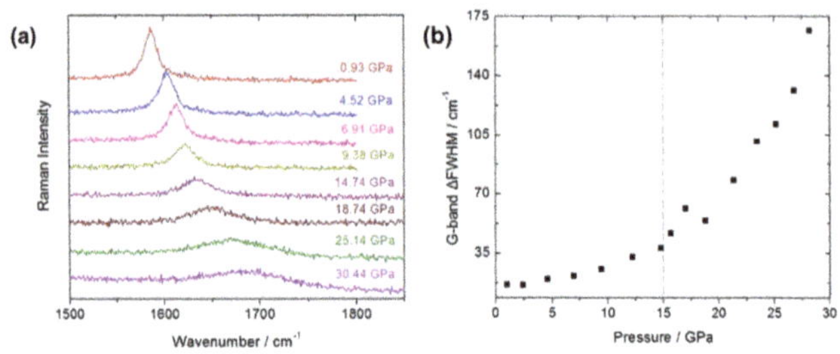

Figure 4.8. (a) G-band of the Raman spectra for graphene nanoplates (1–5 nm) at various pressures, and (b) G-band linewidth variation with pressure. The critical pressure, where the transformation occurs is indicated by the vertical dashed line. Raman analyzes were performed with 633 nm LASER and liquid Ar was used as a phase transition medium. Reprinted from [64], Copyright (2013), with permission from Elsevier.

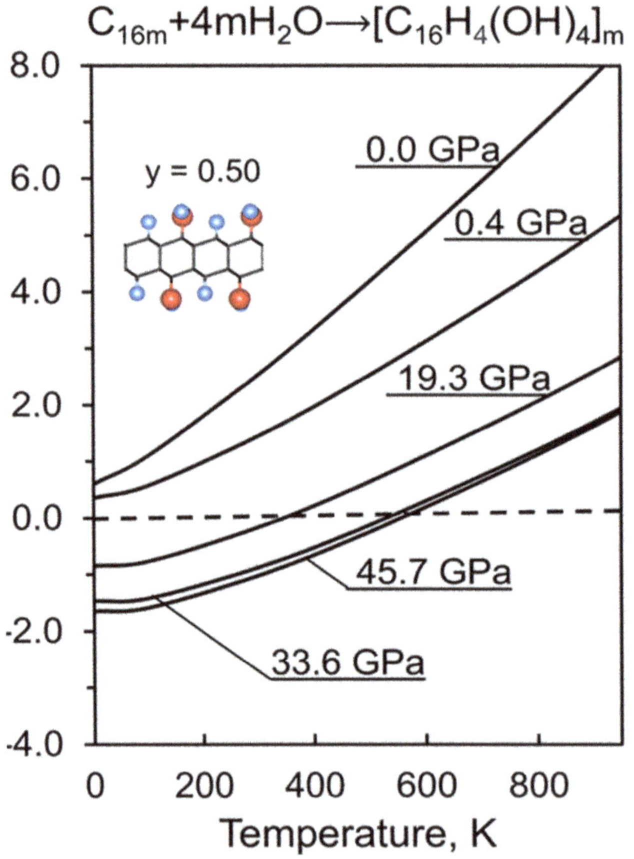

Figure 4.9. Pressure dependence of ΔG of $C_{16}H_4(OH)_4$ film. Reprinted with permission from [65]. Copyright (2022) American Chemical Society.

potential of the bilayer, which is dependent on pressure and temperature of the reaction system, allows the estimation of the stability of the synthesized material. The pressure variation and subsequent variation of ΔG are represented in figure 4.9.

The pressure value has been deduced from the sum of all forces acting on the carbon atoms fixed alongside the *C*-axis of the carbon layer and is represented in the following equation:

$$\Delta G\ (T,\ P) = \Delta U(P) + P\Delta V - \mu \,\mathrm{mol}\ (T,\ P) \tag{4.1}$$

ΔV and $\Delta U(P)$ are the difference of volume and difference of per unit cell internal energy between BLG and diamane, correspondingly. BLG fabricated by CVD method over a CuNi alloy forms a stable structure [66]. This work reports the BLG has variation in electronic properties as compared to the SLG by virtue of its grain size. Here it is reported that the pressure value increase resulted in red shift for the G peak frequencies and blue shift for the reduction in pressure. The pressure range is reported to be from 9 to 40 GPa. However, with water as the phase transition material, blue shift occurs at 37.5 GPa and disappears at 50 GPa pressure. When silicon oil was used as the phase change material, the evolution is nearly reversed, as displayed in figure 4.10.

It is concluded that the existence of water assists the bonding of graphene by acting as the functional groups over the suspended graphene surface. The higher pressure transitioned the graphene layer to a metastable sp^3 bonding and collapses after the loss of pressure. According to the research, the large strains that come along with the formation of diamond nuclei prevent the transformation from happening at the graphite-diamond co-existence pressure. At higher pressures, the nucleation mechanism is favored over the determined transformation. The creation of the metastable hexagonal diamond phase can be attributed to greater distortions of the graphite lattice around cubic diamond nuclei as opposed to those surrounding hexagonal diamond nuclei. Rajasekaran *et al*'s work [51] corroborates Khaliullin *et al*'s [67] and report that phase transition from graphite to diamond directly

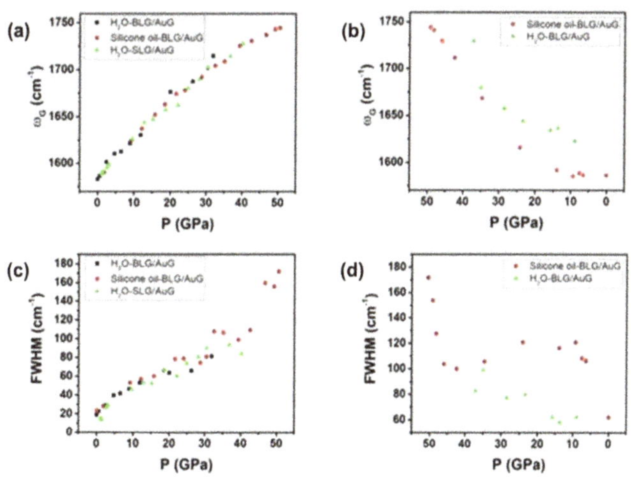

Figure 4.10. (a) The biaxial stress (ω_G) and (c) full width at half maximum (FWHM) of G peak as a function of pressure-up; (b) ω_G and (d) FWHM of G peak as a function of pressure-down. Reprinted from [45], with the permission of AIP Publishing.

Table 4.1. A summary of the production processes and characteristics of diamane. Reproduced from [28]. CC BY 4.0.

Type	Substrate	Precursor	Methods and condition	Product	Retention conditions
HD	Teflon	BLG prepared by CVD method	DAC (the sample is pressurized to 15 GPa and water is used for PTM and the required hydrogen or hydroxyl).	Partially hydrogenated or hydroxylated diamane	High pressure
	Pt (111)	FLG	The process of deuterium adsorption using an atomic cracking source by hot filament.	Partially HD	Ambient pressure
	No substrate	BLG and FLG	The hydrogenation process is 200 s by hot-filament CVD.	Diamond and/or lonsdaleite and/or diamane and/or hybrids of those)	Ambient pressure
FD	No substrate	FGIC (fluorinated graphite intercalated compound)	FGIC is thermal decomposed into FLG and is then fluorinated by FCl_3 to obtain FD.	Partially fluorinated FD	Ambient pressure
	CuNi(111) surface	AB stacking BLG	CVD (BLG is fluorinated by XeF_2 for more than 12 h at 65°C and 50–60 Torr vapor pressure).	Fully fluorinated FD	Ambient pressure
PD	SiC	BLG	Indentation induction by AFM probe.	Partially sp^3 hybrid diamond-like film	High pressure
	No substrate	FLG	High pressure above 19.6 GPa by DAC.	Partially PD	Stable above 1 GPa

corresponds to the kinetic barriers of the diamond nucleation and add that unfavorable amount of strain involved in the distortion of the graphite lattice to promote diamond nucleation is the reason why high pressure is mandated. That being said, the pressure at which these phases may coexist can be very low. This work had proposed an alternate strategy to counter the usage of pressure with hydrogenation as a strategy to incorporate four layers forming a diamond-like structure over a metallic substrate.

Diamane can produce a variety of conformations and functional groups with a wide range of attributes because of the variable stacking order and layers of graphene. Here, as indicated in table 4.1, we include a number of typical diamane production preparation techniques.

4.9 Conclusions

In this chapter, we have focused the synthesis of diamane derivatives in different conditions, and their relative stability and properties. The opening paragraphs of this chapter describe the historical prospects and structural designs of diamane. Then, a brief overview of diamane and diamanoids has been covered. Moreover, the role of pressure and temperature for the synthesis of diamane have been discussed in this chapter. Additionally, the synthesis and design of diamane, as well as the most recent theoretical and experimental findings on HD, FD, and PD have been discussed. Prior to identifying practical uses, sufficient basic and engineering study must be conducted on any newly found material. Since diamanoids and diamane are still relatively new and were only recently discovered, extensive research is urgently needed to properly comprehend them.

Acknowledgments

This research was made possible by funding provided by the Ministry of Science and ICT (MSIT), Government of South Korea, under the National Research Foundation of Korea (2020R1A2C1101561).

Declaration of competing interest

There is no conflict of interest.

References

[1] Chernozatonskii L A, Sorokin P B, Kuzubov A A, Sorokin B P, Kvashnin A G, Kvashnin D G, Avramov P V and Yakobson B I 2011 Influence of size effect on the electronic and elastic properties of diamond films with nanometer thickness *J. Phys. Chem.* C **115** 132–6

[2] Ruoff R S 2012 Personal perspectives on graphene: new graphene-related materials on the horizon *MRS Bull.* **37** 1314–8

[3] Gupta S, Yang J H and Yakobson B I 2018 Two-level quantum systems in two-dimensional materials for single photon emission *Nano Lett.* **19** 408–14

[4] Tiwari S K, Pandey R, Wang N, Kumar V, Sunday O J, Bystrzejewski M, Zhu Y and Mishra Y K 2022 Progress in diamanes and diamanoids nanosystems for emerging technologies *Adv. Sci.* **9** 2105770

[5] Chernozatonskii L A, Sorokin P B, Kvashnin A G and Kvashnin D G E 2009 Diamond-like C 2 H nanolayer, diamane: simulation of the structure and properties *JETP Lett.* **90** 134–8

[6] Maity C K, Sahoo S, Verma K and Nayak G C 2022 SnS_2@Conducting energy level-induced functionalized boron nitride for an asymmetric supercapacitor *Energy Fuels* **36** 2248–59

[7] De S, Acharya S, Maity C K, Sahoo S and Nayak G C 2022 MXene ($Ti_3C_2T_x$)-/amine-functionalized graphene-supported self-assembled Co9S8 nanoflower for ultrastable hybrid supercapacitor *Ind. Eng. Chem. Res.* **61** 7727–38

[8] Maity C K, De S, Acharya S, Siddiki S H, Sahoo S, Verma K, Thakur V K and Nayak G C 2022 Copper oxide stabilized oxy-functionalized boron nitride-carbon nanotube nanohybrid: an ultra-stable electrode for flexible asymmetric supercapacitor device in ionic electrolyte *J. Energy Storage* **56** 105928

[9] De S, Acharya S, Maity C K and Nayak G C 2023 Polyindole-stabilized nanocellulose-wrapped $Ti_3C_2T_x$ (MXene) nanocomposite for asymmetric supercapacitor devices *ACS Appl. Energy Mater.* **6** 969–80

[10] De S, Maity C K, Kim M J and Nayak G C 2023 Tin(IV) selenide anchored-biowaste derived porous carbon-$Ti_3C_2T_x$ (MXene) nanohybrid: an ionic electrolyte enhanced high performing flexible supercapacitor electrode *Electrochim. Acta* **463** 142811

[11] Maity C K, Santra D K, Verma K, Sahoo S, Cotts S, Akinwande D, Berry V and Nayak G C 2021 Induced conducting energy-levels in a boron nitride nano-framework for asymmetric supercapacitors in high charge-mobility ionic electrolytes *Composites* B **212** 108728

[12] Toivola M, Ferenets M, Lund P and Harlin A 2009 Photovoltaic fiber *Thin Solid Films* **517** 2799–802

[13] Tiwari S K, Sahoo S, Wang N and Huczko A 2020 Graphene research and their outputs: status and prospect *J. Sci.: Adv. Mater. Devices* **5** 10–29

[14] Sheehan J E, Buesking K W and Sullivan B J 1994 Carbon–carbon composites *Annu. Rev. Mater. Sci.* **24** 19–44

[15] Maity C K, Hatui G, Tiwari S K, Udayabhanu G, Pathak D D, Nayak G C and Verma K 2019 One pot solvothermal synthesis of novel solid state N-doped TiO_2/n-Gr for efficient energy storage devices *Vacuum* **164** 88–97

[16] Maity C K, Hatui G, Verma K, Udayabhanu G, Pathak D D and Nayak G C 2018 Single pot fabrication of N doped reduced GO (N-rGO)/ZnO–CuO nanocomposite as an efficient electrode material for supercapacitor application *Vacuum* **157** 145–54

[17] Maity C K, Hatui G, Sahoo S, Saren P and Nayak G C 2019 Boron nitride based ternary nanocomposites with different carbonaceous materials decorated by polyaniline for supercapacitor application *ChemistrySelect* **4** 3672–80

[18] Maity C K, Sahoo S, Verma K, Behera A K and Nayak G C 2020 Facile functionalization of boron nitride (BN) for the development of high-performance asymmetric supercapacitors *New J. Chem.* **44** 8106–19

[19] Buckley J D and Edie D D 1993 *Carbon–Carbon Materials and Composites* (Norwich, NY: William Andrew)

[20] Tiwari S K, Mishra R K, Ha S K and Huczko A 2018 Evolution of graphene oxide and graphene: from imagination to industrialization *ChemNanoMat* **4** 598–620

[21] Wallace P R 1947 The band theory of graphite *Phys. Rev.* **71** 622

[22] Elias D C *et al* 2009 Control of graphene's properties by reversible hydrogenation: evidence for graphane *Science* **323** 610–3

[23] Sofo J O, Chaudhari A S and Barber G D 2007 Graphane: a two-dimensional hydrocarbon *Phys. Rev.* B **75** 153401

[24] Clark S M, Jeon K J, Chen J Y and Yoo C S 2013 Few-layer graphene under high pressure: Raman and x-ray diffraction studies *Solid State Commun.* **154** 15–8

[25] Robinson J T *et al* 2010 Properties of fluorinated graphene films *Nano Lett.* **10** 3001–5

[26] Zbořil R, Karlický F, Bourlinos A B, Steriotis T A, Stubos A K, Georgakilas V, Šafářová K, Jančík D, Trapalis C and Otyepka M 2010 Graphene fluoride: a stable stoichiometric graphene derivative and its chemical conversion to graphene *Small* **6** 2885–91

[27] Mathkar A, Tozier D, Cox P, Ong P, Galande C, Balakrishnan K, Leela Mohana Reddy A and Ajayan P M 2012 Controlled, stepwise reduction and band gap manipulation of graphene oxide *J. Phys. Chem. Lett.* **3** 986–91

[28] Qin G, Wu L and Gou H 2022 Diamane: design, synthesis, properties, and challenges *Funct. Diam.* **1** 83–92

[29] Boehm H P, Setton R and Stumpp E 1994 Nomenclature and terminology of graphite intercalation compounds (IUPAC recommendations 1994) *Pure Appl. Chem.* **66** 1893–901

[30] Novoselov K S, Fal′ ko V I, Colombo L, Gellert P R, Schwab M G and Kim K 2012 A roadmap for graphene *Nature* **490** pp 192–200

[31] Piazza F, Monthioux M, Puech P, Gerber I C and Gough K 2021 Progress on diamane and diamanoid thin film pressureless synthesis *C* **7** 9

[32] Piazza F, Gough K, Monthioux M, Puech P, Gerber I, Wiens R, Paredes G and Ozoria C 2019 Low temperature, pressureless sp^2 to sp^3 transformation of ultrathin, crystalline carbon films *Carbon* **145** 10–22

[33] Piazza F, Monthioux M, Puech P and Gerber I C 2020 Towards a better understanding of the structure of diamanoïds and diamanoïd/graphene hybrids *Carbon* **156** 234–41

[34] Li H, Cheng S, Li J and Song J 2015 *J. Nanomater.* **2015** 127235

[35] Fyta M 2013 Nitrogen-vacancy centers and dopants in ultrathin diamond films: electronic structure *J. Phys. Chem.* C **117** 21376–81

[36] Muniz A R and Maroudas D 2012 Opening and tuning of band gap by the formation of diamond superlattices in twisted bilayer graphene *Phys. Rev.* B **86** 075404

[37] Barboza A P, Guimaraes M H, Massote D V, Campos L C, Barbosa Neto N M, Cancado L G, Lacerda R G, Chacham H, Mazzoni M S and Neves B R 2011 Room-temperature compression-induced diamondization of few-layer graphene *Adv. Mater.* **23** 3014–7

[38] Chernozatonskii L A, Demin V A and Kvashnin D G 2020 Ultrawide-bandgap Moiré diamanes based on bigraphenes with the twist angles $\theta \sim 30°$ *Appl. Phys. Lett.* **117** 253104

[39] Chernozatonskii L A, Demin V A and Kvashnin D G 2021 Fully hydrogenated and fluorinated bigraphenes–diamanes: theoretical and experimental studies *C* **7** 17

[40] Bakharev P V *et al* 2020 Chemically induced transformation of chemical vapour deposition grown bilayer graphene into fluorinated single-layer diamond *Nat. Nanotechnol.* **15** 59–66

[41] Piazza F and Monthioux M 2021 Ultra-thin carbon films: the rise of sp^3-C-based 2D materials? *C* **7** 30

[42] Son J *et al* 2020 Tailoring single-and double-sided fluorination of bilayer graphene via substrate interactions *Nano Lett.* **21** 891–8

[43] Felten A, Eckmann A, Pireaux J J, Krupke R and Casiraghi C 2013 Controlled modification of mono-and bilayer graphene in O_2, H_2 and CF_4 plasmas *Nanotechnology* **24** 355705

[44] Sorokin P B and Chernozatonskii L A 2013 Graphene-based semiconductor nanostructures *Phys.-Usp.* **56** 105

[45] Tao Z, Du J, Qi Z, Ni K, Jiang S and Zhu Y 2020 Raman spectroscopy study of sp^2 to sp^3 transition in bilayer graphene under high pressures *Appl. Phys. Lett.* **116** 133101

[46] Cellini F, Lavini F, Cao T, de Heer W, Berger C, Bongiorno A and Riedo E 2018 Epitaxial two-layer graphene under pressure: diamane stiffer than diamond *FlatChem.* **10** 8–13

[47] Ke F *et al* 2020 Synthesis of atomically thin hexagonal diamond with compression *Nano Lett.* **20** 5916–21

[48] Pakornchote T, Ektarawong A, Pinsook U and Bovornratanaraks T 2021 Modifying electronic and elastic properties of 2-dimensional [110] diamond by nitrogen substitution *C* **7** 8

[49] Pakornchote T, Ektarawong A, Busayaporn W, Pinsook U and Bovornratanaraks T 2020 Roles of nitrogen substitution and surface reconstruction in stabilizing nonpassivated single-layer diamond *Phys. Rev.* B **102** 075418

[50] Luo Z, Yu T, Kim K J, Ni Z, You Y, Lim S, Shen Z, Wang S and Lin J 2009 Thickness-dependent reversible hydrogenation of graphene layers *Acs Nano* **3** 1781–8

[51] Rajasekaran S, Abild-Pedersen F, Ogasawara H, Nilsson A and Kaya S 2013 Interlayer carbon bond formation induced by hydrogen adsorption in few-layer supported graphene *Phys. Rev. Lett.* **111** 085503

[52] Rajasekaran S, Kaya S, Abild-Pedersen F, Anniyev T, Yang F, Stacchiola D, Ogasawara H and Nilsson A 2012 Reversible graphene-metal contact through hydrogenation *Phys. Rev.* B **86** 075417

[53] Martins L G P *et al* 2017 Raman evidence for pressure-induced formation of diamondene *Nat. Commun.* **8** 96

[54] Grayfer E D, Makotchenko V G, Kibis L S, Boronin A I, Pazhetnov E M, Zaikovskii V I and Fedorov V E 2013 Synthesis, properties, and dispersion of few-layer graphene fluoride *Chem.–An Asian J.* **8** 2015–22

[55] Jeon K, Lee Z, Pollak E *et al* 2011 Fluorographene: a wide bandgap semiconductor with ultraviolet luminescence *ACS Nano* **5** 1042–6

[56] Gao Y, Cao T, Cellini F, Berger C, De Heer W A, Tosatti E, Riedo E and Bongiorno A 2018 Ultrahard carbon film from epitaxial two-layer graphene *Nat. Nanotechnol.* **13** 133–8

[57] Kvashnin A G, Chernozatonskii L A, Yakobson B I and Sorokin P B 2014 Phase diagram of quasi-two-dimensional carbon, from graphene to diamond *Nano Lett.* **14** 676–81

[58] Li L S and Zhao X 2011 Transformation between different hybridized bonding structures in two-dimensional diamond-based materials *J. Phys. Chem.* C **115** 22168–79

[59] Shang N, Papakonstantinou P, Wang P, Zakharov A, Palnitkar U, Lin I N, Chu M and Stamboulis A 2009 Self-assembled growth, microstructure, and field-emission high-performance of ultrathin diamond nanorods *Acs Nano* **3** 1032–8

[60] Sorokin P B and Yakobson B I 2021 Two-dimensional diamond—diamane: current state and further prospects *Nano Lett.* **21** 5475–84

[61] Piazza F, Cruz K, Monthioux M, Puech P and Gerber I 2020 Raman evidence for the successful synthesis of diamane *Carbon* **169** 129–33

[62] Yagi T, Utsumi W, Yamakata M A, Kikegawa T and Shimomura O 1992 High-pressure *in situ* x-ray-diffraction study of the phase transformation from graphite to hexagonal diamond at room temperature *Phys. Rev.* B **46** 6031

[63] Mao W L, Mao H K, Eng P J, Trainor T P, Newville M, Kao C C, Heinz D L, Shu J, Meng Y and Hemley R J 2003 Bonding changes in compressed superhard graphite *Science* **302** 425–7

[64] Lu S *et al* 2013 High pressure transformation of graphene nanoplates: a Raman study *Chem. Phys. Lett.* **585** 101–6

[65] Varlamova L A, Erohin S V, Larionov K V and Sorokin P B 2022 Diamane oxide. two-dimensional film with mixed coverage and a variety of electronic properties *J. Phys. Chem. Lett.* **13** 11383–90

[66] Wang T, Huang D, Yang Z, Xu S, He G, Li X, Hu N, Yin G, He D and Zhang L 2016 A review on graphene-based gas/vapor sensors with unique properties and potential applications *Nano-Micro Lett.* **8** 95–119

[67] Khaliullin R Z, Eshet H, Kühne T D, Behler J and Parrinello M 2011 Nucleation mechanism for the direct graphite-to-diamond phase transition *Nat. Mater.* **10** 693–7

Chapter 5

Moiré diamanes through fluorination and hydrogenation

Maya Devi, Swetapadma Praharaj and Dibyaranjan Rout

Since the discovery of graphene, new structures have constantly been added to the two-dimensional (2D) structure family. Atomically thin van der Waal crystals stacked on each other create new synthetic materials with unique properties. A twist angle or lattice mismatch between the van der Waal layers gives rise to periodic modulation of atomic stacking, which is called a moiré structure. In bilayer graphene (BLG), the displacement of one layer relative to other by an angle 'θ' leads to formation of super lattice, which is named moiré diamane. Out of the different synthesis methods of diamane, surface functionalization with different adatoms enables chemically-induced phase transition, resulting in diamond structures with sp^3 hybridization. Along with the theoretical studies, experimental methods also confirm the formation of fluorinated diamane (F-diamane). The tunable properties of such diamanes have instigated research of moiré structure with different functional groups, such as hydrogen, chlorine etc. This chapter will briefly discuss the different synthesis methods, the properties, and suitable applications of moiré diamane.

5.1 Introduction

In recent years, carbon-based 2D structures have drawn much attention due to the light atomic mass and strong C–C bonds in different states of hybridization. However, the research area is now focused on the discovery and synthesis of novel 2D carbon allotropes, which are capable of showing a wide range of properties due to their phase transformation. Diamond-like structures have attracted much attention due their unique properties, such as high strength, high thermal conductivity, broad optical transparency [1] etc. After the discovery of graphene (a 2D analog of graphite), the search for production of diamane (a 2D diamond) has continued. For the first time, a single layer diamond structure was predicted in 2009 [2, 3]. It was proposed that due to rehybridization of carbon bonds from sp^2-sp^3, graphene can be converted to

doi:10.1088/978-0-7503-5939-9ch5

diamond phase. Such a transformation in a three-dimensional bulk system from graphite to carbon requires high temperature and pressure because of the high energy barrier of diamond nucleation [4, 5]. This energy barrier in diamond nucleation is significantly reduced by the presence of defects or dangling bonds in a precursor [6]. When a few atomic layers are considered in the 2D phase, many properties (e.g., surface chemistry, interface defect concentration and interlayer concentration) come into play, which gives rise to scope of low energy transformation [7–10]. Both theoretically and experimentally, various methods have proposed to transform sp^2 graphene to sp^3 diamond, such as chemical functionalization, application of mechanical strain, and combination of both external physical and chemical drivers. As a result, various types of 2D sp^3 carbon phases have been proposed, which have diverse properties. These 2D carbon phases can be classified as graphane, diamane, and diamene. Graphane is single layer of sp^3graphene that is fully saturated with hydrogen atoms.

Diamane is an ultrathin single layer of diamond that is obtained by compressing BLG to form an interlayer sp^3 bond. Here the sp^2 hybridization is converted to sp^3 by covalent bonding. This is obtained by chemical functionalization of double-layer graphene. Depending on the functionalizing group, they are classified into (i) pristine diamane, (ii) Diamondol or OH-diamane, (iii) F-diamene, and (iv) H-diamane [11]. In BLG, the two layers can be stacked over each other in different ways. One of them is AB stacking (figure 5.1), where the second layer is shifted relative to the first layer by a vector equal to the edge of the hexagon. This is also called as Bernal stalking. The other type of stacking is AA, where the carbon atoms of both layers have identical lateral coordinates. This is called Lonsdaleite stacking.

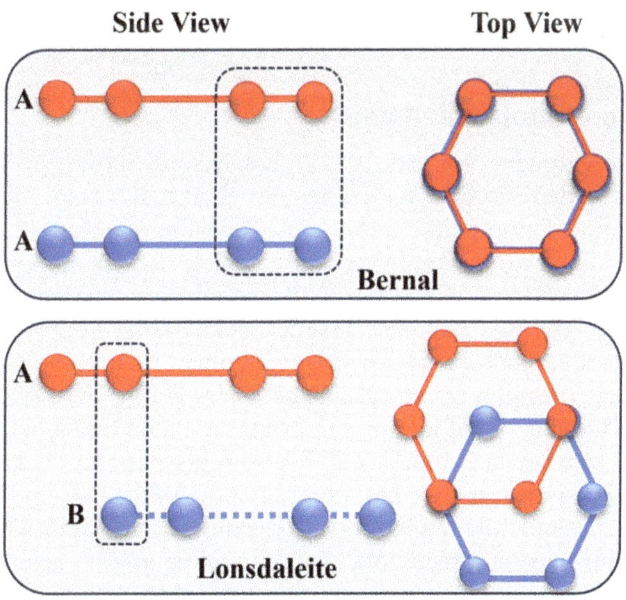

Figure 5.1. Types of stacking in BLG.

Due to lack of thermal stability, atomically thin diamond films cannot be obtained in pristine form [12]. So various functional groups can be added to get a thermodynamically stable structure. In addition to the functional groups, pressure can also be applied for formation of diamanes but on removal of pressure they lose their stability. The theoretical studies on adsorption of hydrogen atom on both types of BLGs were done first and the structures obtained were called diamanes. In 2009, Chernozatonskii used the stacking sequence of both types of BLG to theoretically study the formation of diamane and its isomer diamane II [2]. Many theoretical and experimental works have been conducted to study the formation and various properties of these materials. In 2020, Bakhrav *et al* experimentally synthesized F-diamane [13]. By varying the addition of different functional groups, the properties of diamane are tuned for various applications. The uniqueness of diamane structure depends on (i) the amount of sp^3 carbon and (ii) the area of hydrogenated graphene. For hydrogenation, atomic hydrogen is usually taken because the uniform adsorption of these to BLG will lead to strong chemical bonding between graphene layers, thus overcoming the van der Waal forces [14].

Diamanes can be classified into three types according to structure [15]: (i) complete interlayer bonded configuration from half-hydrogenated stacked bilayers and the non-functionalized carbon atoms bonded to neighboring layers, (ii) hybrid of sp^2 and sp^3system formed from a bilayer stacked with a fully interlayer bonded domain, and (iii) twisted BLG obtained by twisting one layer with respect to the other by a small twisting angle. Hybrid structures with interlayer bonded diamond-like domain along with coexistence of pristine graphene domain and well-defined interface aligned in armchair or zigzag directions of the graphene layer show tunable properties. The twisted BLG is hydrogenated to obtain moiré diamane with unique properties. In this chapter, a brief idea of the synthesis, properties, and applications of moiré diamanes are discussed.

5.2 Structure of moiré diamanes

The monolayer graphene consists of 2D honeycomb lattice. It has hexagonal Brillouin zones with reciprocal lattice vectors K and K′, respectively. The symmetrical position of Dirac points prevents the opening of energy gap in graphene, which limits its application. With vertical stacking of two monolayer graphene, symmetry breaking occurs, which results in inequivalent charge and electrostatic potential between the two graphene layers. So interlayer hopping occurs, which causes hybridization, due to which the bandgap opens up and the Fermi velocity is reduced. The superposition of two graphene layers with a twist angle 'θ' between them gives rise to an array of alternating bright and dark spots, which represent a moiré structure, as shown in figure 5.2. The bright spots represent AA stacking, where every atom in the top layer has a pairing atom in the bottom layer. Similarly, the dark spot in the structure represents the Bernal stacking AB or BA. In AB, the A atoms in each top layer sit directly over an atom in the bottom layer, whereas the B atoms in the top layer do not have any pairing atom in the bottom layer. Exactly the opposite occurs in BA stacking. The superposition of the Brillouin zones of the two

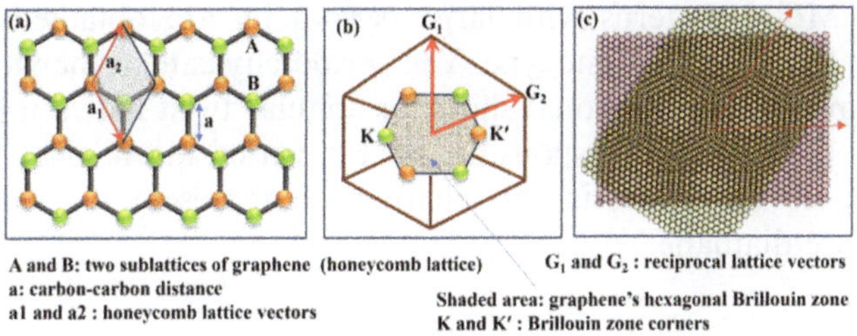

Figure 5.2. Structure of (a) and (b) monolayer graphene and (c) twisted BLG (moiré pattern).

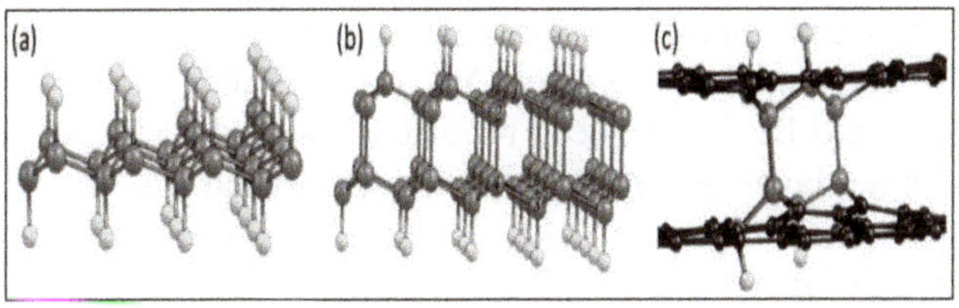

Figure 5.3. Structures of (a) graphane, (b) diamane, and (c) the scheme of the formation of a diamane nucleus on an initial BLG. Reproduced from [3]. CC BY 4.0.

graphene layers gives rise to two mini Brillouin zones. The two layers are coupled together by the sublattice dependent moiré potential.

In the absence of tunneling between layers, the band structure has four Dirac cones from the two layers and two valleys. Due to interlayer tunneling, the crossings at the intersections are avoided, which gives rise to saddle points where the energy band reaches maximum or minimum. These points give rise to enhanced density of state peaks in scanning tunneling spectroscopy, which are termed van Hove singularities (VHS). The twisting angles influence the energy gap between the conduction band and valence band VHSs. For small twisting angles, the VHSs are close to each other, which correspond to strong hybridization between layers. The energy separations of the VHSs' peaks decrease monotonically with the twist angle between the layers, which gives rise to the electronic properties of moiré diamane [16]. Changing the twist angle changes the structure and properties of the parent BLG, and this affects the interlayer bonds during hydrogenation and fluorination. With zero twist angle, AB stacking is more stable than AA stacking [2, 3]. For low twisting angles between 0–16 degree, the formation mechanism of interlayer bonds in structural domain during functionalization is the same as ordinary diamane [2] but moiré diamane is not completely passivated. So a mixture of diamane domain with BLG matrix is obtained. An increase in the twist angle from 16° to 30° significantly changes the stacking pattern. The chemical adsorption of functional groups gives rise to chair like C_6 rings from the pluckered carbon sheets due to pyramidization of hydrogenated or fluorinated carbon atoms. The basic unit of diamane is shown in figure 5.3. The introduction of heteroatom increases the thickness of the material.

The presence of hydrogen on the surface makes the crystal lattice distorted slightly. The structure of the hydrogenated carbon atoms is similar to that of diamond [11]. Moiré patterns with large periodicity are obtained by the superposition of two layers of the same type. The periodicity can be changed by changing the interlayer spacing, translational shifts, or angular twist between the two layers. This change in periodicity is responsible for the varied mechanical, electrical, and optical properties. Therefore, in this chapter we represent the various synthesis methods of moiré diamane.

5.3 Synthesis of moiré diamane

Moiré diamanes are synthesized by stacking of two monolayer graphene with some twisting angle between them. The graphene interlayer connection significantly differs from the connection of AA or AB stalked bilayer. Various methods of synthesis of the twisted BLG have been reported. Twisted BLG can be fabricated by stacking two single layer graphene sheets synthesized by the chemical vapor deposition (CVD) method (figure 5.4(a)). The CVD grown single layer graphene is transferred to SiO_2/Si substrate with a protective layer of poly methyl methacrylate (PMMA). The film is then etched by copper. PMMA is removed by using acetone and annealing. In a similar manner, another layer of graphene is transferred onto the single layer graphene/SiO_2/Si. The sample is baked in air at 150°C for 20 min for proper adhesion between the layers. The removal of PMMA plays a significant role because it can affect the interlayer interaction between graphene. Since the shape of single layer graphene grown by CVD method is random, it is difficult to control the twisting angle in this process [17]. Optical micrographs of the sample show domains of different sizes and shapes (marked in red, yellow and blue), while the atomic force microscopy (AFM) pictures of the single and bilayer depict an increased level of roughness in the bilayer region (figures 5.4(b)–(d)).

The small twist angle BLG can be prepared by using hexagonal boron nitride. In this method, a single graphene flake is divided into two parts. The divided flakes are sequentially picked up by hexagonal boron nitride attached to a hemispherical handle. Between the picking up of two graphene flakes, the substrate is rotated by a

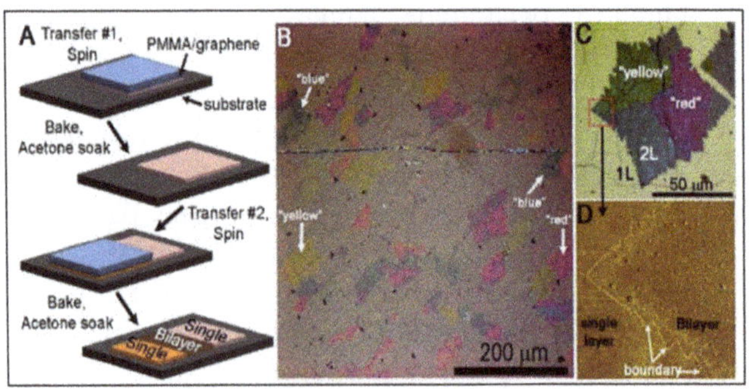

Figure 5.4. (a) Schematic diagram showing the formation of BLG using CVD. (b) Optical micrograph of twisted BLG—domains are labeled in red, yellow and blue. (c) Optical micrograph of sample showing top partial layer (single layer 1L and bilayer 2L). (d) AFM picture showing the interface of single and bilayer. Reprinted with permission from [17]. Copyright (2013) American Chemical Society.

small angle. This gives rise to a twisting angle between the crystal axes of distinctive layers (figure 5.5). A significant control over the twisting angle can be obtained in this method [18].

In another method of synthesis of twisted BLG proposed by Chen *et al* a femtosecond laser was utilized. A single layer graphene sheet exfoliated on SiO_2/Si substrate by use of femtosecond laser was divided by a set of parallel cutting lines. The two sections of graphene sheets were twisted by an angle and stacked together onto a SiO_2/Si substrate by a proper transfer mechanism (figure 5.6). A good control over the twisting angle can be obtained in this method in comparison to other methods [19]. The synthesis method proposed by Wang *et al* [20] controls the folding of single layer graphene. In this method, a large area of monolayer graphene grown on Cu (111) substrate is transferred to a substrate having both hydrophobic and hydrophilic regions with well-defined boundary such that the PMMA layer is in contact with the substrate.

This is then immersed in a solution to etch away the Cu foil. During this process, the PMMA/graphene film that adhered to the hydrophilic part was detached. Then, by replacing the etchant with pure water, the assembly was slowly drawn from the solution. During this process, the floating PMMA/graphene folded over the adhered

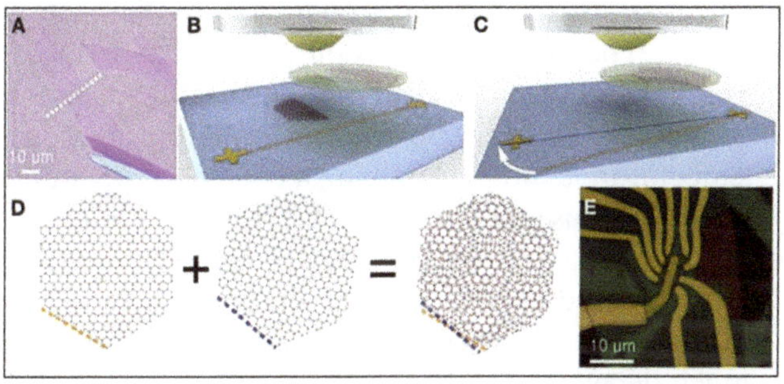

Figure 5.5. (a) Optical micrograph of single flake of graphene: dotted line shows subsequent split into two sections; (b) and (c) two sections detached from the substrate using the same hemispherical handle but in (c) the substrate is twisted by an angle. (d) Schematic diagram of the moiré pattern formed due to twist angle between the two layers. (e) Optical micrograph of small twist angle BLG sample. Reproduced from [18] with permission.

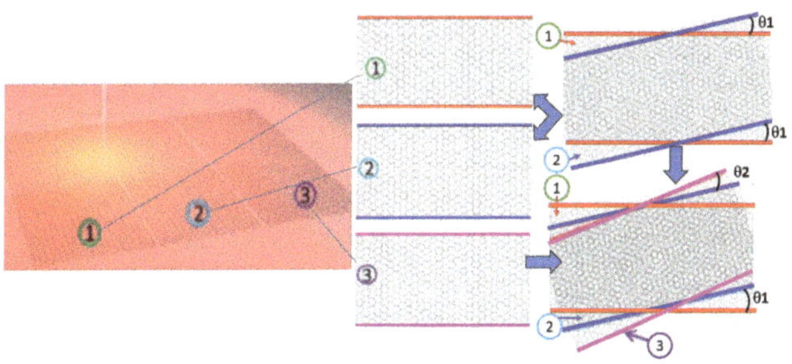

Figure 5.6. Schematic diagram showing the fabrication of twisted BLG double twisted trilayer graphene by cutting-rotation-stacking method. [19] John Wiley & Sons. © 2016 WILEY-VCH Verlag GmbH & Co. KGaA, Weinheim.

layered in the hydrophobic region. The folded film was then transferred to a substrate and the polymer coating was removed by proper washing. By using this method, the required twisting can be obtained by varying the folding angles at the hydrophilic and hydrophobic boundary [20]. Figure 5.7 shows the folding process of monolayer graphene film. Cao *et al* fabricated twisted BLG by vertical stacking of graphene. One of the single layers of graphene is fixed and another layer is stacked vertically by mechanical transfer method. The stacking can be done at various twist angles. Very low twisting angle diamanes are produced in this method, which show the property of superconductivity [21]. Gadelha *et al* presented a preparation of twisted BLG by dry tear and stack method (figure 5.8). In this method, a poly-dimethyl-siloxane (PDMS) semi-pyramidal stamp covered with polycarbonate sheet is used. The PDMS stamp is used to tear the graphene flake into two pieces and they are then stacked. The twisted BLG is then transferred from the stamp to any substrate by detaching the polycarbonate at 180°C [22].

Emelin *et al* reported the formation of diamane from locally tuning BLG. The BLG grown by CVD is transferred to langasite substrate and is then exposed to a focused electron beam through a layer of PMMA. When the electron beam is incident on the substrate, hydrogen and oxygen are released from PMMA and langasite by knock off process. So BLG is converted to diamane by the chemical induction method [23]. Theoretical calculation shows that BLG is difficult to be hydrogenated in comparison to single layer graphene, which makes the synthesis of fully hydrogenated diamane difficult. So the preparation of hydrogenated diamane requires specific temperature control. Applying a controlled pressure on BLG flakes using an electron force microscope tip that is biased by a voltage can produce the moiré structure. Here the tip injects charges on the flakes surface which are covered by water molecules used for controlling the temperature. The water molecules can be controlled by use of resistive heating in dry air [24] but in this process the synthesis of diamane is very much dependent on applied pressure (figure 5.9).

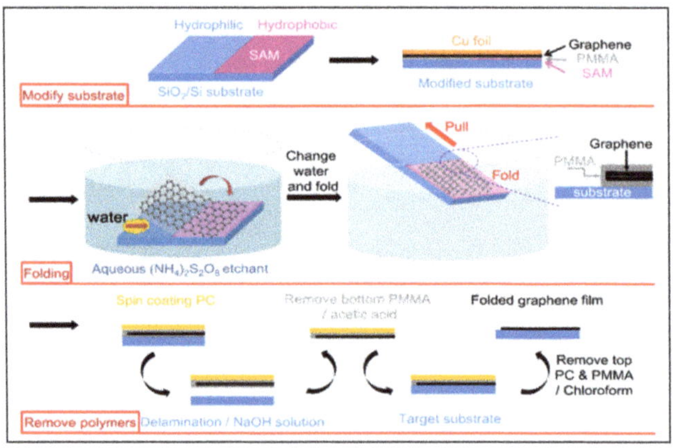

Figure 5.7. Schematic diagram showing the folding process of monolayer graphene film (includes substrate modification, folding regulated by delamination of hydrophilic surface and removal of polymers—PMMA and PC). Reprinted with permission from [20]. Copyright (2017) American Chemical Society.

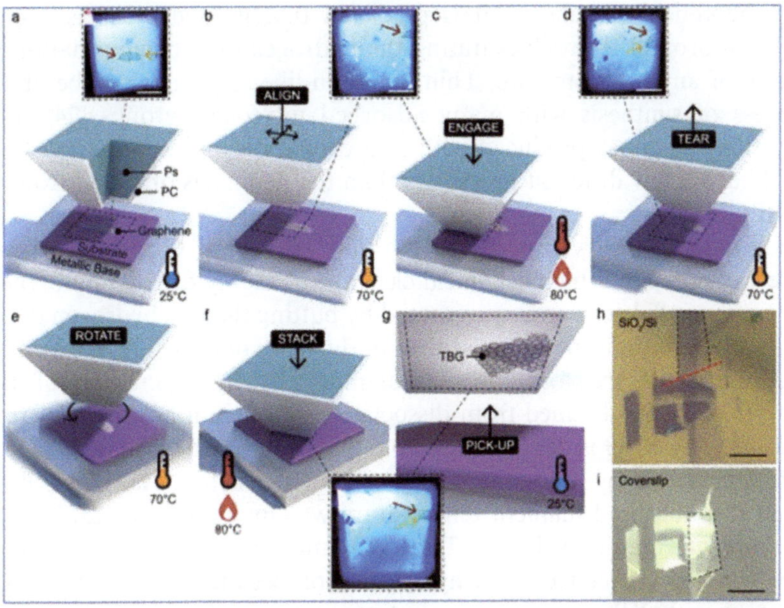

Figure 5.8. Schematic diagram showing tear and stake to build twisted BLG using pyramidal stamp (PS). (a) Visualization of graphene flake through PS. (b) Alignment of PS over a region of the graphene sheet marking the beginning of tearing process. (c) Contact of PS over a part of the graphene sheet while increasing the temperature. (d) Cooling of PS and retraction from the graphene sheet resulting in tearing and picking up of the graphene sheet in contact with the PS. (e) Alignment of PS over the remaining graphene sheet which is being rotated by a desirable angle. (f) Contact of PS with the twisted graphene sheet while increasing the temperature. (g) Separation of PS from the sample while decreasing the temperature. (h) Optical micrograph graphene on SiO_2/Si substrate before the tear and stack process. (i) Optical micrograph of graphene sample transferred to a coverslip after tear and stack process. Reprinted with permission from [22]. Copyright (2021) American Chemical Society.

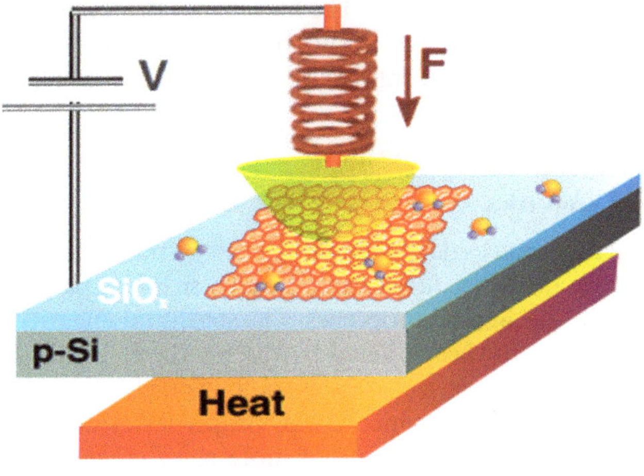

Figure 5.9. Schematic diagram showing the experimental set up of scanning probe microscopy (SPM): the conducting SPM tip applies controlled force during charge injection with bias voltage V onto the graphene layer atop of SiO_x substrate. [24] John Wiley & Sons. Copyright © 2011 WILEY-VCH Verlag GmbH & Co. KGaA, Weinheim

From theoretical calculation it is predicted that graphene sheets can spontaneously twist around themselves during the hydrogenation or fluorination without application of any high pressure. Thin diamond-like structures can be obtained in this process of synthesis with many adsorbed molecular groups [24]. It is also expected that diamanes produced with hydrogenation and fluorination of twisted BLG will have a few different properties than the diamanes obtained from AA/AB stackedBLG.

Hydrogenated BLG was prepared by Luo *et al* [25]. Graphene flakes on Si substrate were prepared by mechanical cleavage of highly ordered pyrolytic graphite. The graphene flakes were hydrogenated by putting them in hydrogen plasma at 1 Torr with different input powers and time of duration of exposure. Recently, Piazza *et al* [26] reported the formation of diamane by hydrogenation of BLG by chemisorptions of H obtained from dissociation of H_2 in a hot filament reactor at low pressure. The BLG is deposited on gold Quantifoil TEM grid by CVD method. It was hydrogenated by a commercial hot filament reactor. The H_2gas flow in an evacuated chamber and filament temperature was maintained at 2350°C–2590°C. The pressure was maintained at 50 Torr. In 6 min 20 s the hydrogenetation occurs [26]. Fluorination is preferred over hydrogenation because it is easy to characterize the structure with XPS due to high electro-negativity of fluorine. Strong binding energy results in shift of C–F peak in comparison to C–C peak. This means that they can easily be distinguished, which cannot be done in hydrogenated graphene. For hydrogenation and hydroxylation, hot filament, plasma, or high pressure are usually required. Therefore, it becomes very difficult to control H/OH coverage but fluorination can be done at room temperature and low pressure by use of XeF_2 as source of fluorine [13]. Fluorination is a lower cost energy process than hydrogenation. The chemical induction method can produce fluorinated moiré diamane from BLG. Bakharev *et al* fluorinated BLG grown by CVD method on a single crystal Cu–Ni. Fluorination was done at 65°C under 50–60 Torr vapor pressure using a home built fluorination system. The fluorinated BLG was transferred to gold TEM grid by using electrochemical bubbling delamination method and the fluorinated diamane structure is confirmed from a HRTEM image [13].

5.4 Properties of moiré diamane

5.4.1 Mechanical properties

Mechanical properties such as deformation, elasticity, strain, tensile strength are key parameters for any nanodevice fabrication. Interlayer bonds play a very significant role in enhancing the mechanical properties. The structural integrity of the system is dependent on resistance to shear offered by the interlayer bonds [27]. Diamane is harder than graphene due to the higher stiffness coefficient, but it is more brittle than graphene. The stiffness and other mechanical properties of diamane is affected by the functional groups [28]. On insertion of more sp^3 C–C interlayer, the uniform load distribution between layers increase, which increases the buckling resistance and reduces the Young's modulus [29]. However, by increasing the Young's modulus of diamane, it can show good stiffness with great flexibility [30]. The addition of selective functional groups can increase the mechanical stability. Functional groups such as H and F increase the mechanical stability but functional groups with bigger atoms like Cl reduce mechanical stability. It is observed that there is more elasticity in fluorinated diamane [31]. The unit area of stalked diamane contains more interlayer C–C bonds than moiré diamane. The corrugated surface of moiré diamane is straightened by the presence of C–C bonds. From calculation, it is found that the moiré diamane has nearly same value of Poisson's ratio as that of diamond. Meanwhile, for stalked diamane the value is very less because of the weak bonding of carbon atoms in the upper and lower layer with hydrogen in comparison to C–C bonding, which confines the structure in adjacent layers. Due to this, more deflection occurs in diamane in comparison to moiré diamane with the application of the same force in a perpendicular direction. All C–C bonds are elongated in daimane but in moiré diamane none of the bonds are significantly elongated under tension. When stress is given, the rupture of the strained diamane occurs at relatively lower strain in comparison to moiré diamane. It is also found that moiré diamane has more rigidity in comparison to untwisted diamane, so a large force is required to breakthrough the membrane with a tip [32]. Moiré diamane has stronger mechanical strength in comparison to untwisted diamane. It is observed that diamane nanoribbons of width less than 100 nm show excellent vibration characteristic with low energy dissipation and high Q factor [33]. The increase in lattice spacing can increase the vibration frequency and lower the stiffness of layer. In moiré diamane, the increased density of the interlayer bonds increases the shear modulus [29]. These exceptional mechanical and vibrational properties of moiré diamane leads to various applications.

5.4.2 Electrical properties

Chemical induction gives rise to phase transition from sp^2 graphene to sp^3 diamond. In this method, surface functionalization of multilayer graphene is done by stacking the layers by different atoms. This phase transition leads to semiconducting properties. These functionalized multilayers show electronic properties of a semimetal. Moiré diamane structure can be fabricated by functionalizing the BLG by hydrogen, fluorine, or chlorine. The transition of electronic properties from metal to semiconductor, or vice versa, is possible by functionalization or defunctionalization [34]. The electronic properties of diamane nanosheets are studied using generalized gradient approximation (GGA) along with Perdew–Bruke–Emzorhof revised for solids (PBEsol) and GGA along with HSE06 [35, 36]. Both calculations predict that two-layer hydrogenated diamane is a semiconductor with wide bandgap. By replacing hydrogen with fluorine, the material shows insulating behavior with a bandgap of 5.6 eV. When comparing the calculated bandgap values with that of bulk diamond it is observed that bilayered hydrogenated and chlorinated diamane have smaller bandgap in comparison to diamond, but the bandgap of bilayered fluorinated diamane is slightly different from that of diamond. The changed bandgap is due to the contribution of surface states to the bandgap edges. The electronic structure of these nanosheets change with the thickness of the layers. As the number of layers changes, the energy barriers of hydrogenation or fluorination decrease. From calculation, it is observed that in hydrogenated and chlorinated diamane the bandgap decreases with an increase in the number of layers. However, in fluorinated diamane the bandgap initially increases with an increase in number of layers but after three layers it decreases. In fluorinated diamane, the bandgap transforms from direct to quasi-direct and then to indirect with an increase in the number of layers. The variation in bandgap with increase in number of layers is due to the formation of surface electronic states in one case and bulk electronic states in the other [37]. By applying Bader theory, it is found that hydrogenated diamane states close to the top of the valence band are bulk electronic states,whereas fluorinated diamanes possess important localized surface electronic states [38]. The projected density of states of bilayer hydrogenated, fluorinated, and chlorinated diamane monolayer, as well as their projected charge density distribution at valence band maxima (VBM) and conduction band minima (CBM), are calculated, which are shown in figure 5.10 [37]. In hydrogenated diamane, the VBM is made up of inplane p-orbitals of carbon atoms (p_x and p_y). Only one lobe of inplane p orbital overlaps on both hydrogenated and non-hydrogenated carbon atoms. CBM consists of antibonding combination of s orbital of hydrogen atom and p_z orbitals of hydrogenated carbon atoms.

In bilayer fluorinated diamane, VBM consists of similar types of bonding as that of hydrogenated diamane along with the existence of two lone pair electrons of

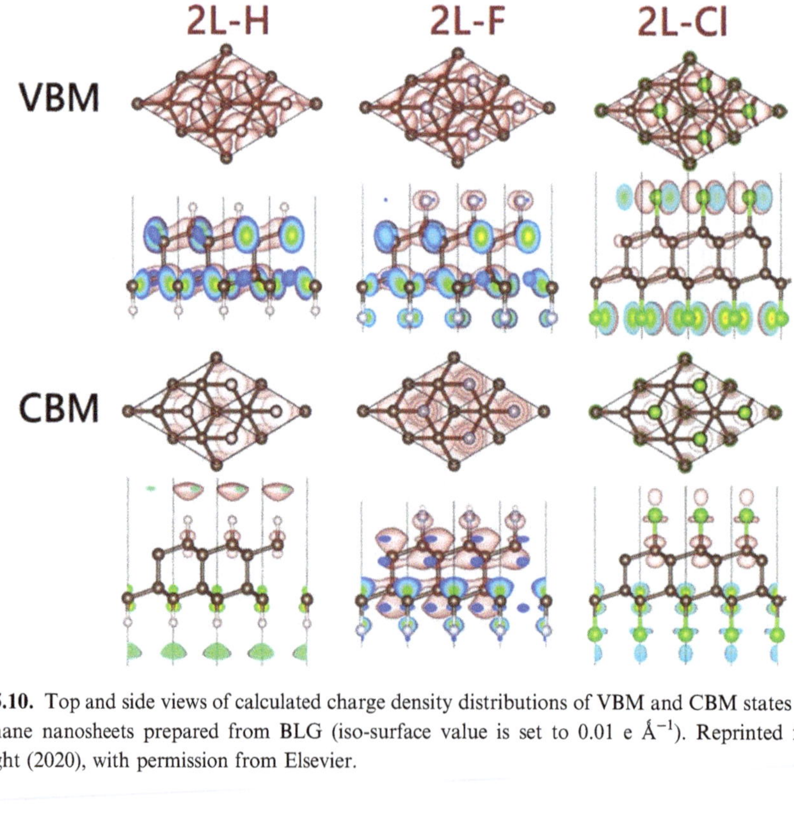

Figure 5.10. Top and side views of calculated charge density distributions of VBM and CBM states of H-, F-, Cl-diamane nanosheets prepared from BLG (iso-surface value is set to 0.01 e Å^{-1}). Reprinted from [37], Copyright (2020), with permission from Elsevier.

fluorine atom. CBM has antibonding C–F states hybridized with large spatial overlap of p_z orbitals of nonflourinated carbon atoms. In bilayer chlorinated diamane, there is a decrease in the contribution of C–C bonds to both VBM and CBM. VBM is dominantly made up of a lone pair of electrons of chlorine, which have much higher energies than fluorine [37]. The absolute energy positions of VBMs are different for hydrogenated, fluorinated, and chlorinated diamanes but the absolute energy positions of CBM of all diamanes are nearly equal. For hydrogenated and chlorinated samples, the quantum confinement effect dominates the bandgap. In fluorinated diamane both the quantum confinement and surface state effects contribute to the bandgap [37]. Strain can affect the electronic bandgap of diamane [39]. A perpendicular electric field can tune the bandgap of diamane. Electric fields monotonically decrease the bandgap of bilayered diamane, modifying the electronic behavior. The application of an electric field induces defects, which can be used to generate a super lattice. Fluorination produces semi-fluorinated diamane with one side conducting and the other side insulating. Such types of diamanes are called Janus-F-diamane. The calculated work function of fluorinated diamane is more than that of hydrogenated diamane. In (110) and (0001) surfaces, the work function shows a similar variation as that of bandgap but in other surfaces the work function decreases with an increase in the number of layers. In these functionalized diamanes, the chemical bonding between carbon and surface adatoms are characterized by the bond-dipole moment. This dipole moment is very much dependent on the electro-negativities of elements. This bond-dipole moment affects the intrinsic electronic property of the material. Calculation shows that the dipole moment is independent of the number of layers. It is also found that the hydrogenated diamanes have the lowest dipole moment, which allows them to saturate faster. However, in fluorinated diamane the strong dipole moment present on the surface does not allow for saturation [34]. These electronic properties of diamane make it suitable for various applications.

5.4.3 Optical properties

From the study of optical properties of diamanes it is found that hydrogenated and fluorinated diamanes do not show absorption peaks in visible region but have large absorption coefficient in the UV region of the light spectrum. However, chlorinated diamanes absorb light over a broad energy range 2–3.2 eV [37]. So the chlorinated diamanes can be suitable for optoelectronic devices. The use of various functional groups could bring improved optical properties.

5.4.4 Magnetic properties

When diamane is substituted/doped with suitable atoms, it gives rise to magnetic properties. Magnetism in diamane is the same as that in graphene doped with boron/nitrogen or suitable dopant. Dopants close the energy gap, narrow the bandgap, and open the magnetic properties. Nitrogen-substituted diamane has a high concentration of holes, which gives rise to strong ferromagnetism. Similarly, creating a silicon vacancy enhances the paramagnetic properties in diamane. With doping, the lowering of the Bohr radius and increasing radiative rate in a single phonon mode makes diamane ferromagnetic. With the application of an electric field, a small to zero energy gap is induced. This small energy gap is different from that in graphene and gives rise to ferromagnetism. Magnetic properties related to both conductive and insulative behavior are shown by diamane. Hydroxylated diamanes usually show insulative behavior and non-hydrogenated diamanes show conductive behavior for ferromagnetism. This important behavior of this type of materials can play an important role in the electromagnetism sector of nanotechnology [15].

5.4.5 Thermal properties

The thermal conductivity of a material has both an electronic and a lattice contribution. Above 300 K, the electronic contribution is insignificant. From calculation it is observed that hydrogenated AB stacked diamane with twist angle zero degree has the highest lattice thermal conductivity (1360 W m^{-1} K^{-1}). This high value of thermal conductivity is due to the ordered lattice structure, with shorter C–C bond length. This provides a short path for heat waves. Since disorderness in twisted diamane give longer paths for heat waves, this in turn reduces the thermal conductivity. Passivation of BLG with heavier fluorine atoms leads to a decrease in lattice thermal conductivity to 361 W m^{-1} K^{-1} in comparison to hydrogenated diamane with 1360 W m^{-1} K^{-1}. In hydrogenated diamane, an increase in twist angle causes the thermal conductivity to decrease drastically to 35 and 32 W m^{-1} K^{-1} in 21.8° and 27.8° twisting angle. Due to passivation by fluorine atoms, the diamane structure is more stabilized because fluorine is heavier in comparison to hydrogen. Consequently, the reduction of thermal conductivity in fluorinated diamane is minimal with an increase in twist angle. The value of thermal conductivity is 97 and 90 W m^{-1} K^{-1} for twist angles 21.8° and 27.8°, respectively [40]. The decrease in thermal conductivity may be attributed to an increase in phonon scattering [41, 42]. Disorder relating to changed twist angle increases the Umklapp scattering [43]. When this scattering dominates the phonon scattering, the thermal conductivity is inversely proportional to temperature. With an increase in structural disorder, thermal conductivity decreases and it also affects temperature dependence. Lattice thermal conductivity is inversely related to lattice anharmonicity. For small twisting angles, the scattering rate remains same for both hydrogenated and fluorinated diamane. However, for higher twisting angles, the scattering rate is more than 10 times that in fluorinated diamane. The higher scattering rates decreases the thermal conductivity. Although fluorine is heavier than hydrogen, at large twisting angles the hydrogenated system becomes more anharmonic, which lowers the value of thermal conductivity for the hydrogenated system [40]. The acoustic phonon branch has the highest contribution to lattice thermal conductivity [43]. It is observed that the contribution from the optical branch of fluorinated moiré diamane decreases monotonically with an increase in twist angle and there is no dependence of phonon contribution to lattice thermal conductivity with increased temperature in the fluorinated system. From all these observations, it is observed that thermal conductivity in moiré diamane is very much dependent on adsorption type and twist angle [40]. Consequently, it can easily be tuned, and therefore diamanes can be considered as prospective candidates for application in thermal management devices.

5.5 Applications of moiré diamane

Diamanes have exceptional electrical, thermal, optical, vibrational, and mechanical qualities that make them valuable for emerging technologies, particularly quantum, photonic, and space-related technologies. Hence, diamanes have a huge potential for application in the rapidly expanding fields of nanotechnology and nanoengineering. Diamanes and diamanoids are also great prospects for nanoelectrical devices, nanoresonators, optical waveguides, and the following generation of thermal management systems because of their properties, such as wide bandgap, tunable thermal conductivity, excellent thermal insulation, etc. Different groups have examined the diamond thin films, or diamanes, under the harsh operating conditions of high temperature, high voltage, and high frequency for electronic applications. In this regard, diamane can be considered to be a good replacement for Si in the semiconductor industry sector because it matched with the efficiency offered by Si. Figure 5.11 shows some important applications and table 5.1 lists various properties of diamanes and their suitability for various applications.

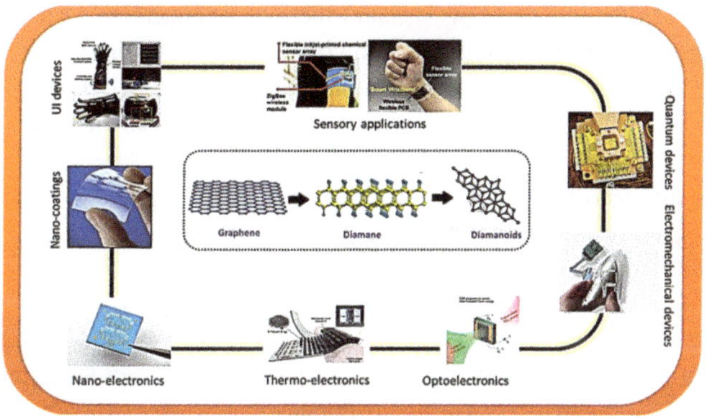

Figure 5.11. Applications of moiré diamanes. Reproduced from [15]. CC BY 4.0.

Table 5.1. Properties of diamanes and their applications.

Sl. No.	Property	Field of application	References
1.	Possibility of engineering bandgap and conductivity by deposition of functional groups and substitution of vacancies.	Field effect transistor. Supercapacitor. Lithium battery and laser medium.	[3, 26, 45]
2.	Possibility of electrical characteristic ranging from metallic to semiconducting. Showing good electron mobility and hole mobility.	Application in semiconducting and nanoelectrical applications.	[27, 37, 46]

3.	Carbon-fluorine dipole formation in layers.	Opportunity for fabrication of electronic and spintronic devices.	[30]
4	Insulating properties of fluorinated diamane.	Can be used in e-beam irradiation for nanopatterning technology.	[13]
5.	Good hydrophobic nature, low dielectric constant, low dissipation factor, and good optical properties.	Suitable for electronic packaging of fan-out wafer-level package.	[47]
6.	Small thickness, excellent electrical, optical and mechanical properties.	Base of touch screen panel.	[28]
7.	Formation of fullerene type of domain from BLG with twist angle of 30 degree.	Suitable for application of semiconducting hybrid film.	[15]
8.	Good electrical conductivity and low thermal conductivity.	Power generating system from thermoelectric material which coverts thermal difference to electrical energy.	[48]
9.	High thermal conductivity in hydrogenated diamane and low thermal conductivity in fluorinated diamane.	Find application in hermos-electric and thermal management systems. Polymer coating to fluorinated diamane enhance chemical stability for harse thermal management systems.	[47, 49]
10.	Wide band-gap spectra with many resonance peaks in valence and conduction bands in density of states.	Essential for nano-optical and nano-photonic applications.	[46]
11.	Coincidence of bandgap with electrical and optical properties of diamane.	Suitable for semiconductor devices and laser media in optoelectronic application.	[26, 28, 50–52]
12.	Fluorine and chlorine doping enhance the bandgap showing thermal and dynamical stability.	Used in photocatalysis.	[37]
13.	Wide bandgap and high mechanical stiffness of fluorinated diamane.	Useful for nano-photonics.	[9]
14.	Light weight, high inplane conductivity and high mechanical strength of fluorinated diamane.	For energy applications.	[47]
15.	Graphite-like diamond structure.	Used in tunnel devices and optical waveguides.	[15]
16.	High hardness, wide bandgap, and tunable thermal conductivity.	Useful for application in electromechanical vibrators.	[46]
17.	Enhancement of electron mobility and hole mobility in hydrogenated and fluorinated diamane due to	Suitable for electromechanical system.	[27, 46]

(*Continued*)

Table 5.1. (*Continued*)

Sl. No.	Property	Field of application	References
	large bandgap opening and elasticity.		
18.	Thermal conductivity tuning as a consequence of hydrodynamic character.	Used in nano-magnetic resonance imaging and sensing.	[53]
19.	Moiré diamane with diamond defects of nitrogen-substituted vacancies.	Application in magnetic, electrical and temperature sensor engineering.	[54]
20.	Hydrogenated, fluorinated, and oxygenated diamane.	Developing chemical and biosensor to detect uric acid, dopamine, and ascorbic acid.	[55]
21.	High hardness and wide bandgap.	Can act as a good mechanical resonator.	[46]
22.	Diamond defects of nitrogen-substituted vacancies and vacancy centers.	Helps in development of quantum information processing.	[54]
23.	Nitrogen-vacancies in diamane.	Gives rise to configuration of q-bits to be used in quantum computing.	[45]
25.	Low coefficient of friction of hydrogenated diamane.	Ultrathin coatings on mechanical parts for improvement in tribological performance.	[15]
24.	Ultra-high strength and improved stiffness in hydrogenated diamane.	Can be used as composite and also finds application in aerospace engineering.	[11, 26, 45]
25.	Nitrogen-substituted diamane can be a super hard coating.	Used for preventing wear and tear and prepare waterproof super hard coating material.	[56]

5.6 Conclusion and future scope

Diamane is a potential 2D structure with various unique properties, which makes it suitable for a wide range of applications. This chapter provides a brief discussion on the structure, synthesis methods, properties, and the predicted applications of diamane in different fields. However, there are still a number of challenges remaining in the synthesis and application of diamane, which needs attention. Out of the reported methods of synthesis, most of them use high pressure or complicated techniques. Consequently, cost-effective synthesis methods using an alternative route should be tried and optimized. The tuning in configuration should be tried to enhance the different properties for wider applications. From theoretical studies, various physical properties of the diamane are predicted. Emphasis must be given to validate these from experiments using different functional groups. Synthesis should also be tried with green synthesis methods. The stable diamane structure should be synthesized and used in the design of various devices with increased efficiency. More effort should be given to study the theory behind the diamane structure, which will help to optimize the synthesis parameters.

References

[1] Moore M 2001 Diamond: natural *Encyclopedia of Materials: Science and Technology* (Amsterdam: Elsevier) 2133–43

[2] Chernozatonskii L A, Sorokin P B, Kvashnin A G and Kvashnin D G E 2009 Diamond-like C 2 H nanolayer, diamane: simulation of the structure and properties *JETP Lett.* **90** 134–8

[3] Chernozatonskii L A, Demin V A and Kvashnin D G 2021 Fully hydrogenated and fluorinated bigraphenes–diamanes: theoretical and experimental studies *C* **7** 17

[4] Bachmann P K, Leers D and Lydtin H 1991 Towards a general concept of diamond chemical vapour deposition *Diam. Relat. Mater.* **1** 1–12

[5] Khaliullin R Z, Eshet H, Kühne T D, Behler J and Parrinello M 2011 Nucleation mechanism for the direct graphite-to-diamond phase transition *Nat. Mater.* **10** 693–7

[6] Boukhvalov D W and Katsnelson M I 2008 Chemical functionalization of graphene with defects *Nano Lett.* **8** 4373–9

[7] Erohin S V, Ruan Q, Sorokin P B and Yakobson B I 2020 Nano-thermodynamics of chemically induced graphene–diamond transformation *Small* **16** 2004782

[8] Rajasekaran S, Abild-Pedersen F, Ogasawara H, Nilsson A and Kaya S 2013 Interlayer carbon bond formation induced by hydrogen adsorption in few-layer supported graphene *Phys. Rev. Lett.* **111** 085503

[9] Kvashnin A G, Avramov P V, Kvashnin D G, Chernozatonskii L A and Sorokin P B 2017 Features of electronic, mechanical, and electromechanical properties of fluorinated diamond films of nanometer thickness *J. Phys. Chem.* C **121** 28484–9

[10] Shul'zhenko A A, Jaworska L, Sokolov A N, Gargin V G and Belyavina N N 2017 Phase transformations of n-layer graphenes into diamond at high pressures and temperatures *J. Superhard Mater.* **39** 75–82

[11] Qin G, Wu L and Gou H 2022 Diamane: design, synthesis, properties, and challenges *Funct. Diam.* **1** 83–92

[12] Ke F, Zhang L, Chen Y, Yin K, Wang C, Tzeng Y K and Chen B 2020 Synthesis of atomically thin hexagonal diamond with compression *Nano Lett.* **20** 5916–21

[13] Bakharev P V, Huang M, Saxena M, Lee S W, Joo S H, Park S O and Ruoff R S 2020 Chemically induced transformation of chemical vapour deposition grown bilayer graphene into fluorinated single-layer diamond *Nat. Nanotechnol.* **15** 59–66

[14] Leenaerts O, Partoens B and Peeters F M 2009 Hydrogenation of bilayer graphene and the formation of bilayer graphane from first principles *Phys. Rev.* B **80** 245422

[15] Tiwari S K, Pandey R, Wang N, Kumar V, Sunday O J, Bystrzejewski M and Mishra Y K 2022 Progress in diamanes and diamanoidsnanosystems for emerging technologies *Adv. Sci.* **9** 2105770

[16] Andrei E Y and MacDonald A H 2020 Graphene bilayers with a twist *Nat. Mater.* **19** 1265–75

[17] Robinson J T, Schmucker S W, Diaconescu C B, Long J P, Culbertson J C, Ohta T and Beechem T E 2013 Electronic hybridization of large-area stacked graphene films *ACS Nano* **7** 637–44

[18] Kim K, DaSilva A, Huang S, Fallahazad B, Larentis S, Taniguchi T and Tutuc E 2017 Tunable moiré bands and strong correlations in small-twist-angle bilayer graphene *Proc. Natl Acad. Sci.* **114** 3364–9

[19] Chen X D, Xin W, Jiang W S, Liu Z B, Chen Y and Tian J G 2016 High-precision twist-controlled bilayer and trilayer graphene *Adv. Mater.* **28** 2563–70
[20] Wang B, Huang M, Kim N Y, Cunning B V, Huang Y, Qu D and Ruoff R S 2017 Controlled folding of single crystal graphene *Nano Lett.* **17** 1467–73
[21] Cao Y, Fatemi V, Demir A, Fang S, Tomarken S L, Luo J Y and Jarillo-Herrero P 2018 Correlated insulator behaviour at half-filling in magic-angle graphene superlattices *Nature* **556** 80–4
[22] Gadelha A C, Ohlberg D A, Santana F C, Eliel G S, Lemos J S, Ornelas V and Campos L C 2021 Twisted bilayer graphene: a versatile fabrication method and the detection of variable nanometric strain caused by twist-angle disorder *ACS Appl. Nano Mater.* **4** 1858–66
[23] Emelin E V, Cho H D, Korepanov V I, Varlamova L A, Erohin S V, Kim D Y and Panin G N 2022 Formation of diamane nanostructures in bilayer graphene on langasite under irradiation with a focused electron beam *Nanomaterials* **12** 4408
[24] Barboza A P, Guimaraes M H, Massote D V, Campos L C, Barbosa Neto N M, Cancado L G and Neves B R 2011 Room-temperature compression-induced diamondization of few-layer graphene *Adv. Mater.* **23** 3014–7
[25] Luo Z, Shang J, Lim S, Li D, Xiong Q, Shen Z and Yu T 2010 Modulating the electronic structures of graphene by controllable hydrogenation *Appl. Phys. Lett.* **97** 233111
[26] Piazza F, Monthioux M, Puech P, Gerber I C and Gough K 2021 Progress on diamane and diamanoid thin film pressureless synthesis *C* **7** 9
[27] Cheng T, Liu Z and Liu Z 2020 High elastic moduli, controllable bandgap and extraordinary carrier mobility in single-layer diamond *J. Mater. Chem.* C **8** 13819–26
[28] Sorokin P B and Chernozatonskii L A 2013 Semiconductor nanostructures based on graphene *Phys. Usp.* **2** 113–32
[29] Machado A S, Maroudas D and Muniz A R 2013 Tunable mechanical properties of diamond superlattices generated by interlayer bonding in twisted bilayer graphene *Appl. Phys. Lett.* **103**
[30] Ahmad Y, Batisse N, Chen X and Dubois M 2021 Preparation and applications of fluorinated graphenes *C* 7 20
[31] Hu Y, Li D, Yin Y, Li S, Ding G, Zhou H and Zhang G 2020 The important role of strain on phonon hydrodynamics in diamond-like bi-layer graphene *Nanotechnology* **31** 335711
[32] Artyukh A A and Chernozatonskii L A 2022 Mechanical characteristics of diamond-like moiré films *JETP Lett.* **116** 737–44
[33] Zheng Z, Zhan H, Nie Y, Xu X, Qi D and Gu Y 2020 Single layer diamond-a new ultrathin 2D carbon nanostructure for mechanical resonator *Carbon* **161** 809–15
[34] Tantardini C, Kvashnin A G, Azizi M, Gonze X, Gatti C, Altalhi T and Yakobson B I 2023 Electronic properties of functionalized diamanes for field-emission displays *ACS Appl. Mater. Interfaces* **15** 16317–26
[35] Csonka G I, Perdew J P, Ruzsinszky A, Philipsen P H, Lebègue S, Paier J and Ángyán J G 2009 Assessing the performance of recent density functionals for bulk solids *Phys. Rev.* B **79** 155107
[36] Brothers E N, Izmaylov A F, Normand J O, Barone V and Scuseria G E 2008 Accurate solid-state band gaps via screened hybrid electronic structure calculations *J. Chem. Phys.* **129**
[37] Mortazavi B, Shojaei F, Javvaji B, Azizi M, Zhan H, Rabczuk T and Zhuang X 2020 First-principles investigation of mechanical, electronic and optical properties of H-, F-and Cl-diamane *Appl. Surf. Sci.* **528** 147035
[38] Bader R F 1991 A quantum theory of molecular structure and its applications *Chem. Rev.* **91** 893–928

[39] Zhang Y, Hu C H, Wen Y H, Wu S Q and Zhu Z Z 2011 Strain-tunable band gap of hydrogenated bilayer graphene *New J. Phys.* **13** 063047

[40] Chowdhury S, Demin V A, Chernozatonskii L A and Kvashnin A G 2022 Ultra-low thermal conductivity of moiré diamanes *Membranes* **12** 925

[41] Wilson J H, Fu Y, Sarma S D and Pixley J H 2020 Disorder in twisted bilayer graphene *Phys. Rev. Res.* **2** 023325

[42] Veeravenkata H P and Jain A 2021 Density functional theory driven phononic thermal conductivity prediction of biphenylene: a comparison with graphene *Carbon* **183** 893–8

[43] Ghosal S, Chowdhury S and Jana D 2021 Impressive thermoelectric figure of merit in two-dimensional tetragonal pnictogens: a combined first-principles and machine-learning approach *ACS Appl. Mater. Interfaces* **13** 59092–103

[44] Piazza F, Gough K, Monthioux M, Puech P, Gerber I, Wiens R and Ozoria C 2019 Low temperature, pressureless sp2 to sp3 transformation of ultrathin, crystalline carbon films *Carbon* **145** 10–22

[45] Ke F, Zhang L, Chen Y, Yin K, Wang C, Zhu W and Chen B 2019 Synthesis and electronic structure characterization of diamane *arXiv preprint* arXiv:1902.01570

[46] Son J, Ryu H, Kwon J, Huang S, Yu J, Xu J and Lee G H 2020 Tailoring single-and double-sided fluorination of bilayer graphene via substrate interactions *Nano Lett.* **21** 891–8

[47] Clark S M, Jeon K J, Chen J Y and Yoo C S 2013 Few-layer graphene under high pressure: Raman and x-ray diffraction studies *Solid State Commun.* **154** 15–8

[48] Zhu L and Zhang T 2019 Suppressed thermal conductivity in fluorinated diamane: optical phonon dominant thermal transport *Appl. Phys. Lett.* **115** 151904

[49] Chernozatonskii L A, Sorokin P B, Kuzubov A A, Sorokin B P, Kvashnin A G, Kvashnin D G and Yakobson B I 2011 Influence of size effect on the electronic and elastic properties of diamond films with nanometer thickness *J. Phys. Chem.* C **115** 132–6

[50] Chernozatonskii L A, Mavrin B N and Sorokin P B 2012 Determination of ultrathin diamond films by Raman spectroscopy *Phys. Status Solidi (b)* **249** 1550–4

[51] Chernozatonskii L A, Sorokin P B and Artukh A A 2014 Novel graphene-based nano-structures: physicochemical properties and applications *Russ. Chem. Rev.* **83** 251

[52] Chernozatonskii L A, Katin K P, Demin V A and Maslov M M 2021 Moiré diamanes based on the hydrogenated or fluorinated twisted bigraphene: the features of atomic and electronic structures, Raman and infrared spectra *Appl. Surf. Sci.* **537** 148011

[53] Chernozatonskii L A, Demin V A and Kvashnin D G 2020 Ultrawide-bandgap Moiré diamanes based on bigraphenes with the twist angles $\Theta \sim 30^\circ$ *Appl. Phys. Lett.* **117** 253104

[54] Felten A, Eckmann A, Pireaux J J, Krupke R and Casiraghi C 2013 Controlled modification of mono-and bilayer graphene in O_2, H_2 and CF_4 plasmas *Nanotechnology* **24** 355705

[55] Pakornchote T, Ektarawong A, Busayaporn W, Pinsook U and Bovornratanaraks T 2020 Roles of nitrogen substitution and surface reconstruction in stabilizing nonpassivated single-layer diamond *Phys. Rev.* B **102** 075418

[56] Tiwari S K, Pandey R, Wang N, Kumar V, Sunday O J, Bystrzejewski M, Zhu Y and Mishra Y K 2022 Progress in diamanes and diamanoids nanosystems for emerging technologies *Adv. Sci.* **9** 2105770

IOP Publishing

Diamane

Fabrication, properties and new advances in 2D diamond

Santosh K Tiwari and Arpan Kumar Nayak

Chapter 6

Properties of diamane anchored with different groups

Ankita Subhrasmita Gadtya, Suresh Sagadevan, Santosh K Tiwari and Srikanta Moharana

Diamane, a diamanoid's prospective two-dimensional analog, has recently received much attention among academic researchers because of its potential unique qualities. The novel nano-allotropes of sp^3 hybridized carbon, diamanes, and diamanoids may be synthesized from a graphitic system by Birch reduction at a regulated pressure with the help of appropriate functionalization and substitution. New technologies, including quantum devices, photonics, and space technologies, may benefit from the exceptional electrical, thermal, optical, vibrational, and mechanical capabilities of these nano-allotropes. Diamanes and diamanoids have great prospects for use in nano-electrical devices, nano-resonators, optical waveguides, and the next generation of thermal management systems due to their broadband gap, adjustable thermal conductivity, superior thermal insulation, etc. The history, synthesis, various properties, structural characteristics, and innovative uses of diamanes and diamanoids are covered in-depth in this typical chapter. We will also emphasize the role that theoretical analysis has played in the in-depth study of diamanes with different anchoring anion groups. This study demonstrates the potential of functionalized diamanes in electronics and engineering applications.

doi:10.1088/978-0-7503-5939-9ch6

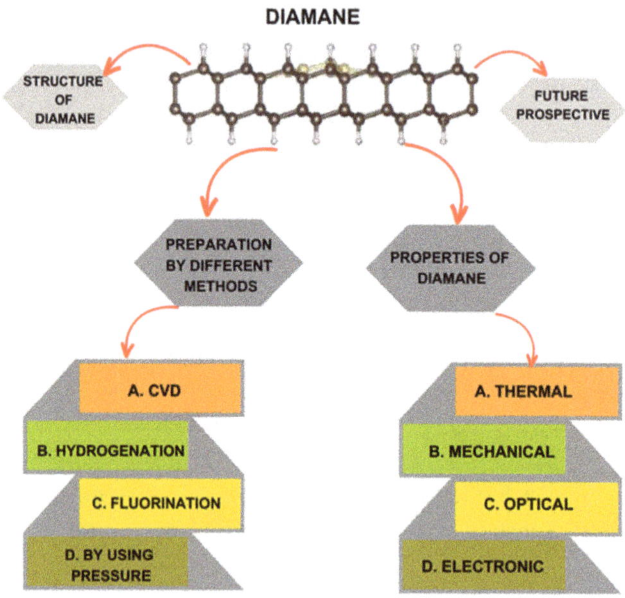

6.1 Introduction

Diamonds and graphite, two allotropes of carbon that have recently been discovered, have remarkably distinct structures and linkages, and better physical, chemical, and mechanical properties [1, 2]. Recently, a lot of theoretical work has been done on a substance called 'diamane.' This material is based on the same carbon atoms as diamond but it has different chemical groups, called functional groups, linked to its surface. The potential for these alterations to provide novel traits and functionalities has broadened the range of possible scientific and industrial uses [3]. Graphite is a soft and opaque semimetal. In contrast, diamond is the hardest known mass material, the best heat conductor, is chemically neutral, and is clear to light. Graphene, or atomically thin graphite, is notable for a number of unusual physical characteristics [3, 4]. These include high carrier mobility, a half-integer quantum Hall effect, and unconventional superconductivity [5]. However, a diamond with a thickness of one atomic layer, or diamane, is expected to behave quite differently from graphene. Diamane is a polymorph that comes in both the cubic (c-diamond) and hexagonal (h-diamond) forms. The plane (100) and (111) are predicted for c-diamane, whereas (0001) and (2110) are predicted for h-diamane [6, 7]. These c- and h-diamonds may have different atomic thicknesses depending on the crystallographic orientations from which they are thinned down. In 2009, Chernozatonskii and his groups [7] made the first effort at modeling bilayer graphene (BLG) with AB and AA stacking with the outer adsorption of hydrogen atoms. They discovered the formation of sp^3-bonded BLG, also known as diamane. Although its Young's modulus is greater than that of bulk diamond, diamane has a narrower bandgap (by 0.8 eV). Moreover, the bandgaps and the effective mass of charge carriers were found to vary with the number of layers of graphene in a fluorinated diamane (FD) sheet, as reported by Kvashnin and his co-workers [8]. Under a perpendicular bias, density functional theory (DFT) simulations performed by Wang and his co-workers have demonstrated a tunable gap and a transition from semiconducting to metal states in hydrogenated diamane (HD) [9]. Additionally, at various pressures and temperatures, the phase diagrams of diamane films adorned by various groups have been demonstrated. Various studies have been reported based on the properties and potential applications of diamane to the diversity of surface functional groups, such as diamane, diamondol, diamondene, and diamine [10]. For instance, Ruoff and his co-workers have reported recent experimental verification of synthesis of the FD under mild conditions, which suggests that the structure's better mechanical qualities may be exploited in practice [11]. There have been many reports on the conversion of BLG into diamane films by attaching fluorine or hydrogen atoms to the outer surface of graphene films [12–14]. In this chapter, we explore diamane anchored with different groups, which represents cutting-edge work in the field of current materials science. By selectively attaching different chemical moieties to the surface of diamond-like carbon structures, researchers have control over the material's behavior, reactivity, and interaction with its surroundings. These results have significantly advanced several fields, from electronics and catalysis to biomedicine and environmental cleaning. However, the six phases, adorned with five functional groups, are explored in terms of their relative thermodynamic and dynamical stability, as well as their geometrical properties [14].

6.2 Overview of diamane

The intriguing family of materials known as 'diamane,' a combination of 'diamond' and 'ane' (representing alkanes), combines the extraordinary characteristics of diamond with the adaptability of organic compounds. Various chemical groups are theoretically attached to the surface of diamond-like carbon structures to make them perform new functions [15, 16]. This functionalization opens up new opportunities for advance in a variety of scientific and industrial domains by allowing the material's qualities to be tailored to suit particular applications.

- **Carbon structure resembling a diamond:** Diamane is composed of the same carbon atoms as a diamond but in a more stable tetrahedral structure. Diamond is one of the most precious minerals on Earth owing to its excellent hardness, thermal conductivity, and optical characteristics, all of which stem from its distinctive crystal structure. The unique characteristics of diamane are inherited from its parent material, diamond, in the form of its carbon skeleton.
- **Modification with the chemical group:** The process of modification of diamane, in which different chemical groups or moieties are covalently bonded to the surface of the diamond-like carbon lattice, is the most innovative aspect of diamane. These functional groups, which can be either organic or inorganic, are crucial in establishing the material's features and behavior.

- **Properties:** Studies have shown that by carefully selecting functional groups and manipulating their density and distribution on the surface of diamane, its characteristics can be adjusted to satisfy a wide range of needs. For instance, a substance can acquire hydrophobic properties by incorporating hydrophobic groups, while electrical conductivity can be increased by incorporating conductive groups.

Diamane has a wide range of applications. Diamane may also be used to develop electronic goods with enhanced functionality and reduced power consumption. For example, certain functional groups can be added to boost catalytic activity and selectivity. Diamane materials have been used in drug delivery and tissue engineering owing to their superior biocompatibility and bioactivity. Furthermore, this theoretical concept has the potential to revolutionize various fields with enhanced material properties [17].

6.2.1 Structure of diamane

Based on BLG, Chernozatonskii *et al* found a new layered structure called C_2H in which each carbon atom is sp^3 hybridized, much like in graphene. One of the graphene layer's two atomic sublattices has covalent bonds with hydrogen atoms, while the other sublattice has carbon atoms, which are situated beyond the carbon atoms of the adjacent graphene, and also have covalent bonds with these atoms. By similarity with hydrocarbon molecules and nanostructures with various hybridizations, we refer to such a C_2H structure as a 'diamane'. Diamane can be produced using discharge hydrogen plasma and BLG. In this case, at the proper conditions (the pressure and temperature of hydrogen gas), hydrogen atoms are chemically adsorbed on both surfaces. A carbon atom from the 'upper' graphene, which is not positioned over a carbon atom from the 'lower' graphene and is hence freer, is displaced from the plane as a result of sp^3 hybridization after a hydrogen atom is joined to a carbon atom. Three of the carbon atoms around it are moved southward here. The 'lower' graphene's atoms undergo a similar rearrangement that results in the sp^3 hybridization of carbon atoms from adjacent layers that are positioned on top of one another. As a result, the diamane nucleus is created. Diamane is stronger than graphane but declines at lower applied loads, according to a molecular dynamics simulation of a membrane being punched via a nanotip. It has a bigger effective stiffness coefficient than graphane [7, 18].

Figure 6.1 shows that the basic unit of graphane (1(a)) and diamane (1(b, c)) with the chair-like C6 rings creates carbon sheets that are a bit wrinkled due to the pyramidalization of hydrogenated (fluorinated) carbon atoms, in contrast to the flat honeycomb pattern in graphene. Meanwhile, as heteroatoms contribute extra atomic layers, their presence makes the material thicker [7]. The extremely symmetric top, hollow, and bridge stackings successively arise on the path of the outer layer of the diamane as it glides along the diagonal of two primitive lattice vectors onto the lower one. Diamane's characteristics and stability boundaries can be drastically altered by its structure. The latter is principally brought about by the graphitization that occurs as a result of graphite's affinity for energy in comparison

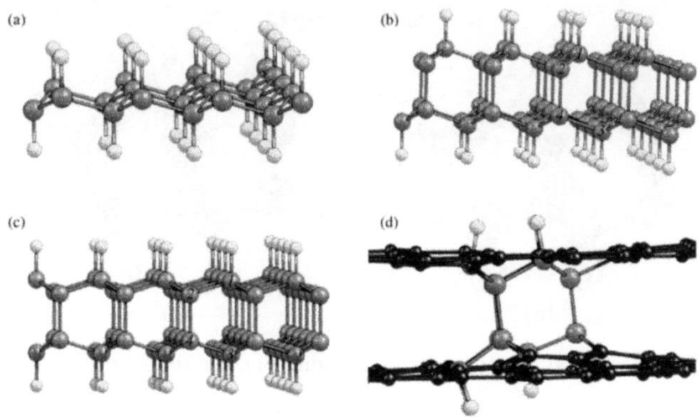

Figure 6.1. Schematic illustration of atomic compositions of (a) graphane, (b) diamane with AB stacking, (c) diamane with AA stacking, and (d) interlayer C–C bond formation scheme on double-layer graphene, where hydrogen atoms (light color) settle on both sides of the graphene and cause carbon atoms (dark color) to create bonds between neighboring layers. Reprinted from [7], Copyright (2009), with permission from Springer Nature.

to diamond. When the dimensions are reduced, surface processes take center stage. However, in the bulk, a sizable transformation barrier keeps the diamond metastable and able to endure across geological periods. When the BLG's rotation angle exceeds 60°, AB stacking tends to prefer coherent stacking sequences in terms of energy. BLG is simply piled with AA stacking without rotations. The cubic diamond (c-D) and hexagonal diamond (h-D), which approached the (111) crystal plane of c-D and the (0001) crystal plane of h-D, are created when the various stacked graphenes come into contact with the surface functional groups. Diamane having Bernal stacking has the lowest formation energy and also the greatest bandgap of all the stacking kinds having the same number of layers. Less than 0.02 eV/atom separates the formation energies of HD with various stacking configurations in the same number of layers. The possibility of all types of stacked diamanes exists given the smaller energy difference. Additionally, a discovery was made that the band-gap energy of a diamond alongside a higher count of carbon layers remains lower than graphane (3.5 eV), and the bandgap continuously reduces as the growth of the number of carbon layers increases. Even five-layer HD could exhibit the quantum confinement effect comparable to the one above [19].

Recently, several study teams have conducted experimental creations of h- and f-diamane (HD and FD). The remaining C atoms that are not linked make covalent bonds with the sublattices of other C atoms in the surrounding graphene, forming a new carbon system. The two different C atoms from two different subatomic lattices with H (F) atoms construct a multilayer unit in h-diamane (f-diamane). The computed in-plane lattice constants for HD and FD respectively, are 2.52 Å and 2.56 Å, which in fact remain next to other computational and experimental values. The interlayer C–C linkages in the two systems are almost identical at 1.56 Å, which is very close to the value of bulk diamonds (as shown in figure 6.2). In accordance with the atomic roughness hypothesis of friction, a geometric fold of 0.5 Å emerges in the graphene as a result of interactions between the carbon atoms in the

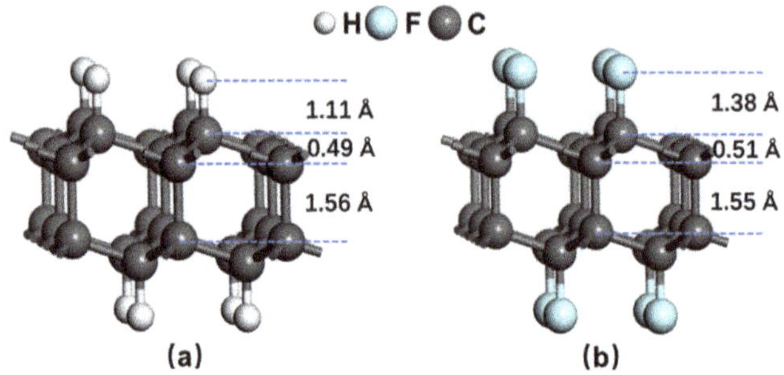

Figure 6.2. The improved geometrical structures of (a) h-diamane and (b) f-diamane. Reproduced from [23]. CC BY 4.0.

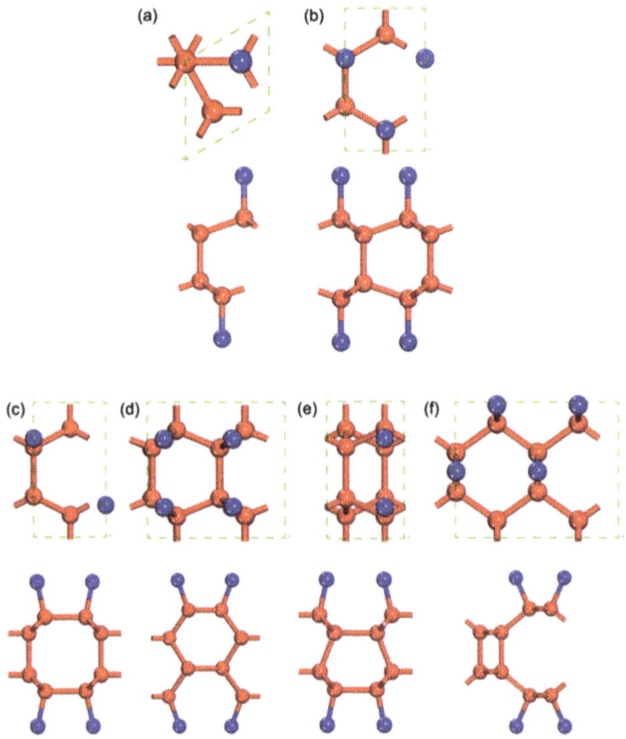

Figure 6.3. Six-stage plans are shown schematically. First chair (a), second chair (b), third boat (e), fourth boat (f), and so on. Carbon nuclei are depicted by the red spheres, while the other groups—H, F, Cl, OH, and NH_2—are shown by the blue spheres. Reproduced from [10] with permission from the Royal Society of Chemistry.

surrounding layers and the passivated atoms [20–23]. Fully HD preparation is still challenging and necessitates careful temperature control. Compared to HD, FD has much superior thermal stability. Fully FD is easier to manufacture experimentally than fully HD, and FD may be stabilized at 400°C, even in air [23–26].

The six phases are schematically represented in figures 6.3(a)–(f), which are identified as chair1, chair2, boat1, boat2, boat3, and boat4, respectively. Ge and his co-workers analyzed six distinct phases ornamented by five groups, i.e., a total of thirty structures.

6.3 Brief synthesis techniques of diamane

6.3.1 Chemical vapor deposition technique

The production of diamane and its analogs typically uses the chemical vapor deposition (CVD) technique. The stability of hybridized sp^3 structures formed from the pure sp^2 graphene is highly important in the formation of diamane, and this stability can be attained using a high-quality CVD approach in which the vapor is solely deposited onto the external side of the BLG. However, a little amount of edge deformation is frequently observed during CVD deposition, which identifies the technique's disadvantage. The peak of the nucleation barriers, which controls the maximum development of functional groups chemisorbed onto the layers of diamanes, can have faults if functional groups are restricted to the surface of the diamanes. The functional groups in the chemical vapor source start to determine whether the phase shift from graphene to diamanes is thermodynamically feasible or not. The frequently employed functional group sources are ammonia, oxygen, water, hydroxyl group, and chlorine for the CVD method, which are used to synthesize diamanes. Chemisorption, which is ultimately responsible for the stabilization of diamane, is an important step in CVD. The virgin BLG's dangling bonds can be passivated via a sophisticated CVD method, allowing for the creation of diamane, which are stable embedded nanostructures. As initial components for the manufacture of diamane by CVD method, suspended BLG sheets from graphene were deposited on 3 mm diameter gold Quantifoil TEM grids. Nearly 98% of the grid was covered. The graphene stacking order produced by the BLG films results from the sequential relocation of two multi-domain mono-layered graphenes (1LGs) generated by CVD, which is why it changes randomly across the film. However, because the 1LG layers are polycrystalline (20-m-large grains for the most part) and rather thin (3 mm), there is a good likelihood that the layers will locally have the favorable AB or AA stacking arrangements at the micrometer scale [27].

6.3.2 Hydrogenation

A commercial hot filament reactor was used during the preparation procedure of diamane via hydrogenation. Two tungsten wires were used as heating elements. To prevent carbon contamination, the filaments were washed with acetone before mounting. The hydrogenation process for carburization or conditioning, which is normally necessary for the traditional generation of diamonds by hot filament CVD, was not performed on the filaments because of the same reason. The stainless-steel substrate holder was mobile and fluid-cooled, and the grids were set up vertically on it. A thermocouple tip that is situated on the surface of the substrate holder, directly beneath the filaments, was used to estimate the temperatures of the grid and substrate holder. Before injecting a controlled amount of ultrahigh purity H_2 gas from the top, the chamber was at least 10^{-1} Torr evacuated. An automated valve that was situated below the substrate holder controlled the pressure. The temperature of the filament was improved and held within the parameter of 2350°C –2590°C, when H2 gas was further added to the chamber and criteria of steady-state, was set up. Along with this, the flow and pressure inside the chamber were maintained at 1 cm and 50 Torr respectively. The substrate holder and the filaments were spaced away from each other at 22.8 mm. As a result, for few layer graphene (FLG) and BLG, the maximum temperature for substrate holding was 325°C and 400°C, respectively. These estimates consider a 50°C difference in substrate holding temperature across the grid. It took the hydrogenation process around 10 min to finish. Recent progress in the fabrication of broad-area and superior-quality graphene films with a precisely controlled number of layers has revealed promising pathways for the experimental conversion of graphene into diamane films. Experimental evidence of the partial phase shift of a FLG to a diamond-like structure on Pt (111) was given by Rajasekaran *et al* [22]. In order to create diamane films, BLG has been exposed to hydrogen radicals using the hot filament procedure. The authors tracked the structural conversion using UV Raman spectroscopy, which revealed the existence of a distinct sp^3-bonded carbon in the stretching mode and the absence of a sp^2-bonded carbon peak. DFT simulations have been used to investigate the stability, mechanical, electrical, and optical aspects of diamane nanosheets, based on experimental studies into the substance [20, 28, 29].

6.3.3 Fluorination

Fluorination is a technique that uses materials to react with compounds containing fluorine to yield the desired result. Under ambient circumstances, fluorination is more stable than hydrogenation. According to Bakharev *et al* [11], an f-diamane film was produced when fluorine was chemically absorbed into BLG that had undergone CVD growth. In their experimental section, BLG films developed on the surface of CuNi (111), which were again fluorinated at 65°C under a vapor pressure of 50–60 Torr of XeF_2. It was discovered that the fluorination parameters (temperature, XeF_2 partial pressure, and exposure duration) could be altered to control the fluorinated graphene structures' C/F ratios. In addition, they found that the FD film acts as an extremely thin wide bandgap semiconductor. In another study, f-diamane nanosheets were prepared using liquid-phase exfoliation. A suspension (0.5 mg ml^{-1}) of each sample in chloroform was sonicated for 5 min to produce stable dispersions of exfoliated fluorinated expandable graphite and fluorinated expanded graphite (F-Exp-able and F-Exp-ed) (as shown in figure 6.4). The highly expanded structure aids in achieving effective exfoliation. In contrast, the previously reported method for exfoliating unexpanded (C2F)n required a highly intense process and engaged the use of probe sonication with greater power input as well as further treatment (ice bath and pulse effect) to prevent overheating and defluorination [28].

6.3.4 By using pressure

In recent times, Gao and his co-workers [30] used SiO2 as the substrate and in situ examined the pressure's impact on the BLG sample using an atomic force microscope studied at ambient temperature. The findings of the experiment show that the resulting

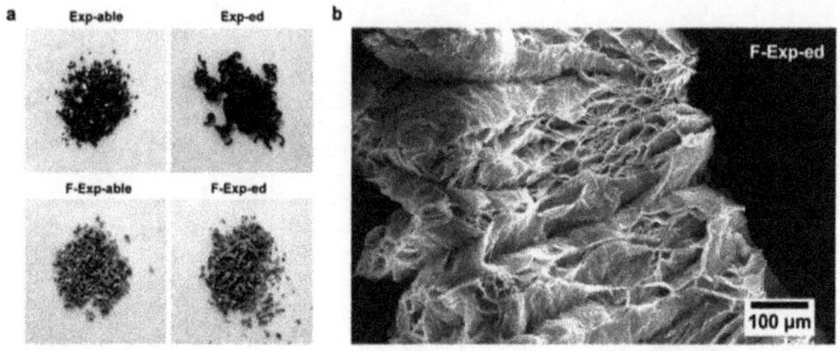

Figure 6.4. Images of expandable graphite (Exp-able), expanded graphite (Exp-ed), fluorinated expandable graphite (F-Exp-able), and fluorinated expanded graphite (F-Exp-ed). (b) SEM image of highly expanded F-Exp-ed. Reprinted from [28], Copyright (2022), with permission from Elsevier.

sample's hardness was approximately identical to that of diamond and that indenting caused a reversible decrease in electrical conductivity. It was thought that a carbon film like a diamond was produced in this experiment. However, when graphene thickness increased (3–5 layers), its performance was limited by the emergence of an unfavorable layered stacking structure (AB stacking). Earlier, the conditions including hydrostatic and non-hydrostatic in a diamond anvil cell (DAC) were used for pure FLG to study the high-pressure phase transformation behavior. The findings demonstrated that at 16GPa, seven-layered graphene generated interlayer sp^3 C–C bonds with partial sp^2 bonds. Unfortunately, after the pressure was released, the graphene structure was still intact. Recently, the measurement of resistance was used to estimate the pressure required for the phase change of pure graphene. As a function of pressure, the sheet resistance of samples with different layers. Combining XRD and theoretical simulation, it is found that the hexagonal pristine diamane (PD) is produced during compression with three-layer or thicker graphene. According to earlier theoretical research, the phase transition pressure gradually reduced as the number of graphene layers increased. Contrary to earlier theoretical claims, the hexagonal PD created under enhanced pressure could be preserved to 1 GPa [25, 30–32].

To functionalize the top of BLG, Martins *et al* conducted high-pressure tests in a DAC. In the experiment, water served as a pressure transmission medium, as well as a source of hydroxyl or hydrogen atoms (figure 6.5(a)). Transferring the BLG created by CVD to a Teflon substrate. The substrate could stop the bottom carbon atoms from getting functionalized because it did not react with the precursor graphene. Therefore, chemical adsorption only took place on top of graphene. Raman spectroscopy measurements (figures 6.5(b) and (c)) during the pressurization process revealed that the carbon layer underwent sp^3 hybridization, resulting in a mixed sp^2 and sp^3 and a broadening of the G band, as the pressure increased. According to calculations made using DFT, diamane was created at 4–5GPa [33].

6.4 Properties of diamane

6.4.1 Thermal and mechanical properties

In addition to being tunable by diamane's thickness, functional group type, and conformation, diamane has a shorter bandgap than diamond. It has a significantly

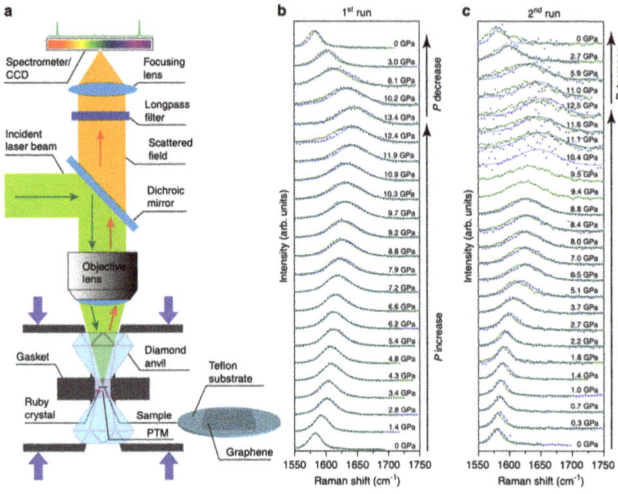

Figure 6.5. Illustration of (a) experimental setup and (b) and (c) Raman spectra with the increase in pressure. Reproduced from [33]. CC BY 4.0.

higher stiffness coefficient than graphene and is more brittle, diamane gets destroyed at lower applied stresses despite appearing to be stronger than that material. The electronic states are also changed by dopants or imperfections to narrow the diamane bandgap. Selective FD conversion to graphene is possible with electron beam irradiation. Diamane has potential uses in the realm of nano-optoelectronics due to these characteristics. Due to the minimal coefficient of friction of surfaces that have been hydrogenated, diamane can also be used to strengthen the wear resistance of coated mechanical parts. The arrangement of the stacking and the functional groups considerably affect thermal conductivity, according to contemporary theoretical studies on diamane. HD demonstrates better heat conductivity than FD. The AA stacking sequence has a superior thermal conductivity value than the AB stacking sequence, which is a significant distinction to make. Out-of-plane acoustic phonon approaches and optical modes in HD and FD, respectively, are the key contributors to thermal conductivity. Diamane also has excellent rigidity, making it particularly desirable for ultrathin protective coatings, such as those used on composite materials in flight equipment. Recent research on the vibrational characteristics of diamane nanoribbons (DNRs) further revealed that the DNR resonator has a significantly high natural inherent frequency and a significant quality factor (*Q*-factor). The stacking order of a diamond has no effect on its vibration qualities, unlike heat conductivity. The ultrasensitive resonator-based sensors are anticipated to utilize high frequency in diamane [34, 35].

The first principal method based on DFT was used to investigate the frictional properties of diamonds by Wang *et al.* Because the folded carbon in f-diamane can be hidden by the packed charges of the passivated fluoride layer, it has essentially the same value as flat graphene. They also compared the frictional properties of HD/HD and FD/FD and found that HD/HD has twice the value of μ than the latter, indicating an electronic structural difference. Corresponding to the charge density distribution, the charged layer of FD has more compact packing than that of HD, which can effectively prevent contact between two carbon films and intensify electron–electron repulsion between both of the F layers. Greater interlayer spacing and reduced binding energy in FD/FD show a close relationship between the electrical configuration of passivated atoms and the frictional characteristics of diamane. Their study demonstrates that the thinnest sp^3-hybridized diamond films show admirable friction properties as compared to graphene. Additionally, in previous reports the film also showed a favorable bandgap, satisfactory heat conducting properties, and immense hardness properties. The properties related to friction can be changed by altering the passivated atoms. The findings add to our understanding of diamane in tribology and offer a tribological foundation for its widespread use in micro- and nano-devices [23].

The intriguing physical and electrical characteristics of diamane, which have been discovered through both experimental and numerical analyzes, include high electron carrier mobility, a huge bandgap, and mechanical rigidity that is comparable to bulk diamonds. Additionally, the range of thermal conductivity was estimated from 1960 to 2440 W m^{-1} for various allotropes at ambient temperature, which is significantly higher than that of the majority of 2D materials alike MoS_2, $MoSe_2$, hexagonal boron nitride, phosphorene, etc. In-depth research by Hu *et al* revealed that tensile strain significantly affects the in-plane thermal conductivity (κ) of single-layer diamane. This research used first-principles calculations based on DFT. Under 5% tensile strain, it was discovered that the strength of the in-plane reduced by 50%-1081 W m^{-1} K^{-1}. The computed strength decreased to only 226 W m^{-1} K^{-1}, which was only about 11% of its initial value when the strain was further increased to 20%. The authors discovered that with a 10% strain, the phonon relaxation time weakened by a relatively array of magnitude, although the group velocity value was only marginally affected, which they used to explain this phenomenon [36]. Zhu and his co-workers have studied the thermal properties of diamane with the help of the Peierls–Boltzmann transport equation method in both AA and AB stacking sequences of diamane and called them D-AA and D-AB, respectively. The symmetries of D-AB and D-AA are considerably distinct—although D-AB has an inversion symmetry, D-AA has special horizontal reflection symmetry. Despite the fact that phonon dispersions are identical, the two allotropes may be distinguished thanks to a significant variation in Raman intensity at 1345 cm^{-1}. Most intriguingly, at 300 K, the thermal conductivities of D-AB and D-AA, two diamane allotropes, are 1960 and 2240 W m^{-1} K^{-1}, respectively. Even at normal temperatures, the phonons in diamond correspond to a hydrodynamic regime that is relatively close to that of graphene, and the ZA phonons (>40%). Meanwhile, the material's extraordinarily high thermal conductivity is mostly due to the other two acoustic modes. Due to the reflection symmetry of D-AA, the amount of thermal conductivity of D-AA is approximately 15% greater than that of D-AB. In particular, reflection symmetry is sustained at all *k*-points in the Brillouin zone. The extremely high thermal conductivity indicates that wide bandgap semiconductors in thin-diamane films have a lot of potential for use in electrical applications [37].

Cheng *et al* have discovered that FD has a broad bandgap but a significant curvature was found at the conduction band minimum (CBM), indicating that its effective mass of conduction electrons is somewhat low. Carbon-based materials such as bulk diamond and graphene typically have extremely high mechanical strengths. In order to further confirm the accuracy of the effective mass, they computed the effective masses of the minimum conduction band and the doubly degenerated maximum valence band using the G_0W_0 and PBE methods. We discovered a few differences between the two methods. Given the difference in thickness, the values of FD (480 J m^{-2}) and HD (459 J m^{-2}) are comparable to (and somewhat lesser than) that of graphene (340.3 J m^{-2}). Additionally, the shear modulus of G is comparable to that of graphene under shear strain (144 J m^{-2}) (217 for FD and 226 J m^{-2} for HD). These larger elastic moduli than the majority of other 2D materials point to the exceptional mechanical capabilities of diamane materials, which are comparable to those of graphene and bulk diamond. They have also observed that FD is a direct-broad bandgap material with a 7.96 eV bandgap at the Γ point, consisting of doubly degraded bands at the valence band maximum (VBM) and two branches for hole mobility. These parameters are comparable to HD, with the exception that HD has a lower bandgap using the G_0W_0 technique, at 4.10 eV. In terms of mechanical strength, Young's moduli of 480 and 459 J m^{-2} for FD and HD, respectively, are discovered to be comparable to graphene and also superior to the majority of 2D materials. Since they apply applied biaxial and uniaxial loads, they further investigated the optimum strength and failure mechanism, which also involved elastic and phonon instability. It has been determined that the critical uniaxial strain for FD is $\varepsilon_x = 8\%$ and that the critical biaxial strain is $\varepsilon_x = 15.5\%$ due to elastic instability and soft-mode phonon instability, respectively. The critical uniaxial strain value was found to be $\varepsilon_x = 13\%$ for HD which is caused by the phonon instability failure mechanism. Furthermore, it appears that the effective mass of conduction electrons and valence holes is fairly low given the unusually huge curvature at the VBM in HD and the CBM in FD. The electron mobility of around 2732 $cm^2\ V^{-1}\ s^{-1}$ in FD and the hole mobility of around 1565 $cm^2\ V^{-1}\ s^{-1}$ in HD are both substantially higher than bulk silicon and MoS_2 and also competitive with standard III–IV (as shown in figure 6.6) semiconductor materials, which are both the result of outstanding elastic modulus and modest effective mass. These two diamond-like materials, which could possibly be a suitable substrate for nano-electromechanical systems, are anticipated to have significant potential for use in nano-optics and nanoelectronics in the future [38–41].

Similar to other 2D materials, phonon dispersions in diamane always have three acoustic phases at the Γ point, two of which are linear and one of which has a quadratic connection. The absence of hypothetical frequencies in the phonon dispersion relations for FD is evidence of the dynamical and structural stability. For Cl- and HD, however, tiny imaginary frequencies in the shape of pits can be seen at the Γ point. The flexural acoustic modes are characterized by patterns in the U-shape, which can be challenging for 2D materials to converge and are typically caused by computational artefacts. In fact, they discovered that the small imaginary frequencies in the achieved phonon dispersion equations for the 2L-H and 2L-Cl

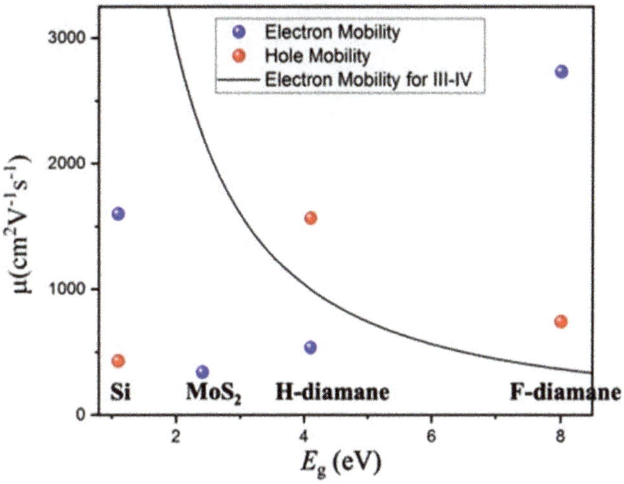

Figure 6.6. The relationship between carrier mobility and bandgap E_g for FD and HD in bulk III–IV semiconductor materials (solid line, represented by $\mu = 0.83 \times 10^4\ E_g^{-3/2}$). Reproduced from [38] with permission from the Royal Society of Chemistry.

lattices fade out when utilizing bigger supercells of 5 × 5 × 1. Around the frequency of 86 THz, relatively flat bands form as a result of the enhanced vibrational frequencies of the hydrogen atoms. One might draw the conclusion that the diamane nanosheets under consideration are dynamically stable based on the phonon dispersion measurements. Additionally, we estimated the Br-diamane phonon dispersions. The results show significant imaginary frequencies, which point to dynamical instability. The sheets of diamane which are constructed of BLG demonstrate the lowest mechanical strength, according to an analysis of mechanical properties. At a high temperature of 1000 K, HD and FD nanomembranes could maintain their entire integrity. Nevertheless, Cl-diamane can be entirely stable at 500 K but disintegrates above 1000 K. At a high temperature of 1000 K, the diamane's surface first begins to gradually lose Cl atoms, and eventually, as more Cl atoms leave, the structure changes to become BLG [39]. Using atomistic simulations, Wu *et al* have investigated the mechanical properties of the single-layer diamond, or diamane, following tensile and bending distortion. Their findings indicated an enhanced Young's modulus value in the diamane which is proportionate to graphene's and that this modulus is independent of the stacking order and also quite similar along both the zigzag and armchair directions [40, 41]. For instance, when we take the AB-stacked diamond, its Young's modulus in the zigzag direction was found to be 730 GPa. Contrary to Young's modulus, the crystallographic direction has a major influence on the fracture strain and strength. In particular, the fracture strain/strength value was found to be significantly higher in the zigzag direction than in the armchair direction. Stress–strain curves for H-, F-, and Cl-diamane nanosheets under uniaxial loading in the armchair and zigzag configurations are shown in figure 6.7 [42]. The cracks spread in zigzag paths in accordance with the atomic arrangements, just as was seen in graphene. This deformation

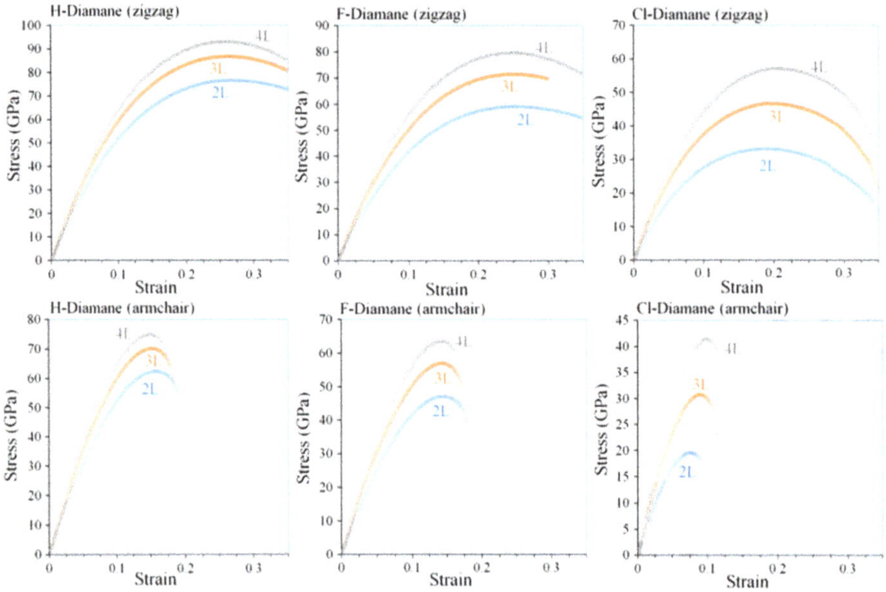

Figure 6.7. Diamane nanosheets' reactions to uniaxial stress and strain vary by functional group, carbon layer count (thickness), and loading direction (armchair, zigzag). Reprinted from [42], Copyright (2020), with permission from Elsevier.

process is continuously visible in tensile deformation in the zigzag and armchair orientations. More significantly, the temperature affects the tensile mechanical characteristics of diamonds. While Young's modulus and fracture strength demonstrate a linear dropping relationship with temperature, the fracture strain displays a nonlinear decreasing tendency. In particular, there is good agreement between the kinetic fracture hypothesis and the association between fracture strain/strength and temperature. At about 3600 eV Å, it is also found that the high bending stiffness of diamane is comparable to the zigzag and armchair directions [42, 43].

In figure 6.8(a) Young's moduli show a similar shifting tendency in both zigzag and armchair directions for the D-AA (AA-stacked diamane) and D-AB (AB-stacked diamane) samples. In the armchair and zigzag orientations, Young's moduli of D-AB at 300 K are 687 and 689 GPa, comparable to those of 2D diamond-like films with totally and partially covered surfaces in hydrogen. In the zigzag and armchair directions, the predicted Young's modulus is found to be greater than graphane, which has a value of roughly 610 and 590 GPa, respectively. In figure 6.8(b), a diminishing tendency is seen when the fracture strength of the D-AB sample is compared along the armchair direction at various temperatures. When the temperature rises, the fracture strength diminishes more quickly than in other carbon nanostructures, such as graphene and CNT. According to calculations, diamane tends to be unstable or break down at temperatures higher than 1000 K [40, 41]. Overall, our findings show that diamane's mechanical characteristics are strongly temperature-dependent. Ultrathin protective coatings, such as those used to

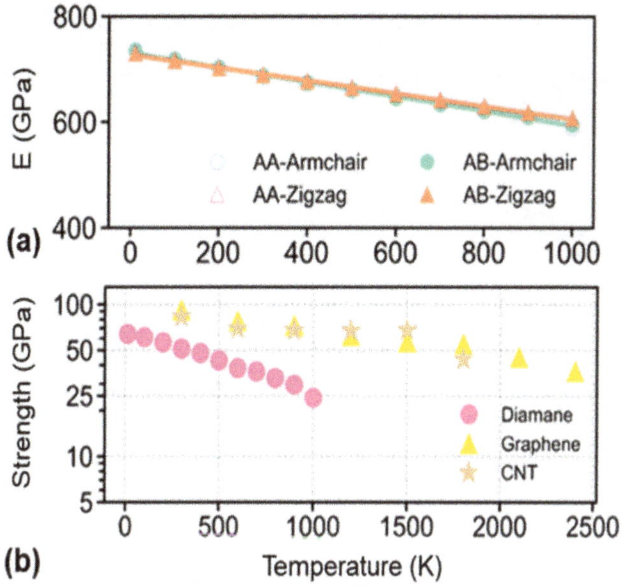

Figure 6.8. Schematic diagram of (a) D-AB and D-AA samples' Young's moduli in the zigzag and armchair directions and (b) temperature-dependent fracture strength of the D-AB sample in the armchair orientation. Reprinted with permission from [41]. Copyright (2020) American Chemical Society.

protect aircraft equipment and other ultrahigh-strength composite components, may find diamane to be a desirable material due to its high stiffness [43].

6.4.2 Optical properties

Recent studies of the vibrational properties of diamane nanoribbons have demonstrated that DNR resonators have a high natural frequency and a substantial quality factor (Q-factor). Diamane's vibration qualities are less sensitive to its stacking sequence than its heat conductivity. It is anticipated that the high frequency of diamane will be put to use in ultrasensitive resonator-based sensors [13].

To investigate the potential use as sensors based on first-principles analysis, the adsorption behaviors and sensing performances of six types of typical volatile organic compounds (VOCs), i.e., C_6H_6, $CHCl_3$, C_2HCl_3, CH_2O, C_2H_6O, and C_3H_6O, on pure and Li-doped f-diamane are systematically investigated. The molecules of VOCs are discovered to be poorly physiosorbed on pure f-diamane that has low adsorption energies (−0.224 to −0.553 eV), and the molecules would desorb as the temperature rose above a crucial range (136–250 K). The stronger VOC–substrate interactions following Li doping are indicated by the increased adsorption energy, shorter interlayer distance, greater charge transfer, and more hybridized peaks in the density of states (DOS), which can be seen in figure 6.9. The critical temperature for the desorption of VOCs (292–504 K) rises concurrently [44].

In another study, Fu and his co-workers gave the optical and electronic properties of h-diamane and f-diamane. Degenerate p-type and n-type doping features for

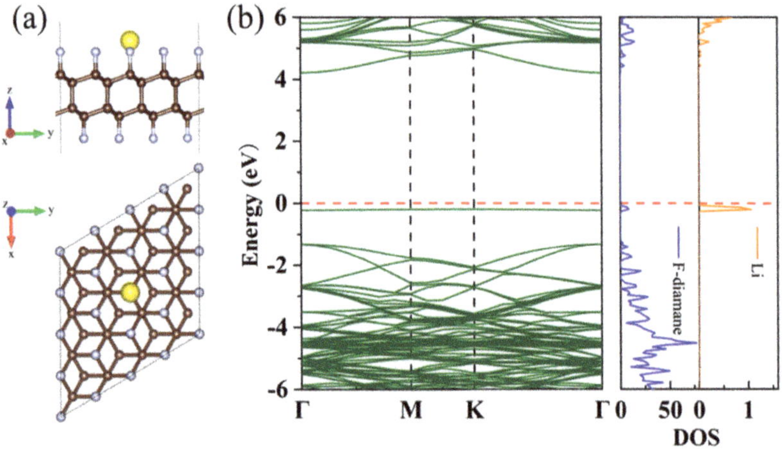

Figure 6.9. Schematic diagram of (a) Li-doped F-diamane (side view and top view) and (b) band structure and DOS of Li-doped F-diamane. Reprinted from [44], Copyright (2023), with permission from Elsevier.

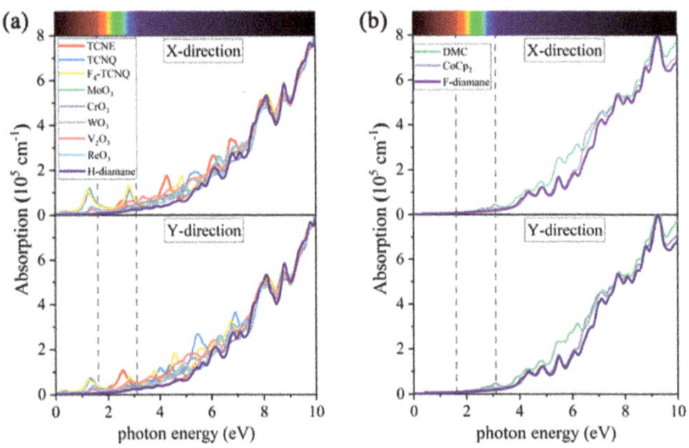

Figure 6.10. Absorption curves of (a) H-diamane and (b) F-diamane with molecule adsorption as a function of photon energy. Reprinted with permission from [45]. Copyright (2023) American Chemical Society.

h-diamane and f-diamane, respectively, are produced on the basis of the surface transfer doping mechanism. They found that degenerate DOS with carrier densities of 3.38×10^{13} cm^{-2} and 1.96×10^{13} cm^{-2} are discovered when significant donors of DMC and $CoCp_2$ adsorb on f-diamane, which is associated with significant charge transfer. With molecular adsorption, the optical absorption for diamanes increases in the visible-light domain. From figure 6.10 we can see that the adsorption peaks that have elevated absorption intensities for pure h- and f-diamane are primarily in the UV range. Following molecular adsorption, fresh absorption peaks are seen in the *x*-direction in the infrared (IR) and visible regions, i.e., ∼1.1–1.4 and ∼2.5–2.9 eV, for h-diamane and for f-diamane it is found to be near visible region, i.e., ∼2.7–3.1 eV. So, it was concluded that the optical absorption has significantly enhanced in the visible-light region [45].

6.4.3 Electronic properties

The abundance of free electrons and their capacity to conduct have an impact on the electronic characteristics of molecular systems. Diamanes with various surface functionalizations could make excellent FED candidates. In fact, the ability to dope the surface of 2D materials in a reversible manner through functionalization/functionalization makes them noteworthy. As a result, chemical or photochemical reactions cause the shift of electronic characteristics from metals to semiconductors, and vice versa. In contrast to the insulating properties of diamond and lonsdaleite, the 2D analog of these materials, graphene, is a semimetal because it is made up of sp^2-hybridized carbon atoms. The chemical and electrical properties of carbon compounds rely heavily on the hybridization of carbon atoms [46]. It was recently discovered by Gierster *et al* [47] that the $(10\bar{1}0)$ surface of ZnO is capable of reversible photodoping, wherein a phase transition is generated by photoinduced downward surface band bending due to photodepletion of donor-type deep surface defects. Diamane is a special multilayer that, depending on the kind of functionalization (fluorine or hydrogen), displays two completely different band-gap behaviors, as shown in figure 6.11. Band structures predicted by TB09 for hydrogenated and FDs, as well as KS band structures $(\bar{2}110)$, are presented in figures 6.11(a) and (b), revealing both bulk and surface electronic states [48]. Table 6.1 summarizes the different properties of diamane-diamanoids when compared to other 2D systems, such as graphene [49].

6.5 Future perspectives

There is huge promise for future revolutionary advances in materials science and beyond by researching the characteristics of diamane anchored with diverse groups. We should expect a multitude of promising new possibilities as research continues to expand our understanding of functionalization approaches. Customizing diamane to meet the requirements of various industries is a promising area of research. Functionalization allows for the tailoring of the material's properties, leading to potential electronic applications where the material's improved conductivity and unique optoelectronic features could result in state-of-the-art devices with unrivaled

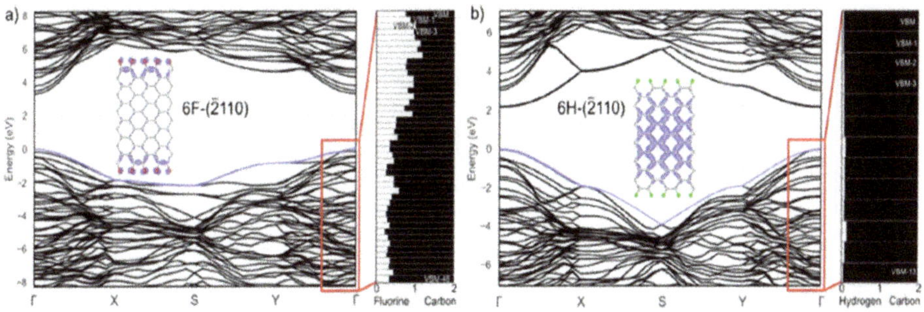

Figure 6.11. Shows the estimated electronic band structures of fluorine- and hydrogen-functionalized six-layer $(\bar{2}110)$ diamanes using TB09. Reproduced from [48]. CC BY 4.0.

Table 6.1. Property comparison of graphene, hexagonal boron nitride, molybdenum sulfide, phospherene, diamane, and diamanoids [49].

2D materials	Graphene	Hexagonal boron nitride	Molybdenum sulfide	Phospherene	Diamane	Diamanoid
Properties	Two distinct varieties of new (chiral) quantum Hall phenomena, with almost no quantum conductivity. The quantum suppression of interference effects seems to be robust. A semimetal with minimal overlap and high electronic conductivity, ideal for applications requiring reliable electrical conductivity.	The enormous bandgap and the surface optical phonon modes are responsible for excellent electrical insulation. The value of the bandgap grows with the amount of B and N present. Enhanced mobility of the field effect. Resistive switching behavior and a larger on/off current ratio.	Thicker objects have different electrical characteristics. Both double-layered materials with an indirect bandgap and single-layer materials with a wide gap may be considered semiconductors. The effect of doping on the charge density is weak. Electrical structures are influenced by defects, electric fields, mechanical strain, and surface adsorption.	The bandgap is direct, narrow, and potentially significant depending on thickness. Band dispersion is highly anisotropic in the Brillouin zone close to the Fermi energy. Group of functional and essential elements that may alter electrical characteristics. The armchair direction has better mobility and conductivity.	Having a high bandgap after being functionalized, results in good electrical characteristics. The variable bandgap in relation to the thickness of the material. Band gap tuning using atom-substitution, passivation, and functionalization. Functionalization enables superconductivity and insulative properties. Due to the fluorination process, diamane now has the characteristics of both a conductor and an insulator.	The film's crystalline structure and thickness affect its bandgap and conductivity. The quantum confinement effect and the direct bandgap are both diminished when the number of layers is increased.

Thermal properties	The highest thermal conductivity in the plane of the carbon atom.	High stability and thermal conductivity. Optimal over 1600°. Excellent stability at very high temperatures. Superior dielectric performance.	Very little heat is conducted out of planes. High-quality thermoelectric properties. Modifiable thermal conductivity to a range of grain sizes.	Biaxial strain and doping cause a material to be superconductive between 3 and 16 Kelvin, although it has poor thermal conductivity.	At room temperature, there is hardly little thermal expansion. Due to its horizontal stacking symmetry, AA-stacked diamane displays greater thermal conductivity than AB-stacked diamane. Diamanes with fluorine and chlorine atoms are more effective thermal insulators than diamanes with hydrogen atoms.	High thermal conductivity; additional enhancement possible by dehydrogenation.
Mechanical properties	Superior mechanical strength, toughness, and rigidity. Highly effective substance.	Durability under mechanical stress, proportional to dimensions and wall thickness. Minimal resistance to sliding.	Superior mechanical robustness. The Young modulus is comparable to that of graphene oxide and steel. The strain-induced phase change from	Superior adaptability in the face of extreme stress and strain. The mechanical response is anisotropic. Strong anisotropy in plane.	Material having a hardness greater than graphene and strength that decreases with the number of layers. Highly elastic and tensile.	Strong, biocompatible, and with a low coefficient of friction. More study is needed.

(*Continued*)

Table 6.1. (*Continued*)

2D materials	Graphene	Hexagonal boron nitride	Molybdenum sulfide	Phospherene	Diamane	Diamanoid
		Having piezoelectric properties in the plane.	semiconductor to metal.		Large modulus of elasticity. Mechanical characteristics can be further improved through functionalization and replacement.	
Optical properties	Fine structural constant, infrared optical absorbance and high-frequency photo response are only a few of the remarkable optical features exhibited by these materials. Depending on the photon energy, the doping agent, the doping amount, and the absorption peak, the properties of the material will change.	Transparency of 99.9%, a wave length range of 250–900 nm, and minimal optical absorption. Sharp absorption in the deep UV range is provided by a wide direct bandgap and UV luminescent characteristics. Planar defects influence the intensity of the deep UV photoluminescence and catho doluminescence emissions.	Superior to other 2D materials in terms of photoresponse. Improved photodetection is achieved by a switching behavior. Mobility and optical features like band gaps can be tuned by adjusting the sheet thickness.	Band topology that varies with thickness. Strong influences on photoresponse, Raman spectra, photoluminescence emission yield, screening, and plasmonic effects come from linear dichronism, optical absorption, and relative orientation. The high degree of polarization in the armchair direction.	Exceptional optical qualities are provided by the direct energy gap, the wide bandgap, the optical pecu liarities, the resonance peak in the valence band-conductive band, and the optical peculiarities. To improve their optical properties, heterostructures of hydrogenated graphene and twisted BLG have been developed.	For investigation and evidence is being collected.

performance. In addition, catalysis can gain from carefully designed active sites that boost reaction selectivity and efficiency, paving the door for greener chemical processes. Furthermore, researchers have actively investigated biocompatible and bioactive changes to diamane that have the potential to revolutionize drug delivery systems, tissue engineering, and medical imaging. Diamane's malleability makes it possible to incorporate targeting ligands and therapeutic payloads, which bodes well for the development of more potent and individualized treatments. Functionalized diamane also has environmental uses that could be very useful, such as carbon capture and storage, water purification, and pollution remediation. Prospects for diamane characteristics connected with various groups have the ability to significantly impact a variety of businesses and help solve some of the world's most urgent problems. Through functionalization, scientists may engineer and fine-tune diamane's properties, ushering in a new era of materials with personalized qualities that will pave the way for revolutionary breakthroughs in electronics, catalysis, biomedicine, and environmental sustainability. Functionalized diamane has the potential to significantly advance several fields of study and improve the quality of life for people everywhere if more study and careful planning go into its creation.

Despite its limitations, a diamane anchored with diverse groups has the potential to revolutionize several disciplines, including materials research, electronics, medicine, and more. Researchers will be able to realize diamane's full potential and usher in an era where custom-made materials revolutionize technology while also providing answers to pressing global problems if they take the initiative to solve these obstacles.

6.6 Conclusions

In summary, diamane's characteristics when anchored with various groups constitute an intriguing new frontier in materials research, opening up a wealth of opportunities for game-changing developments and applications. Strategic functionalization allows researchers to alter diamane's qualities to meet a variety of needs beyond its natural diamond-like attributes. Functionalized diamane materials are on the cusp of bringing about a revolution in the biomedical arena. Creating a biocompatible and bioactive material has the potential to revolutionize healthcare in many ways, including drug delivery, tissue engineering, and medical imaging. Additionally, there is hope that functionalized diamane can help solve environmental problems. Sustainable solutions for a cleaner and greener future may involve its usage in carbon capture and storage, water purification, and pollutant remediation. However, various obstacles must be cleared in order to fully exploit functionalized diamane. The stability and durability under different conditions, as well as the scalability of the synthesis processes, need to be carefully evaluated. Advanced characterization tools must be developed to comprehend the nanoscale behavior, and quantifying the connection between functional groups and material properties is essential for focused design. The potential to reimagine materials science and a wide range of industries is represented by the features of diamane anchored with various groups. Functionalized diamane has the potential to have far-reaching effects on technology, medicine, the environment, and society as a whole if more study, collaboration, and attention to ethics and sustainability are given to its development.

References

[1] Garg R, Dutta N K and Roy Choudhury N 2014 Work function engineering of graphene *Nanomaterials* **4** 267–300

[2] Hirsch A 2010 The era of carbon allotropes *Nat. Mater.* **9** 868–71

[3] Georgakilas V, Perman J A, Tucek J and Zboril R 2015 Broad family of carbon nanoallotropes: classification, chemistry, and applications of fullerenes, carbon dots, nanotubes, graphene, nanodiamonds, and combined superstructures *Chem. Rev.* **115** 4744–822

[4] Tiwari S K, Kumar V, Huczko A, Oraon R, Adhikari A D and Nayak G C 2016 Magical allotropes of carbon: prospects and applications *Crit. Rev. Solid State Mater. Sci.* **41** 257–317

[5] Zhang Y, Tan Y W, Stormer H L and Kim P 2005 Experimental observation of the quantum Hall effect and Berry's phase in graphene *Nature* **438** 201–4

[6] Kvashnin A G and Sorokin P B 2014 Lonsdaleite films with nanometer thickness *J. Phys. Chem. Lett.* **5** 541–8

[7] Chernozatonskii L A, Sorokin P B, Kvashnin A G and Kvashnin D G E 2009 Diamond-like C 2 H nanolayer, diamane: simulation of the structure and properties *JETP Lett.* **90** 134–8

[8] Kvashnin A G, Avramov P V, Kvashnin D G, Chernozatonskii L A and Sorokin P B 2017 Features of electronic, mechanical, and electromechanical properties of fluorinated diamond films of nanometer thickness *J. Phys. Chem.* C **121** 28484–9

[9] Samarakoon D K and Wang X Q 2010 Tunable band gap in hydrogenated bilayer graphene *ACS Nano* **4** 4126–30

[10] Ge L, Liu H, Wang J, Huang H, Cui Z, Huang Q and Lu Y 2021 Properties of diamane anchored with different groups *Phys. Chem. Chem. Phys.* **23** 14195–204

[11] Bakharev P V, Huang M, Saxena M, Lee S W, Joo S H, Park S O and Ruoff R S 2020 Chemically induced transformation of chemical vapour deposition grown bilayer graphene into fluorinated single-layer diamond *Nat. Nanotechnol.* **15** 59–66

[12] Pakornchote T, Ektarawong A, Alling B, Pinsook U, Tancharakorn S, Busayaporn W and Bovornratanaraks T 2019 Phase stabilities and vibrational analysis of hydrogenated diamondized bilayer graphenes: a first-principles investigation *Carbon* **146** 468–75

[13] Zheng Z, Zhan H, Nie Y, Xu X, Qi D and Gu Y 2020 Single layer diamond-a new ultrathin 2D carbon nanostructure for mechanical resonator *Carbon* **161** 809–15

[14] Hu C H, Zhang P, Liu H Y, Wu S Q, Yang Y and Zhu Z Z 2013 Structural stability and electronic and magnetic properties of fluorinated bilayer graphene *J. Phys. Chem.* C **117** 3572–9

[15] Oganov A R, Kvashnin A G and Saleh G 2018 Computational materials discovery: dream or reality? (London: Royal Society of Chemistry) 1–14 DOI: https://doi.org/10.1039/9781788010122-00001

[16] Pinto A K M, Pontes J M, Matos M J, Mazzoni M S and Azevedo S 2022 BCN diamondol-like compounds: stability trends and electronic properties *Comput. Mater. Sci.* **215** 111737

[17] Capek I 2009 Dispersions, novel nanomaterial sensors and nanoconjugates based on carbon nanotubes *Adv. Colloid Interface Sci.* **150** 63–89

[18] Qin G, Wu L and Gou H 2022 Diamane: design, synthesis, properties, and challenges *Funct. Diam.* **1** 83–92

[19] Zhu L, Hu H, Chen Q, Wang S, Wang J and Ding F 2011 Formation and electronic properties of hydrogenated few-layer graphene *Nanotechnology* **22** 185202

[20] Chernozatonskii L A, Demin V A and Kvashnin D G 2021 Fully hydrogenated and fluorinated bigraphenes–diamanes: theoretical and experimental studies *C* **7** 17

[21] Dong Y, Wu X and Martini A 2013 Atomic roughness enhanced friction on hydrogenated graphene *Nanotechnology* **24** 375701

[22] Rajasekaran S, Abild-Pedersen F, Ogasawara H, Nilsson A and Kaya S 2013 Interlayer carbon bond formation induced by hydrogen adsorption in few-layer supported graphene *Phys. Rev. Lett.* **111** 085503

[23] Wang J, Li L, Wang J, Yang W, Guo P, Li M and Zhao B 2022 First-principles study on the nanofriction properties of diamane: the thinnest diamond film *Nanomaterials* **12** 2939

[24] Elias D C, Nair R R, Mohiuddin T M G, Morozov S V, Blake P, Halsall M P and Novoselov K S 2009 Control of graphene's properties by reversible hydrogenation: evidence for graphane *Science* **323** 610–3

[25] Clark S M, Jeon K J, Chen J Y and Yoo C S 2013 Few-layer graphene under high pressure: Raman and x-ray diffraction studies *Solid State Commun.* **154** 15–8

[26] Smith D, Howie R T, Crowe I F, Simionescu C L, Muryn C, Vishnyakov V and Proctor J E 2015 Hydrogenation of graphene by reaction at high pressure and high temperature *ACS Nano* **9** 8279–83

[27] Piazza F, Cruz K, Monthioux M, Puech P and Gerber I 2020 Raman evidence for the successful synthesis of diamane *Carbon* **169** 129–33

[28] Colin M, Chen X, Dubois M, Rawal A and Kim D J 2022 F-diamane-like nanosheets from expanded fluorinated graphite *Appl. Surf. Sci.* **583** 152534

[29] Ahmad Y, Batisse N, Chen X and Dubois M 2021 Preparation and applications of fluorinated graphenes *C* **7** 20

[30] Gao Y, Cao T, Cellini F, Berger C, De Heer W A, Tosatti E and Bongiorno A 2018 Ultrahard carbon film from epitaxial two-layer graphene *Nat. Nanotechnol.* **13** 133–8

[31] Paul S and Momeni K 2019 Mechanochemistry of stable diamane and atomically thin diamond films synthesis from bi-and multilayer graphene: a computational study *J. Phys. Chem.* C **123** 15751–60

[32] Kvashnin A G, Chernozatonskii L A, Yakobson B I and Sorokin P B 2014 Phase diagram of quasi-two-dimensional carbon, from graphene to diamond *Nano Lett.* **14** 676–81

[33] Martins L G P, Matos M J, Paschoal A R, Freire P T, Andrade N F, Aguiar A L and Cançado L G 2017 Raman evidence for pressure-induced formation of diamondene *Nat. Commun.* **8** 96

[34] Niu C, Zhang J, Zhang H, Zhao J, Xia W, Zeng Z and Wang X 2022 Configuration stability and electronic properties of diamane with boron and nitrogen dopants *Phys. Rev.* B **105** 174106

[35] Piazza F, Gough K, Monthioux M, Puech P, Gerber I, Wiens R and Ozoria C 2019 Low temperature, pressureless sp2 to sp3 transformation of ultrathin, crystalline carbon films *Carbon* **145** 10–22

[36] Hu Y, Li D, Yin Y, Li S, Ding G, Zhou H and Zhang G 2020 The important role of strain on phonon hydrodynamics in diamond-like bi-layer graphene *Nanotechnology* **31** 335711

[37] Zhu L, Li W and Ding F 2019 Giant thermal conductivity in diamane and the influence of horizontal reflection symmetry on phonon scattering *Nanoscale* **11** 4248–57

[38] Cheng T, Liu Z and Liu Z 2020 High elastic moduli, controllable bandgap and extraordinary carrier mobility in single-layer diamond *J. Mater. Chem.* C **8** 13819–26

[39] Lee C, Wei X, Kysar J W and Hone J 2008 Measurement of the elastic properties and intrinsic strength of monolayer graphene *Science* **321** 385–8

[40] Wort C J and Balmer R S 2008 Diamond as an electronic material *Mater. Today* **11** 22–8

[41] Wu Y C, Shao J L, Zheng Z and Zhan H 2020 Mechanical properties of a single-layer diamane under tension and bending *J. Phys. Chem.* C **125** 915–22

[42] Mortazavi B, Shojaei F, Javvaji B, Azizi M, Zhan H, Rabczuk T and Zhuang X 2020 First-principles investigation of mechanical, electronic and optical properties of H-, F-and Cl-diamane *Appl. Surf. Sci.* **528** 147035

[43] Wei Q and Peng X 2014 Superior mechanical flexibility of phosphorene and few-layer black phosphorus *Appl. Phys. Lett.* **104** 251915

[44] Liu Y, Gao L, Fu S, Cheng S, Gao N and Li H 2023 Highly efficient VOC gas sensors based on Li-doped diamane *Appl. Surf. Sci.* **611** 155694

[45] Fu S, Liu Y, Li J, Wan L, Gao N and Li H 2023 Degenerate p-type and n-type doping of diamane by molecular adsorption *J. Phys. Chem.* C **127** 9939–46

[46] Novoselov K S, Andreeva D V, Ren W and Shan G 2019 Graphene and other two-dimensional materials *Front. Phys.* **14** 1–4

[47] Gierster L, Vempati S and Stähler J 2021 Ultrafast generation and decay of a surface metal *Nat. Commun.* **12** 978

[48] Tantardini C, Kvashnin A G, Azizi M, Gonze X, Gatti C, Altalhi T and Yakobson B I 2023 Electronic properties of functionalized diamanes for field-emission displays *ACS Appl. Mater. Interfaces* **15** 16317–26

[49] Tiwari S K, Pandey R, Wang N, Kumar V, Sunday O J, Bystrzejewski M and Mishra Y K 2022 Progress in diamanes and diamanoids nanosystems for emerging technologies *Adv. Sci.* **9** 2105770

Chapter 7

Diamanoids and diamane oxide synthesis and properties: an overview

Sanglap Mondal, Swapnadipa Patra, S K Rozi Unnesa Alam, Soumyadeep Mitra and Subhasis Roy

Carbon-based structures are recent developments of material that have gained a lot of research attention and serve as the cornerstone of a technologically sophisticated contemporary civilization. Comprehensive band-gap semiconducting materials are now available in nanometer-thick, crystalline, and sp3-bonded carbon sheets for use in electronics, optoelectronics, and medical devices. This chapter provides a full examination of diamanoids and diamane oxides, taking into account their historical potential, method of synthesis, structural properties, inherent traits, and state-of-the-art applications. Diamanoids and diamane oxide are both active research topics, and therefore the production methods and characteristics of these materials may change as new discoveries are found. Based on the detailed reporting of the potential for novel applications of dimanes and diamanoid oxide derivatives, this chapter intends to introduce essential upgrading principles for a new generation of diamanes and diamanoid oxide-based quantum devices to advance carbon nanotechnology.

7.1 Introduction

Due to its exceptional features, graphene [1], which replaced carbon nanotubes [2] in 2004, has a very wide range of applications (including electronics, optics, photovoltaics, spintronics, etc) [3]. However, graphene does not have an electrical bandgap. Due mainly to this constraint, innovative two-dimensional (2D) materials development for use in electronics and optoelectronics has been encouraged and a wide range of new materials have been created from mono-elemental materials. Sub-10 nm sp3-C-rich carbon films, labeled as sp3C2D in the subsequent discussion, have been relatively overlooked, despite possessing several notable properties. These properties, either anticipated through analogies with diamond in computational models or discovered experimentally, have not garnered significant attention. These

doi:10.1088/978-0-7503-5939-9ch7

qualities include: (i) diamond-like mechanical properties, (ii) a wide electronic bandgap, (iii) chemical inertness as well as biocmopatibility, and (iv) they work by turning different gaseous forms of carbon into materials using one of Earth's most abundant elements, which is a win-win situation [3].

Massive developments in carbon materials have been observed during the last 5 years. Several experimental approaches for synthesizing diamane and diamanoids have been developed based on the findings of computational and theoretic investigations. Scientists discovered 2D graphene-based materials known as 'diamane' and 'diamanoid' [4] while researching new 2D structures, which have tremendous potential for applications in diverse fields, such as of quantum computing technology.

More recent developments in fabrication methods have made it possible to produce diamanoids with several layers. Due to the structural instability of diamanoid, Bernal stacking is not a common practice. Diamanoids will be produced more readily when other stacking sequences, such as (ABB) diamanoids, (ABBA) diamanoids, and (AABBCC) diamanoids, are used. Due to thermodynamic constraints, the scientific community has not given the recent updates of many diamanoids with fluorine and hydrogen the attention that they deserve. Additionally, most current research on dimanes, diamanoid compounds, and thin films resembling carbon uses theoretical and simulation-based computational methodologies. Therefore, there is a great deal of room for experimenting. Considerable experimental studies on diamane and diamanoids, including their synthesis, characterization, and applications, are covered in the following sections [5].

7.2 Diamanoids

7.2.1 Early works

Diamanoids and diamane oxide are indeed specific terms used in the context of carbon-based materials. Diamanoids are a class of diamond-like carbon (DLC) structures that are composed of smaller diamond cages fused together. These structures can be thought of as 'nano-diamonds' with complex shapes and architectures. Diamanoids exhibit properties similar to both diamond and amorphous carbon, making them intriguing for various applications, including electronics, catalysis, and energy storage. Diamanoids can be synthesized through different methods, such as molecular dynamics simulations or experimental techniques. One common approach is to use hydrocarbon precursors and subject them to high temperatures and pressures, similar to the conditions required for natural diamond formation. This process enables the growth of small diamond clusters, which can then aggregate and rearrange to form larger diamonoid structures and diamane oxide, also known as diamond oxide or diamantane oxide, which is a specific derivative of diamantane that is a hydrocarbon compound with a diamond-like structure. Diamantane consists of four fused cyclohexane rings arranged in a diamond lattice configuration. The most promising work is diamane and diamanoids formation. Several diamond production techniques have been developed, the most popular being HPHT, chemical vapor deposition (CVD), and PVD [6]. The initial investigation of carbon nanomaterials in 1985 and later saw the emergence of experimental nanotechnology and carbon materials engineering. The International Union of Pure and Applied Chemistry (IUPAC) coined the word 'graphene' for the first time in 1994. This provided vocabulary, terminology, and reactions for graphene and related nanosystems [5]. Even though initial research on graphene was started in 1853, the name was only a brief introduction because a proper synthesis had not yet been achieved. The IUPAC nomenclature increased scientists' and researchers' interest in creating 2D structures. Although the theoretical discovery of graphene was made in 1947, it took some years of experimentation to identify and isolate the material. Graphene was synthesized and extracted for the first time in 2004. Single-layer graphene was isolated in 2007 through several studies, hydrogenated the single-layered structure, and then assessed the characteristics of these novel materials [7]. In 2009, single-layer graphene was hydrogenated to provide a reversible structure with better electrical and transport characteristics [8]. In this work, diamane or hydrogenated bilayer graphene (BLG) was created by hydrogenating BLG. BLG stacking sequences were hydrogenated to produce diamane and its isomer diamane II, respectively, in the AB and AA orientations. These preliminary results stimulated investigation into the different chemical, physical, mechanical, and thermal properties of diamanes. In 2012, trilayer graphene underwent hydrogenation, resulting in the production of triamane, which was a brand-new structure [9]. By fluorination bilayer, graphene created a structure with the first-ever carbon–carbon interlayer bonding in 2020 after researching the feasibility of fluorinated diamond (F-diamane) [4]. Unfortunately, the experimental diamane formation was not done until 2018 [10]. To adjust diamane's characteristics, other researchers added additional functional groups such as chlorine and hydroxyl. BLG has been interestingly swapped with atoms of nitrogen, boron, lithium, etc over the past 5–6 years to adjust the characteristics for a variety of uses. To create Moire's pattern on the diamond with unique features, twisted BLG was additionally hydrogenated. Until 2019, it was assumed that diamanoids might develop in cold hydrogen plasma from unsupported few-layer graphene. To create a stable sp3 structure, researchers produced diamanoids, or hydrogenated few-layer graphene, inside a hot-filament reactor.

In conclusion, it can be said that a lot of theoretical as well as experimental research was done on graphene and its derivatives from 2000 to 2010, much of which was devoted to evaluating diamane and diamanoids [5].

7.2.2 Structure of diamanoids

Diamanoids are a class of nanostructures that are composed of fused diamond cages. These structures are similar to diamond in terms of their atomic arrangement but differ in size and connectivity. While the exact structure of diamanoids may vary depending on their specific configuration, they generally possess a core composed of smaller diamond cages that are interconnected to form a larger and more complex structure. A general description of the structure of diamanoids follows:

Diamond cages: The fundamental building blocks of diamanoids are diamond cages. These cages are three-dimensional (3D) structures formed by a network of carbon atoms arranged in a diamond-like lattice. Each carbon atom within a diamond cage is bonded to four neighboring carbon atoms, resulting in a highly stable and rigid structure.

Fused structure: Diamanoids consist of multiple diamond cages that are fused together. The fusion occurs when carbon atoms from neighboring cages form additional covalent bonds, connecting the cages and creating a larger, interconnected structure. This fusion leads to the formation of a continuous network of diamond-like cages within the diamanoid.

Nanostructure: Diamanoids are typically nanoscale in size, meaning that their dimensions are on the order of nanometers. They can range from relatively small diamanoids composed of a few fused diamond cages to more complex and larger diamanoid structures.

Structural variations: The specific arrangement and connectivity of the fused diamond cages in diamanoids can vary, leading to structural variations and different properties. The fusion patterns may determine the overall shape, symmetry, and stability of the diamanoid structure.

Potential internal cavities: The fused diamond cage structure of diamanoids can create internal cavities or channels within the nanostructure. These cavities may offer space for encapsulating other molecules or nanoparticles, potentially enabling applications in areas such as drug delivery or catalysis. It is important to note that while the concept of diamanoids as fused diamond cage structures is plausible, their synthesis and characterization are still largely theoretical at this point.

A new name has been suggested for diamane, even though it was previously recognized as a large class of compounds generated structurally from the hydrogenated graphene layer. Two structural configurations are viable depending on whether the first graphene layer's stacking sequence is AB or AA [11]. An ABA-stacked graphene layer, in contrast to an AB-stacked graphene layer, cannot form a diamond structure because half of layer B's carbon atoms are situated in front of this layer's hexagonal center as opposed to another atom, preventing them from bonding with layer A's second layer. While stacking sequences ABCA can create the diamond structure, stacking sequences AAA can create the lonsdaleite structure [12]. When constructed from more than two graphenes, lonsdaleite and diamond can also unite to produce the structure of a diamanoid. Therefore, utilizing the stacking sequence of the initial graphenes, such as (ABBA)D, (ABB)D, (AABBCC)D, (ABBC)D, and so on (here 'diamanoid' has been abbreviated as 'D'), subsequent hydrogenated multilayer diamanoid structures with layer counts of 3, 4, and so forth up to 'few' should be found. This partial change from sp^2 to sp^3-carbon atoms may occur during commensurate stacking when carbon atoms can superimpose locally. This and the tension in graphene and diamanoid hybrids was anticipated by scientists [10].

7.2.2.1 Diamanoid/graphene hybrid model

A graphene hybrid model refers to a system or device that combines graphene with other materials or structures to enhance its properties or create new functionalities. Graphene, a 2D material composed of a single layer of carbon atoms arranged in a hexagonal lattice, possesses exceptional mechanical, electrical, thermal, and optical properties. By incorporating graphene into hybrid structures, researchers aim to leverage its unique characteristics while addressing certain limitations.

Figure 7.1(a) shows example systems that were partially hydrogenated and converted, as well as four unique carbon layers were simulated. All of the C atoms in the first layer are sp^3-hybridized, with half of them related to one H atom and the other half to carbon atoms from the second layer. This first layer is totally hydrogenated, which is typical of diamane and diamanoid molecules. Half of the C atoms in layer two are sp^3-C and connected to those in layer one because layer two is not hydrogenated. Figure 7.1(c) displays an identically sp^3-hybridized free orbital with one unpaired pz. The third and fourth layers are both still pure sp^2-C, but the third layer interacts with the second layer more forcefully than the fourth because it contains more unmet valence bonds. As a result, the second-third interlayer distance (between 0.3037 and 0.3308 nm) is less than the third-fourth interlayer distance (0.3512 nm). The L1 combination of the first, second, and third layers may repeat itself when dealing with larger systems [10]. A few examples of graphene hybrid models follow:

Graphene-based composites: Graphene can be combined with polymers, metals, ceramics, or other materials to form composites. The addition of graphene enhances the mechanical strength, electrical conductivity, and thermal properties of the composite material. These composites find applications in various industries, including aerospace, automotive, and electronics.

Graphene transistors: Graphene has excellent electrical properties, such as high carrier mobility, which makes it a promising material for transistors. However, pure graphene lacks an energy bandgap necessary for digital applications. To overcome this limitation, researchers have explored hybrid models where graphene is combined with other semiconducting materials, such as transition metal dichalcogenides or boron nitride. This hybridization enables the creation of graphene-based transistors with tunable bandgaps and improved performance.

Graphene-based sensors: Graphene's high sensitivity to various stimuli, such as gases, chemicals, or biological molecules, makes it an attractive material for sensors.

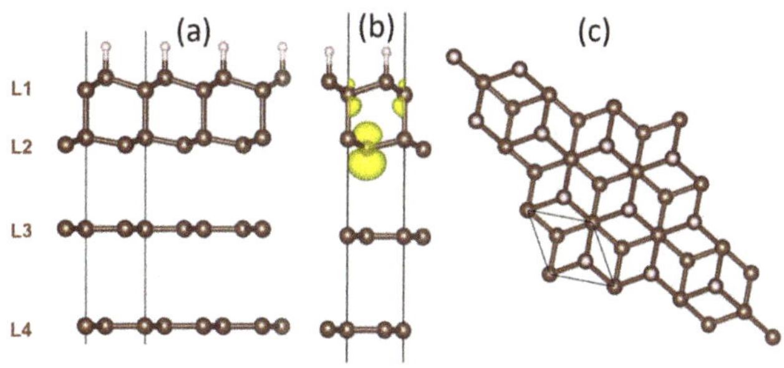

Figure 7.1. (a and b) Two side-views and (c) the top image of FLG, that was used in calculations using ABBA stacking and density functional theory. The predicted structure is a face-centered diamond cubic. Panel (b) shows the favored orientation of the p_z orbitals in the direction of the layer below (L3). The black thin lines show the primitive cell. Reprinted from [13], Copyright (2020), with permission from Elsevier.

Graphene can be integrated with other materials to develop hybrid sensors that exhibit enhanced sensitivity, selectivity, and response times. For example, graphene can be combined with metal nanoparticles or metal oxides to create highly sensitive gas sensors or biosensors.

Graphene-based energy storage: Graphene's large surface area and high electrical conductivity make it suitable for energy storage applications. By combining graphene with other materials, such as metal oxides or polymers, researchers have developed hybrid supercapacitors or batteries with improved energy storage capacity, faster charging rates, and longer cycle life.

Graphene-based optoelectronics: Graphene's excellent optical properties, such as high transparency and broadband absorption, make it an ideal candidate for optoelectronic devices. Hybrid models involving graphene and other semiconducting materials enable the creation of photodetectors, solar cells, or light-emitting devices with enhanced performance. These examples demonstrate how graphene can be combined with other materials to create hybrid models that capitalize on the unique properties of graphene and address specific application requirements. The field of graphene hybrid models is still evolving, and ongoing research aims to explore new combinations and applications to unlock further possibilities.

7.2.2.2 UV Raman spectroscopy analysis

UV Raman spectroscopy is a useful technique for analyzing the structural and vibrational properties of materials, including diamanoids. Although they are diamanoids materials, we discuss the potential analysis of diamanoids using UV Raman spectroscopy based on the assumption that they share some similarities with diamond-like structures. An overview of how UV Raman spectroscopy tools applied to diamanoids follows:

Vibrational modes: Diamanoids, being composed primarily of carbon atoms in a diamond-like lattice, are expected to exhibit vibrational modes similar to diamond. Diamond has a well-known Raman spectrum with prominent peaks associated with different vibrational modes. These include the sp^3-hybridized C–C stretching mode (approximately 1332 cm^{-1}), the sp^3 C–H stretching mode (approximately 2900 cm^{-1}), and the combination modes involving the bending and stretching motions of carbon atoms.

Shifts in Raman peaks: The Raman peaks in diamanoids might exhibit slight shifts compared to diamond due to potential structural variations, defects, or impurities. These shifts provide information about the specific bonding and local environment within the diamanoid lattice. However, the magnitude and nature of these shifts would depend on the specific structural modifications present in the diamanoid.

Intensity and line broadening: The intensity and line broadening of Raman peaks in diamanoids vary depending on their size, crystallinity, and structural disorder. Smaller diamanoids might show broader peaks due to size confinement effects, while larger diamanoids with better crystallinity may exhibit sharper peaks. Structural disorder, such as lattice defects or irregularities in the fused diamond cages, lead to additional broadening or intensity variations in the Raman spectra.

Surface effects: The surface properties of diamanoids could influence their UV Raman spectra. Surface reconstructions, steps, or other surface features may affect the surface vibrational modes, leading to changes in the observed Raman peaks or additional features in the spectrum.

Chemical modifications: If diamanoids contain impurities or functional groups, such as dopants or surface adsorbates, then they could introduce additional Raman-active modes in the spectrum. These modes might arise from vibrational motions involving the impurity or functional groups, providing information about their presence and interactions within the diamanoid structure.

The actual UV Raman spectroscopy analysis of diamanoids would require their successful synthesis and characterization, which remains a subject of ongoing research and technological development.

Like double graphene layers, the spectra are distinguished and have a strong G peak at approximately 1582 cm^{-1} [14]. Following hydrogenation, the Raman spectra exhibit significant alterations across the board. The comparable UV Raman spectra are shown in figure 7.2 [10].

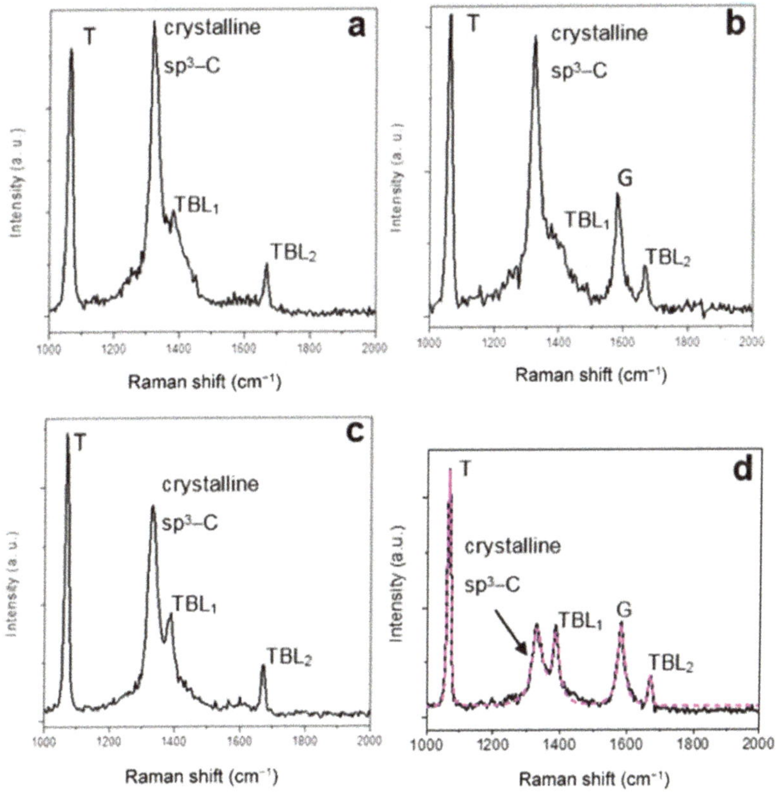

Figure 7.2. UV Raman spectra of FLG following exposure to the hydrogenation process driven by hot filament (a–d). The crystallized sp^3-C peak is visible in the spectra at 1319–1337 cm^{-1}. Only the partial conversion is seen by the G band. (b and d) Two peaks at 1385 cm^{-1} and 1669 cm^{-1} are representative of twisted 2LG and designated TBL1 and TBL2 (twisted bilayer). This is an illustration of a fitting that uses symmetrical Voigt functions to estimate peak positions (c–d). Reproduced from [10]. CC BY 4.0.

7.2.2.3 Electronic diffraction analysis

Electron diffraction patterns in a few layered graphene regions undergoing hydrogenation differ from those in more typical layered graphene and crystalline sp^3-C sheet regions. Hence, the patterns show the occurrence of coherent domains stacked and somewhat twisted and comparable periodicity [10].

7.2.2.4 FTIR microscopy analysis

The symmetrical sp^3-C–H_2 and symmetrical sp^3-C–H_3 modes are disregarded because their antisymmetric counterparts would also be identified simultaneously with the detected wave number 2950 cm^{-1}. Olefinic and aromatic sp^2-C–H modes are disregarded because, in the case of unbound molecules, they are frequently observed at wavenumbers over 2975 cm^{-1} [10].

FTIR microscopy analysis on carbon-based materials might also be applicable to these materials.

Sample preparation: Prepare the diamanoid or diamane oxide sample for FTIR microscopy analysis. This may involve mounting or positioning the sample appropriately, considering factors such as sample thickness, transparency, and compatibility with the infrared beam.

Instrument calibration: Perform calibration of the FTIR microscopy instrument by acquiring spectra from reference materials with known infrared absorption characteristics. This helps ensure accurate spectral measurements and alignment of the instrument.

Spectral acquisition: Collect FTIR spectra from the diamanoid or diamane oxide sample using FTIR microscopy. Select specific regions of interest for analysis, ensuring that the infrared beam is focused on these areas. Acquire a series of interferograms and Fourier transform them to obtain the infrared spectra.

Data analysis: Process and analyze the acquired spectra using suitable software. Identify characteristic absorption bands or peaks in the spectra that correspond to specific functional groups or chemical bonds present in diamanoids or diamane oxide. Compare the obtained spectra with reference spectra or databases, if available, to aid in identification.

Chemical mapping: Generate chemical maps based on specific spectral features of interest. Use the FTIR microscopy instrument to scan the sample and collect spectra from different spatial locations. Analyze the variation in spectral intensities or absorption bands across the sample to visualize the distribution of chemical components or functional groups.

Data interpretation: Interpret the FTIR microscopy data in the context of diamanoids or diamane oxide. Correlate the observed spectral features with the expected chemical composition or structural characteristics of these materials. Consider other complementary characterization techniques, such as Raman spectroscopy or electron microscopy, to support the analysis and provide a comprehensive understanding.

It is important to note that specific considerations and analysis strategies may vary depending on the particular properties and structures of diamanoids and diamane oxide.

7.3 Diamane oxide

7.3.1 Early works

Diamane oxide refers to diamantane that has been functionalized with oxygen atoms. The oxygen atoms are introduced to the molecule through oxidation reactions, where specific positions in the diamantane structure are modified by replacing carbon atoms with oxygen atoms. The synthesis of diamane oxide would involve the functionalization of diamantane using appropriate oxidizing agents or reactions, similar to the general principles of organic synthesis. The specific methods and conditions used for diamane oxide synthesis may depend on the desired positional substitution and the choice of oxidizing reagents. Diamane oxide is a 2D film with variable electrical characteristics and heterogeneous coverage. The endurance of the diamane oxide films, a recently developed technology, is being studied utilizing a variety of components depending on the predecessors, temperature, and pressure. The fabrication of the film using the frequently utilized oxygen supply in the H_2O form demands pressures of GPa, and the electrical characteristics of bilayer diamane customized using various functional groups. With its electrical characteristics dependent on the layer thickness, detailed research on diamane oxide film has been conducted to clarify its relationship with surface states. The current study discusses the variable-composition convex hulls of diamane oxide and how the proportion of OH-groups and precursor chemical potential affect the breakdown enthalpy. In order to enhance films, researchers employ the Gibbs free energy difference of diamane films with stoichiometry and the dependency of diamane oxide films' formation energy on pressure [15].

7.3.2 Structure

Diamane oxide exhibits variable coordination and film structure in the presence of varying fractions of OH functional groups, as seen in figures 7.3 and 7.4.

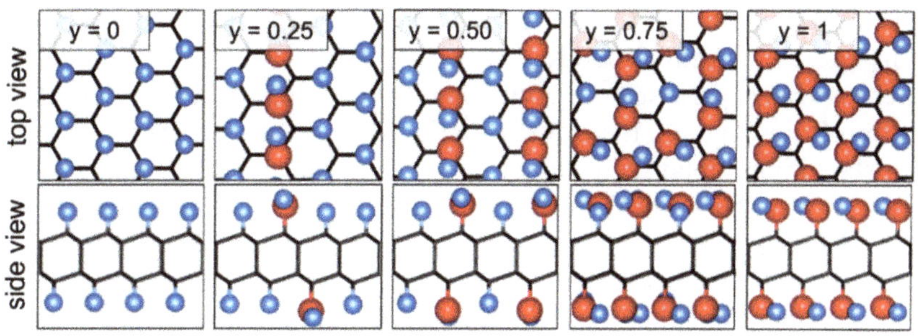

Figure 7.3. Views of the diamane oxide atomic structure from above and from the side, showing various OH-group percentages. Sticks denote carbon atoms, while red and blue balls depict oxygen and hydrogen atoms, respectively. Reprinted with permission from [15]. Copyright (2022) American Chemical Society.

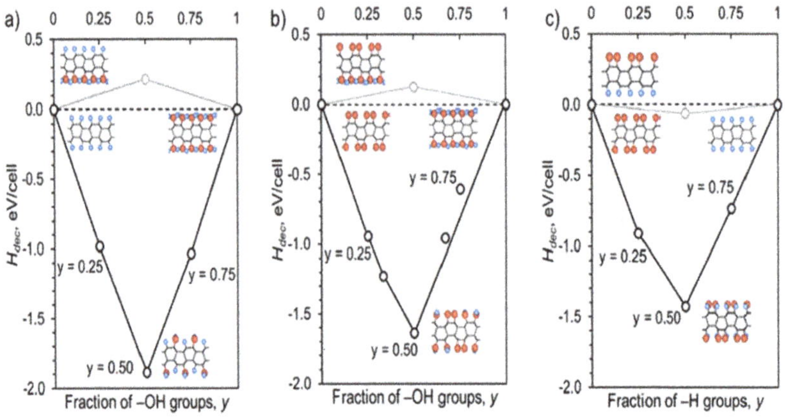

Figure 7.4. The link between the decomposition enthalpy Hdec and the proportion of OH-groups in (a) C4H2*x*(OH)2*y*, (b) C4O2*x*(OH)2*y*, and the diamane oxide convex hulls of varying composition show (c) C4O2*x*H2*y*. The color gray is used to represent Hdec for Janus-type structures. Reprinted with permission from [15]. Copyright (2022) American Chemical Society.

7.4 Properties

7.4.1 Properties of diamanoids

A unique 2D carbon structure that is known as a diamanoid, or partially hydrogenated few layers graphene, is generated as the number of hydrogenated graphene layers is increased from two to a few layers (3–5). Diamanoid has several properties, which include: (i) electronic, (ii) magnetic, (iii) optical, (iv) thermal, and (v) mechanical.

7.4.1.1 Electronic properties

Diamanoids are preferred for ultrathin protective coatings, such as those used on aircraft equipment as ultrahigh-strength components in composite materials, because of their high stiffness. DNR resonators have a high natural frequency and a considerable quality factor (*Q*-factor), according to recent studies into the vibrational properties of diamanoids nanoribbon (DNR) [16]. The uppermost graphene layer in diamanoids has the capacity to functionalize, tuning the electrons across the entire chemical structure.

Numerous techniques, such as functionalization, doping with heteroatoms, the production of diamanoids, the transformation of graphene into graphene ribbons, and even breaking sublattice symmetry and periodic potential, can improve the electrical characteristics and bandgap of graphene [17]. Diamanoids are substantially more manipulable than diamond, graphite, and graphene in terms of the electrical bandgap. When we analyze the stacking pattern for electrical properties, Bernal's graphene stacking has the lowest formation energy and the largest bandgap. However, the quantum confinement effect and bandgap gets narrower as the number of layers rises. Once the number of layers increases, the direct bandgap transforms into an indirect bandgap, which substantially affects the mobility of the electron carriers and quantum tunneling. Recent theoretical revisions show that the type of surface modification and film thickness affect where the valence band is located at the top of the graphene layer.

Diamond is known for its wide bandgap, high electrical resistivity, and excellent electrical insulating properties. Diamanoids, being derived from diamond, could exhibit similar characteristics. Diamane oxide, or diamond oxide, is a derivative of diamantane where specific positions in the diamantane structure are functionalized with oxygen atoms. The electronic properties of diamane oxide would be influenced by the introduction of oxygen atoms and their effect on the electronic structure. The exact impact on the electronic properties would depend on the specific positions and concentrations of the oxygen atoms and their interaction with the carbon framework. The bandgap is an important electronic property of diamanoids that determines whether a material behaves as a conductor, semiconductor, or insulator. Carbon-based materials, including diamond and diamond-like structures, typically have wide bandgaps, making them excellent insulators or wide-bandgap semiconductors. The presence of dopants or functional groups can modify the bandgap and introduce additional electronic states, potentially affecting the conductivity of the material. Defects, impurities, or intentional doping can significantly impact the electronic properties of carbon-based materials. For example, vacancies, substitutional impurities, or functional groups can introduce localized electronic states within the bandgap, altering the electrical conductivity and optical properties. The presence of such defects or dopants are relevant to diamanoids and diamane oxide because they could influence the electronic behavior of these materials. In nanoscale carbon-based materials, such as carbon nanotubes or graphene quantum dots, quantum confinement effects can arise due to the confinement of charge carriers in a reduced dimensional space. These effects can lead to size-dependent changes in the electronic band structure and properties. While diamanoids and diamane oxide may not exhibit the same level of quantum confinement as these nanoscale materials, it is possible that their electronic properties could be influenced by their specific structures and dimensions. It is important to note that the electronic properties of diamanoids and diamane oxide are areas of ongoing research, and more comprehensive studies may have been conducted since my knowledge cutoff.

In addition, it has been demonstrated that interlayer hybridization and pressure can be used to adjust the electrical structure of the pure graphene layer seen in diamanoids. The type of graphene (the beginning material) employed in producing dimanes directly affects their electronic characteristics [18].

7.4.1.2 Magnetic properties

Diamanoids and their derivatives exhibit fascinating magnetic behavior, opening up brand-new theoretical and experimental research topics with tremendous potential. Instead of functionalization, diamanoids are typically replaced or doped with atoms to produce magnetic behavior. Diamanoids have magnetism comparable to graphene doped with nitrogen, boron, sulfur, or other appropriate dopants to generate vacancies in various charged states. These dopants, like graphene, close the energy gap, narrower the bandgap, and open the magnetic properties of diamanoids instead of the electronic properties. While nitrogen replacement in graphene showed very mild ferromagnetic activity, nitrogen-vacancy on the surface functioned paramagnetically. Graphitic carbon edges have other flaws. Intriguingly, nitrogen-substituted diamanoids exhibit new magnetic behavior in quasiparticles at large hole concentrations, where ferromagnetic behavior becomes dominant. The paramagnetism and spin ordering state in diamanoids are also slightly boosted by creating a silicon (Si) vacancy.

As previously mentioned, the single-photon emission properties of dimanes differ from those of typical non-doped dimanes when the Bohr radius is diminished and the radiative rate is increased. Its magnetic characteristics are then affected by this occurrence. Decreases in the bandgap of diamanoids due to the application of an electric field or the desorption of hydrogen from the surface of graphene also imply the presence of a 2D spin-polarized semiconductor. Semi-hydrogenation encourages a small indirect bandgap distinct from that of graphene and displays ferromagnetic property while combining magnetic forces. An energy gap between low to zero is found when an electric field is given. The top three layers of un-hydrogenated sides of semi-hydrogenated diamond film structures have also displayed ferromagnetic properties. For ferromagnetism, hydroxylated diamanoids displayed insulative activity. As a result, the magnetic characteristics of diamanoids relate to conductive and insulative behavior. These characteristics distinguish these 2D nanostructures from others and extremely important in the nanotechnology field of electromagnetism [5].

7.4.1.3 Optical properties

Transparency and absorption: Carbon-based materials, such as diamond and graphene, can exhibit exceptional transparency in the visible and ultraviolet regions of the electromagnetic spectrum. This transparency is due to their wide bandgap, which results in minimal absorption of photons in these energy ranges. Diamanoids, being derived from DLC structures, could potentially exhibit similar optical transparency properties.

Refractive index: The refractive index of a material describes how light propagates through it and determines the bending or refraction of light rays. Carbon-based materials, including diamond, generally have high refractive indices due to their high density and high refractive power. The refractive index of diamanoids could depend on their specific structure, composition, and any additional functional groups or dopants present.

Luminescence and photoluminescence: Some carbon-based materials, such as diamond nanoparticles or carbon dots, can exhibit luminescent or photoluminescent properties. This means they can emit light when excited by photons or other energy sources. The luminescence or photoluminescence behavior of diamanoids would depend on their specific structure, surface functionalization, defects, and impurities.

Raman scattering: Raman scattering is an optical phenomenon where light interacts with the vibrational modes of a material, resulting in a shift in its energy. Carbon-based materials, including diamond and graphene, exhibit strong Raman scattering signals due to their unique vibrational properties. Diamanoids may also exhibit characteristic Raman scattering spectra that can provide valuable information about their structural and electronic properties.

Optical bandgap: The optical bandgap of a material determines the range of energies at which it absorbs or transmits light. Carbon-based materials, including diamond, typically have wide bandgaps in the ultraviolet or visible energy range. The specific optical bandgap of diamanoids would depend on their structure, composition, and any additional functionalization or doping. Diamanoids are supposed to have increased optical capacities due to their ocular peculiarities, large gap spectra, and many resonance peaks in valence bands and conductive bands in the density of states (DoS). These properties are essential for the effective operation of various optical applications [19]. According to reports, fluorinated and hydrogenated diamanoids cannot absorb visible light due to a significant shift in their electronic structures. Still, chlorinated diamanoids can absorb light across a broad energy range within the visible regions. During this time, an insulating diamond layer known as fluorinated diamanoids can irradiate electron beams. This property can be utilized in nanopatterning processes and the large bandgap and direct energy gap conditions permit for optical applications. The silicon vacancy in diamanoids offers the single-photon emission at near-visible wavelengths with limited bandwidth. These photon sources could provide updated optical characteristics by emitting photoluminescence from single-photon sources. The optical features of the structure could be improved by heterostructures of hydrogenated graphene [20] and TBG that generate Moire's diamanoids by narrowing the small bands inside the ultrawide 2D bandgap. Due to dielectrics and broad bandgap, substituting atoms for functionalization of UTHD films modifies the optical characteristics [21].

7.4.1.4 Thermal properties

Heat transfer, consequently impacted by the acoustic and optical modes, impacts thermal conductivity. The acoustical way should be dominant and overlapping for a high heat conductivity. Group velocities affect thermal conductivity and heat transport. The rise in group velocity is associated with increased thermal velocity and the thickness of the designated frequency for acoustic bands.

As a result, the thermal conductivity increases when the optical mode's low frequency results in a larger group velocity. The thermal expansion of graphite, pure AB BLG (diamanoids), graphene, and diamond is almost nil at ordinary temperatures. According to several excellent works, interlayer bonding reflection can be used to design the heat conductivity of these 2D systems. The symmetry of horizontal reflection constrains the phonon scattering channels, which affects structural configurations. As a result, the lonsdaleite (AA) stacking sequence has a higher conductivity than Bernal (AB). On account of impediments at the edge borders of conduction channels, the thermal conductivity varies with temperature and decreases as temperature rises. Hence, all of the mentioned elements may impact the thermal conductivity of diamonds, either directly or indirectly.

BLG has undergone fluorination, hydrogenation, and other processes to create diamanoids with better thermal characteristics. Under carefully regulated experimental circumstances, diamanoids hydrogenation has been shown to produce materials with extraordinarily high thermal conductivity. However, hydrogenated diamanoids have high thermal conductivity and are highly intriguing to high-temperature applications. Due to horizontal mirror symmetry, the AA-stacked hydrogenated diamanoids exhibit better heat conductivity than the AB-stacked hydrogenated diamanoids. Hence, depending on the functionalization technique utilized, diamonds' thermal characteristics and modified forms offer both good heat conductivity and great insulating performance, making them distinct [22, 23]

7.4.1.5 Mechanical properties

Diamanoids and their derivatives are more brittle than graphene, but they are also harder and stronger than graphene because of their higher stiffness coefficient. The functional moieties affect the coefficient of the finite interlayer-linked domain of carbon by varying the sp^2–sp^3 hybridization state ratio. Finally, this affects the stiffness and other mechanical properties of the diamanoid. The nano-sheet of diamanoids has the lowest elasticity and tensile strength when isotropic elastic properties are dominant because of functionalization and symmetry disruption. These features are consistent with graphene's greater thickness and absence of functional groups. In functionalized diamanoids, adding more sp^3-hybridized carbon atoms would boost tensile strength while preserving suitable application flexibility, as evidenced by the constrained hole mobility in hydrogenated diamanoids and the constrained electron mobility for fluorinated diamanoids.

Recent research indicates that there has not been much impact on the bandgaps of hydrogenated and lonsdaleite diamanoids. Diamanoids nano-ribbons' high natural frequency, minimal energy loss, and high Q-factor value provide good vibrational characteristics. With a width of under 100 nm, these nano-ribbons resemble graphene nano-ribbons. In diamanoids and diamanoids nano-ribbons, increasing the lattice spacing exponentially increases the vibrational frequency and lowers layer stiffness [24, 25]. To change TBG's mechanical characteristics with the symmetry's functionalization, the dimer's periodicity and the Moire's pattern must be kept [26].

7.4.2 Diamane oxide

7.4.2.1 Dependence of diamane oxide films formation energy on pressure

Diamane oxide exhibits inverse proportionality for the Gibbs free energy difference regarding pressure and proportionality with temperature in the case of pressure dependence for formation energy. For different diamane oxide compositions, a decrease in the proportion of hydroxyl groups indicates a more minor Gibbs free energy difference [15, 27] (figure 7.5).

7.4.2.2 Electronic properties of multilayer H-diamane and OH-diamane

Regarding the DoS over a wide range of electron volt energies, various layers of diamane oxide exhibit various electrical features. The electronic properties of multilayer H-diamane (hydrogenated diamane) and OH-diamane (hydroxylated diamane) are specifically limited in the scientific literature. Some general insights into the electronic properties of DLC materials and their potential implications for multilayer H-diamane and OH-diamane follow.

Bandgap: DLC materials, including multilayer H-diamane and OH-diamane, generally exhibit a wide bandgap. The bandgap refers to the energy range in which

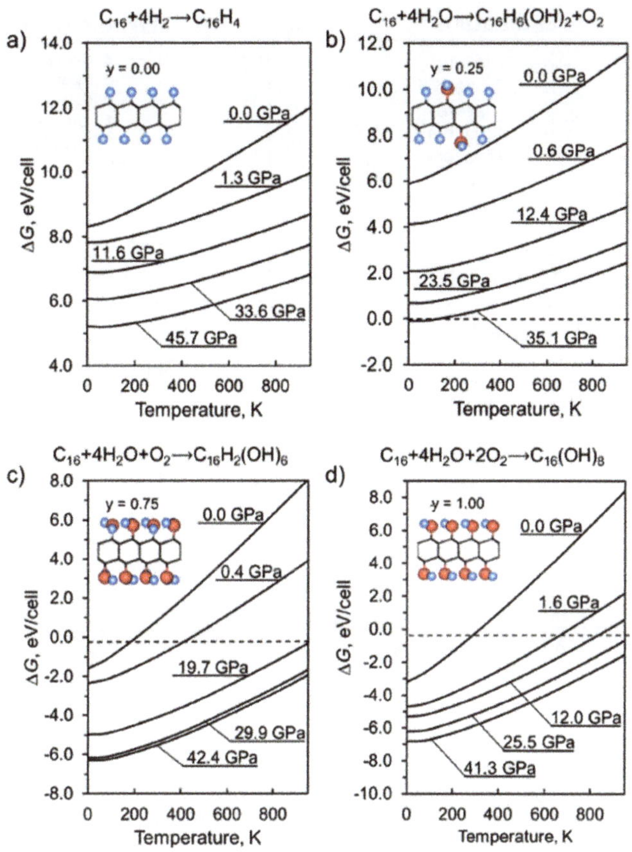

Figure 7.5. The dependence of Gibbs free energy difference of diamane films with stoichiometry (a) C16H4, (b) C16H6(OH)2, (c) C16H6(OH)6 and (d) C16(OH)8 on pressure. Reprinted with permission from [15]. Copyright (2022) American Chemical Society.

electrons are not allowed to exist. The presence of a bandgap indicates that these materials are typically insulating or have a high electrical resistivity. The specific bandgap of multilayer H-diamane and OH-diamane would depend on their structural characteristics and the specific arrangement of hydrogen or hydroxyl groups.

Density of states: DOS is a measure of the number of electronic states available at each energy level in a material. In the case of multilayer H-diamane and OH-diamane, the presence of additional hydrogen or hydroxyl groups may modify the DOS compared to pristine diamond or graphene. The DOS profile can affect the material's electrical and optical properties.

Carrier mobility: Carrier mobility refers to the ease with which charge carriers (electrons or holes) can move through a material in response to an electric field. The carrier mobility in multilayer H-diamane and OH-diamane may be influenced by various factors, including the presence of hydrogen or hydroxyl groups and the structural arrangement of the layers. Understanding the carrier mobility is crucial for assessing the material's potential for electronic applications.

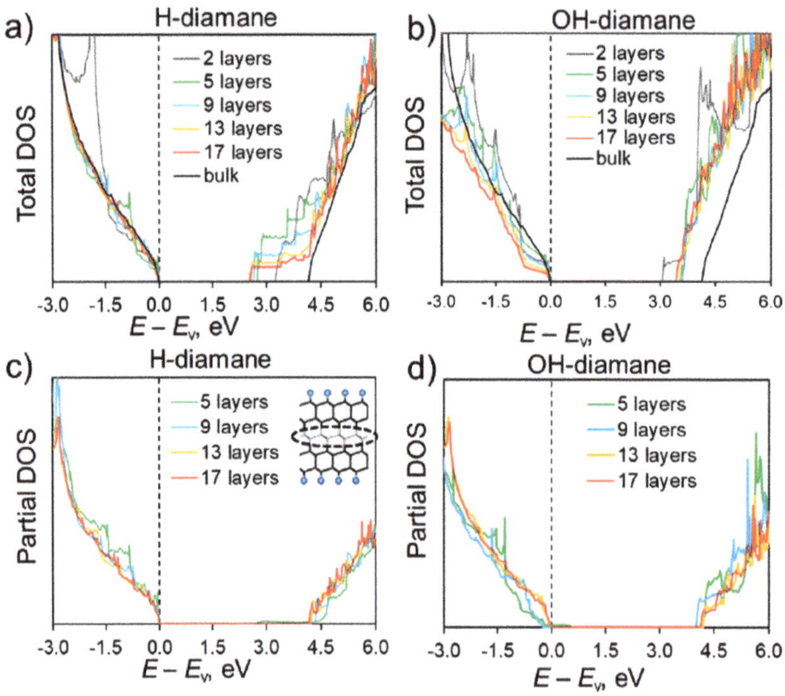

Figure 7.6. (a and b) Total DoS per atom in the structures calculated for different number of layers of H-diamane and OH-diamane. (c and d) Partial DoS for inner carbon atoms in the central layer of the diamanes. The inset in (c) displays a 5-layered diamane with central atoms marked by the dashed ellipse. Energy is shifted to the VBM. Reprinted with permission from [15]. Copyright (2022) American Chemical Society.

Surface states: The surface properties of multilayer H-diamane and OH-diamane, including the presence of hydrogen or hydroxyl groups on the surface, can influence the electronic properties. Surface states can affect surface reactivity, charge transfer, and the interaction of the material with other substances. These surface states may have implications for applications such as catalysis or sensing.

It is important to note that the electronic properties of multilayer H-diamane and OH-diamane may be influenced by their specific structure, layer arrangement, doping level, and the nature of the hydrogen or hydroxyl functionalization. Additionally, further research and experimentation are necessary to fully understand and characterize the electronic properties of these specific materials (figure 7.6).

7.5 Synthesis

7.5.1 Method of diamanoid synthesis

Selection of precursor molecules: Diamanoids are DLC structures composed of smaller diamond cages fused together. The first step in diamanoid synthesis would involve selecting suitable precursor molecules that can undergo reactions to form the desired diamond-like structure. The precursor molecules should have the necessary carbon–carbon bonds and connectivity to enable the fusion of smaller diamond cages.

Molecular assembly and fusion: Once the precursor molecules are selected, the next step would involve the assembly and fusion of these molecules to form larger diamond cages. This process may require controlled heating, catalysis, or other techniques to promote the bonding and rearrangement of carbon atoms into the desired diamond-like structure.

Purification and characterization: After the synthesis step, the resulting diamanoid material would need to be purified to remove any unreacted precursor molecules or byproducts. Various purification methods may be employed, such as solvent extraction, filtration, or high-temperature treatments. The purified diamanoid material would then be characterized using techniques such as microscopy, spectroscopy, or diffraction to verify its structural properties.

Further modifications: Depending on the specific application or desired properties, additional modifications may be carried out on the diamanoid material. This could include functionalization with different atoms or molecules, doping with impurities to alter its electronic properties, or surface modifications to enhance its compatibility with other materials or interfaces. It is important to note that the above process description is a generalized overview and may not capture the intricacies and specific details of diamanoid synthesis. The field of diamanoid synthesis is an active area of research, and scientists are continually exploring and developing new methods to create these structures.

7.5.1.1 Pristine graphene materials

Pristine graphene refers to graphene materials that are in their purest form, consisting of a single layer of carbon atoms arranged in a 2D honeycomb lattice. Pristine graphene possesses remarkable properties due to its unique structure, including exceptional mechanical strength, high electrical conductivity, thermal conductivity, and optical transparency. Here are some key aspects of pristine graphene materials [10].

Structure: Pristine graphene is composed of a single layer of carbon atoms tightly bonded together in a hexagonal lattice. Each carbon atom forms three covalent bonds with its neighboring atoms, creating a sp^2-hybridized structure. The resulting lattice structure gives graphene its extraordinary properties.

Mechanical properties: Pristine graphene is known for its exceptional mechanical strength and flexibility. It has a tensile strength of around 130 GPa, making it one of the strongest known materials. It also exhibits excellent elasticity, allowing it to stretch significantly without breaking.

Electrical conductivity: Pristine graphene exhibits exceptional electrical conductivity, allowing the efficient flow of electric charge. Its unique electronic structure, with delocalized π electrons, enables high electron mobility and conductivity. This property makes graphene suitable for various electronic and optoelectronic applications.

Thermal conductivity: Pristine graphene also demonstrates excellent thermal conductivity. Its 2D structure allows for efficient heat transfer, making it a promising material for thermal management applications, such as heat sinks and thermal interface materials.

Optical transparency: Pristine graphene is optically transparent and can absorb and transmit light across a broad spectrum, ranging from ultraviolet to infrared. It absorbs around 2.3% of incident white light due to its unique electronic band structure. This property makes graphene attractive for applications in transparent conductive films, photovoltaics, and sensors.

Other properties: Pristine graphene exhibits several other fascinating properties, including impermeability to gases and liquids, high surface area, and chemical stability. These properties make it suitable for various applications in fields such as energy storage, filtration, catalysis, and biomedicine. It is important to note that pristine graphene is typically synthesized using methods such as mechanical exfoliation of graphite, CVD, or epitaxial growth techniques. The synthesis method can influence the quality, size, and properties of the graphene material. Pristine graphene has attracted significant research interest and holds immense potential for a wide range of applications. Ongoing research continues to explore and harness its unique properties, as well as develop scalable and cost-effective synthesis methods for its production.

As prepared FLG films placed on the 3 mm TEM grids (2000 Mesh copper grids) from graphene were employed as graphene made materials to fabricate diamanoid/graphene hybrids. CVD produced the FLG films from CH4 at 1000°C on Ni substrate and then transferred onto a commercial TEM grid using a polymer-free transfer technique to minimize contamination. Nevertheless, larger films, up to 20 layers (or 10, depending on the folding position), were seen in TEM. FLG thickness is commonly 0.3–2 nm (for 1–6 monolayers) [10].

7.5.1.1.1 Hydrogenation process

Hydrogenation is more cost-effective compared to high-temperature and high-pressure procedures. 'Hydrogenation' typically refers to reacting hydrogen with specific materials to produce desired compounds. There is the possibility of adsorption, irreversibility, and other forms of reactions. Adsorption of hydrogen causes hybridization, which results in sp^2 to sp^3 spreading across graphene layers via interlayer bonds. Top carbon layer in the subpoint lattice A forms hydrogen bonds during the hydrogenation process, whereas the lower subpoint lattice A carbon layer forms bonds with the upper subpoint lattice B carbon layer. The sp^3-hybridized state is created when the bottom subpoint lattice B carbon layer comprises hydrogen bonds. While hydrogen atoms adsorbed on the outer layer of multilayered graphite stabilize the films that give rise to diamane, they drain electrons to generate sp^3 hybridization, and the absence of electrons gives rise to BLG. Complete hydrogenation is feasible, but additional alterations are needed to stabilize the structure. The quantity of hydrogen coverage would depend on the processing subunits, as hydrogenation process is a layer-making system regulated by hydrogenation energy boundaries [28]. To create diamane nanopatterns, the swelling regions of the BLG absorb hydrogen with poor coverage. Dehydrogenation, which occurs due to annealing, can be used to begin electrical and chemical functionalization. As a result, this reversible technique may produce BLG and dimanes. By joining carbon atoms from the top layers with hydrogen and carbon atoms from the bottom with carbon atoms from nearby layers, diamanes are created when two-layered graphene is hydrogenated. When graphene is hydrogenated in a single layer, one hydrogen atom is connected to each carbon atom above and below the layer [5].

7.5.1.1.2 Hot-filament process

The hot-filament process, also known as the hot-filament chemical vapor deposition process, is a technique that is used for the deposition of thin films or coatings onto substrates. It involves the use of a heated filament, typically made of tungsten or another refractory metal, to decompose precursor gases and enable the growth of desired materials. Some potential applications of the hot-filament process for diamanoids and diamane oxide follow:

Diamond film deposition: The hot-filament process is commonly employed for the deposition of diamond films. By introducing hydrocarbon gases, such as methane or acetylene, onto the heated filament, carbon atoms are dissociated and deposited onto a substrate, resulting in the growth of diamond films. This process may potentially be adapted to synthesize diamanoids or diamane oxide films using appropriate precursor gases.

Surface modification: The hot-filament process can also be utilized for surface modification of carbon-based materials. By controlling the gas composition and process conditions, it may be possible to introduce functional groups or modify the surface chemistry of diamanoids or diamane oxide, thereby tailoring their surface properties for specific applications.

Doping and alloying: The hot-filament process can facilitate the incorporation of dopant atoms or the formation of alloyed structures in carbon-based materials. By introducing appropriate precursor gases containing the desired dopant or alloying elements, it may be possible to modify the properties of diamanoids or diamane oxide, such as their electrical conductivity or optical properties.

Nanostructured growth: The hot-filament process can be used to grow nanostructured materials, including nanocrystalline or nanocomposite films. By controlling the process parameters and the growth conditions, it may be possible to synthesize nanostructured diamanoids or diamane oxide with unique properties and applications.

It is important to note that the specific process parameters, precursor gases, filament temperature, and growth conditions would need to be optimized for the synthesis or modification of diamanoids and diamane oxide using the hot-filament process. Additionally, further research and exploration may be required to determine the suitability and effectiveness of the hot-filament process for these specific materials.

Diamanoid can be produced effectively because hot filament has a characteristic quality of efficient hydrogen generation process and ability to form materials at low temperatures. This method accelerates high-kinetic-energy ions toward the substrate keeping the graphene unaffected. Under different temperatures and pressures, the hydrogenation of BLG to produce hydrogenated diamane has proven effective in fabricating thin films at the molecular level. To prevent carbon contamination, the filaments were washed with acetone before mounting. Using a thermocouple tip placed on top of the substrate holder's surface, just below the filaments, the temperatures of the grid and substrate holder were determined. The chamber's top was used to admit a controlled flow of ultra-pure H_2 gas once the chamber had been evacuated to at least 10^{-1} Torr. The resulting maximum substrate holding temperature for FLG and 2LG was 325°C and 400°C, respectively [10].

7.5.1.1.3 ***Plasma approach***

Plasma is a partially ionized gas that is composed of charged particles, neutral species, and free electrons. It can be used as a versatile tool for various processes in material synthesis and modification. Some potential applications of the plasma approach for diamanoids and diamane oxide follow [5]:

Plasma-enhanced chemical vapor deposition (PECVD): PECVD is a widely used technique for depositing thin films of materials onto substrates. It involves introducing reactive gases into a plasma discharge to induce chemical reactions and the formation of desired materials. Plasma can be utilized to deposit carbon-based films, potentially including diamanoids or diamane oxide, onto suitable substrates.

Plasma surface modification: Plasma treatment can be employed to modify the surface properties of materials, including carbon-based materials. Plasma can introduce functional groups or alter the surface chemistry, thereby enhancing the desired characteristics of diamanoids or diamane oxide. Surface functionalization can influence properties such as wettability, adhesion, and biocompatibility.

Plasma etching: Plasma etching is a process used to selectively remove material from a substrate. Plasma can be employed to etch carbon-based materials, potentially enabling the fabrication and patterning of diamanoids or diamane oxide structures with high precision. Plasma etching techniques can involve reactive ions or reactive radicals to remove specific areas or layers.

Plasma activation: Plasma activation can be used to improve the bonding between different materials or interfaces. By exposing the surfaces of diamanoids or diamane oxide to plasma, the surface energy and reactivity can be increased, facilitating better adhesion or bonding with other materials.

It is important to note that the specific plasma parameters, gas mixtures, power levels, and process conditions would need to be optimized for the synthesis or modification of diamanoids and diamane oxide. Additionally, further research and exploration may be required to determine the suitability and effectiveness of the plasma approach for these specific materials.

A functional group is added in carbon lattices using the straightforward, quick, and scalable plasma treatment method. The fused hydrogen plasma in the graphene is collected for its positive ions, which are essential in functionalization. The double-bonded carbon atoms in this hydrogen plasma are broken, and the C–C sp^3 hybridization is increased, forming diamanes. Graphene nano-circuitry can be patterned using these plasma irradiation [29] and high thermal stability diamanes created. Cold hydrogen plasma for hydrogenation is created by employing a hydrogen and argon combination under low pressure. Between two electrodes made of aluminum, DC plasma is ignited to produce hydrogen plasma. In BLG, hydrogen atoms are adsorbed chemically on both sides, creating a potential barrier to hydrogenation. This hydrogen plasma discharge results in the diamane formation. Graphene's top and bottom layers both undergo sp3 hybridization. However, hydrogenation only occurs at the top and bottom of the top and bottom layers, respectively [5].

By chemisorbing hydrogen on the top and bottom layers of a two-layer graphene, where conversion can also occur on metal for hydrogenation, cold plasma will make it possible for carbon atoms to link between their layers. The technique of naturally physisorbed water [25], followed by further (H_2O) upper and lower surface of trilayer graphene irradiation in a hydrogen plasma, produces a novel structure known as triamane [9]. Moreover, backscattered electrons may start the hydrogenation process by influencing the disintegration of physisorbed H_2O. When twisted graphene is exposed to hydrogen plasma, diamanes can also be seen to develop. Pressure and temperature are used to help hydrogen chemisorb on both surfaces of graphene [30]. Table 7.1 shows the contemporary literature delving into diamanoid synthesis methods highlights the latest strategies and techniques employed by researchers.

Oxidative pits were formed when hydrogen plasma was injected into hydrogenated BLG and monolayered graphene; however, they were not obvious without hydrogenation. Following that, the hydrogenation process keeps causing the creation of diamane oxide and diamanoid from diamane [5].

Table 7.1. Recent reports on synthesis process and methodology on diamanoid synthesis.

Sl. No.	Process	Methodology	References
1.	Pristine graphene materials	FLG films as graphene materials. The less FLG films were transferred onto a commercial TEM grid using a polymer-free transfer procedure.	[10]
2.	Hydrogenation process	Adsorption of hydrogen causes hybridization, which results in sp^2 to sp^3 spreading across the graphene layers via interlayer bonds. The top carbon layer of sublattice A form bonds with hydrogen atoms during the hydrogenation process, whereas the lower carbon layer of sublattice A forms bonds with the upper carbon layer of sublattice B.	[1, 5]
3.	Plasma approach	A functional group is added to the carbon lattices using the straightforward, quick, and scalable plasma treatment method. The double-bonded carbon atoms in this hydrogen plasma are broken, and the C–C sp^3 hybridization is increased.	[5, 9, 25, 29, 30]
4.	Hot-filament process	The method accelerates high-kinetic-energy ions toward the substrate. The chemisorption produces hydrogenated diamane.	[5, 10]

7.5.2 Diamane oxide

The term 'diamane' is used to describe a stratified diamond or graphene exhibiting the characteristics of an exceptionally thin and durable substance. Qualities of the thinnest hard material. Covalent bonding techniques transform bilayer sp^2-hybridized graphene into sp^3-hybridized structure. This diamond oxide,also known as DLC, can be synthesized using various methods, including PECVD, sputtering, and combustion-based techniques. Table 7.2 depicts the current publications addressing the synthesis process and methodology of diamane oxide. A brief overview of the PECVD involves depositing a thin film of DLC onto a substrate by exposing a hydrocarbon gas (such as methane, ethane, or acetylene) to a plasma. The plasma breaks down the gas molecules into their constituent ions, depositing them onto the substrate. The substrate is typically made of a material that can withstand the high temperatures and energetic ions the plasma generates, such as silicon or quartz. The substrate is also usually cleaned and etched before the deposition process to ensure good adhesion of the DLC film. The plasma is generated by applying a high voltage between two electrodes in a vacuum chamber filled with hydrocarbon gas. The resulting discharge ionizes the gas molecules and generates a plasma of reactive species, including ions, radicals, and excited atoms. The hydrocarbon gas is introduced into the plasma chamber and undergoes a series of chemical reactions, forming a DLC. The exact reaction mechanism is complex and depends on various factors, including gas composition, pressure, temperature, and power input. Depending on the deposition conditions, the resulting DLC film can have multiple properties. It can be amorphous or crystalline, with varying sp^2 and sp^3 hybridization degrees. The film can also be doped with different elements to modify its electronic, optical, and mechanical properties. PECVD is a versatile and widely used method for synthesizing DLC films, with applications in electronics, coatings, and biomedical devices. However, it requires careful control of the deposition conditions to achieve the desired film properties and avoid unwanted side reactions.

Table 7.2. Recent report on synthesis process and methodology of diamane oxide.

Sl. No.	Process	Methodology	References
1.	PECVD	PECVD involves depositing a thin DLC film onto a substrate by exposing a hydrocarbon gas to a plasma. The plasma breaks down the gas molecules into their constituent ions, depositing them onto the substrate.	[16]
2.	Irradiation with a focused electron beam	The process described involves transferring a bilayer CVD graphene onto a langasite substrate and subsequently irradiating it with a focused electron beam through a layer of polymethyl methacrylate.	[6]

3.	CVD approach	Passivating graphene involves the passivation of dangling bonds found on the edges or defects of the graphene lattice and embedded nanostructures. These structures act as a template or host for the synthesis production of diverse embedded nanostructures. By introducing foreign atoms or molecules into the graphene lattice, it is possible to create localized modifications or structures within the graphene sheet. These embedded nanostructures can exhibit unique properties and functionalities	[6, 16]
4.	Hot-filament process	In order to traditionally create metastable diamond at low pressure from a diluted mixture of a hydrocarbon in H_2, hot-filament CVD has been used. Since using the hydrocarbon for carburization or conditioning should prevent carbon contamination, these processes were carried out without its usage.	[3]

7.5.2.1 Different methods for method of diamane oxide synthesis

Diamane oxide, also known as diamond oxide or diamantane oxide, is a synthetic compound with the molecular formula $C_{14}H_{20}O$. One possible method for synthesizing diamane oxide involves the following steps:

(i) Start with commercially available starting materials: Appropriate starting materials can be used to construct the desired molecular structure of diamane oxide. These starting materials may include cyclic hydrocarbons or other suitable precursors.

(ii) Functionalize the starting materials: Introduce specific functional groups to the starting materials that are necessary for further reactions and to achieve the desired structure. This can be done through various chemical transformations such as substitution, addition, or oxidation reactions.

(iii) Perform cyclization reactions: Carry out cyclization reactions to form the ring structures present in diamane oxide. Cyclization can be achieved through a variety of methods, including thermal or photochemical reactions, catalytic processes, or using specific reagents.

(iv) Oxidation: Once the cyclization has occurred, you will need to oxidize specific positions in the molecule to introduce the oxygen atom. Oxidation reactions can be performed using oxidizing agents such as hydrogen peroxide (H_2O_2), potassium permanganate ($KMnO_4$), or other suitable reagents.

(v) Purification and characterization: After the synthesis, the resulting diamane oxide needs to be purified to remove any impurities and isolated as a pure compound. Purification techniques may include recrystallization, column chromatography, or other methods. The final product should be characterized using spectroscopic techniques like nuclear magnetic resonance, infrared spectroscopy (IR), and mass spectrometry to confirm its identity.

It is important to note that the specific reaction conditions, reagents, and techniques used in the synthesis may vary depending on the availability of starting materials, desired yield, and other factors. It is recommended to consult the scientific literature or seek guidance from an expert in organic synthesis for a more detailed and precise method for synthesizing diamane oxide.

7.6 Future scope

The future scope of diamanoids and diamane oxide, as emerging carbon-based materials, holds potential for various technological applications and scientific advancements. Some potential areas where these materials could find future applications follow:

(i) Electronics and optoelectronics: Diamanoids and diamane oxide explored for their electronic and optical properties. Their unique structures and potential semiconducting properties could make them valuable for developing novel electronic devices, sensors, and optoelectronic components.

(ii) Energy storage: Carbon-based materials, including diamanoids, have been investigated for energy storage applications such as batteries and supercapacitors. Diamanoids' high surface area, stability, and potential tunability may make them suitable for improving energy storage device performance.

(iii) Catalysis: Diamanoids, with their complex structures and surface functionalities, could exhibit catalytic properties that could be harnessed for various chemical reactions. Their potential applications in catalysis could include environmental remediation, chemical synthesis, and energy conversion processes.

(iv) Biomedical applications: Carbon-based materials have shown promise in biomedical applications such as drug delivery, bioimaging, and tissue engineering. Diamanoids and diamane oxide, with their unique structures and potential biocompatibility, could be explored for these purposes.

(v) Nanoelectromechanical systems (NEMS): Diamanoids may have applications in the field of NEMS, which involves the integration of nanoscale mechanical systems into electronic devices. The unique structural properties of diamanoids could be utilized for developing high-performance nanomechanical devices and sensors.

(vi) Quantum technologies: The controlled fabrication of nanoscale diamond structures, including diamanoids, could contribute to the development of quantum technologies. Diamond-based materials have shown promise in quantum computing, quantum sensing, and quantum communication due to their excellent quantum coherence properties.

It is important to note that the future scope of diamanoids and diamane oxide is still being explored, and further research and development are necessary to fully understand their properties and unlock their potential in various applications. As scientific understanding advances and new synthesis and fabrication techniques are developed, additional opportunities for these materials are likely to emerge.

7.7 Concluding remarks

Nanometer-thick sp^3-bonded-carbon-rich films (sp^3C^2D) have undergone several notable advancements over the previous 4 years, including the creation of genuine diamanes and diamanoids, monolayer amorphous carbon, and sub-10 nm nanocrystalline diamond. These developments outperform earlier developments in sub-10 nm tetrahedral amorphous carbon films, which have been researched as potential low-*k* dielectric materials for VLS and have long been employed in hard drive technology. The synthesis of diamanes and diamanoids, their atomic structure, physical characteristics (including unique electrical, optical, thermal, mechanical, and vibrational characteristics), and cutting-edge technological applications are all covered in-depth in this extensive book. The first section of this review discussed historical perspectives, structural issues, and physical characteristics. Particular focus was given to the diamanes' bandgap engineering, surface engineering, hetero atom/metal doping, and quantum mechanical properties. Nanoelectronics, semiconducting, optoelectronics, quantum electronics, and other fields are among the applications for dimanes and dimanoid materials. The third section briefly covered the creation of diamanoids and a few real-world point-by-point applications of diamane in various nanodevices.

Users are able to convert a predesigned van der Waals layered precursor into a distinctive covalently linked layered nanosystem with required properties, for applications, with the aid of the current research reports in this article, with clear explanations and a useful bibliography. In the final section, future technology based on dimanes and dimanoid has been described. The goal of this work is to inspire more research into the creation of diamanes, diamanoids, and other ultrathin films.

In conclusion, information specifically focused on diamanoids and diamane oxide was limited in the scientific literature. There were no extensive studies or well-established research on their synthesis methods, structural characteristics, general qualities, cutting-edge applications, electronic properties, optical properties, or mechanical properties. However, it is worth noting that scientific knowledge is constantly evolving, and new research and developments may have emerged since then. It is possible that more recent studies have provided deeper insights into diamanoids and diamane oxide, exploring their potential, properties, and applications.

References

[1] Pumera M and Wong C H A 2013 Graphane and hydrogenated graphene *Chem. Soc. Rev.* **42** 5987

[2] Paul S, Mondal S, Saha A and Roy S 2023 Fundamentals and functionalization of CNTs and other carbon nanomaterials *Functionalized Carbon Nanomaterials for Theranostic Applications* (Amsterdam: Elsevier) 77–90

[3] Piazza F, Monthioux M and Puech P 2022 2D (<10 nm) sp^3-C-rich carbon materials, possibly hydrogenated: a review *Carbon Trends* **9** 100219
[4] Bakharev P V *et al* 2019 Chemically induced transformation of chemical vapour deposition grown bilayer graphene into fluorinated single-layer diamond *Nat. Nanotechnol.* **15** 59–66
[5] Tiwari S K, Pandey R, Wang N, Kumar V, Sunday O J, Bystrzejewski M, Zhu Y and Mishra Y K 2022 Progress in diamanes and diamanoids nanosystems for emerging technologies *Adv. Sci.* **9** 2105770
[6] Emelin E V, Cho H D, Korepanov V I, Varlamova L A, Erohin S V, Kim D Y, Sorokin P B and Panin G N 2022 Formation of diamane nanostructures in bilayer graphene on langasite under irradiation with a focused electron beam *Nanomaterials* **12** 4408
[7] Sofo J O, Chaudhari A S and Barber G D 2007 Graphane: a two-dimensional hydrocarbon *Phys. Rev.* B **75**
[8] Elias D C *et al* 2009 Control of graphene's properties by reversible hydrogenation: evidence for graphane *Science* **323** 610–3
[9] Nascimento A J M and Nunes R W 2013 Lubrication of Stone–Wales transformations in graphene by hydrogen and hydroxyl functional groups *Nanotechnology* **24** 435707
[10] Piazza F, Monthioux M, Puech P, Gerber I C and Gough K 2021 Progress on diamane and diamanoid thin film pressure less synthesis *C* **7** 9
[11] Sun X, Kiang C H, Endo M, Takeuchi K, Furuta T and Dresselhaus M S 1996 Stacking characteristics of graphene shells in carbon nanotubes *Phys. Rev.* B **54** R12629–32
[12] Endo M, Takeuchi K, Hiraoka T, Furuta T, Kasai T, Sun X, Kiang C H and Dresselhaus M 1997 Stacking nature of graphene layers in carbon nanotubes and nanofibres *J. Phys. Chem. Solids* **58** 1707–12
[13] Piazza F, Monthioux M, Puech P and Gerber I C 2020 Towards a better understanding of the structure of diamanoïds and diamanoïd/graphene hybrids *Carbon* **156** 234–41
[14] Saito R, Hofmann M, Dresselhaus G, Jorio A and Dresselhaus M S 2011 Raman spectroscopy of graphene and carbon nanotubes *Adv. Phys.* **60** 413–550
[15] Varlamova L A, Erohin S V, Larionov K V and Sorokin P B 2022 Diamane oxide. two-dimensional film with mixed coverage and a variety of electronic properties *J. Phys. Chem. Lett.* **13** 11383–90
[16] Chen C, Guo D, Fan D, Lu S, Jiang M, Li X and Hu X 2022 Semiconductive vertical graphene nanoribbons self-assembled on diamond (1 0 0) surface by oxidation: a DFT study *Appl. Surf. Sci.* **604** 154646
[17] Dey A, Das B C, Biswas A B, Biswas P, Dhar A, Roy S and Moyez S A 2017 Graphene co-doped TiO_2 nanocomposites for photocatalysis and photovoltaics applications *Indian J. Sci. Technol.* **10** 1–6
[18] Qiu D, Wang Q, Cheng S, Gao N and Li H 2019 Electronic structures of two-dimensional hydrogenated bilayer diamond films with Si dopant and Si–V center *Res. Phys.* **13** 102240
[19] Li J, Li H, Wang Z and Zou G 2013 Structure, magnetic, and electronic properties of hydrogenated two-dimensional diamond films *Appl. Phys. Lett.* **102** 073114
[20] Gao H, Wang L, Zhao J, Ding F and Lu J 2011 Band gap tuning of hydrogenated graphene: H coverage and configuration dependence *J. Phys. Chem.* C **115** 3236–42
[21] Wang J, Li L, Wang J, Yang W, Guo P, Li M, Liu D, Zeng H and Zhao B 2022 First-principles study on the nanofriction properties of diamane: the thinnest diamond film *Nanomaterials (Basel)* **12** 2939
[22] Cheng T, Liu Z and Liu Z 2020 High elastic moduli, controllable bandgap and extraordinary carrier mobility in a single-layer diamond *J. Mater. Chem.* C **8** 13819–26

[23] Clark S, Jeon K J, Chen J Y and Yoo C S 2013 Few-layer graphene under high pressure: Raman and X-ray diffraction studies *Solid State Commun.* **154** 15–8

[24] Wang H, Wang Y, Cao X, Feng M and Lan G 2009 Vibrational properties of graphene and graphene layers *J. Raman Spectrosc.* **40** 1791–6

[25] Zhou H, Xie T, Taniguchi T, Watanabe K and Young A F 2021 Superconductivity in rhombohedral trilayer graphene *Nature* **598** 434–8

[26] Mori Y, Takeda K, Charbonneau M and Refetoff S 1990 Replacement of Leu227 by Pro in thyroxine-binding globulin (TBG) is associated with complete TBG deficiency in three of eight families with this inherited defect *J. Clin. Endocrinol. Metab.* **70** 804–9

[27] Dong J, Zhang K and Ding F 2015 Hydrogenation of bilayer graphene: a small twist makes a big difference *Nano Res.* **8** 3887–97

[28] Wang H *et al* 2014 Carbon dioxide hydrogenation to aromatic hydrocarbons by using an iron/iron oxide nanocatalyst *Beilstein J. Nanotechnol.* **5** 760–9

[29] Sarinont T, Amano T, Attri P, Koga K, Hayashi N and Shiratani M 2016 Effects of plasma irradiation using various feeding gases on growth of *Raphanus sativus* L. *Arch. Biochem. Biophys.* **605** 129–40

[30] Sorokin P B and Chernozatonskii L A 2013 Graphene-based semiconductor nanostructures *Phys.-Usp.* **56** 105

Chapter 8

Synthesis, properties, and overview of nanodiamonds

Rakesh Shrestha, Sagar Ban, Blessy Thomas Baby, Abdul Mumin and Rajendra Joshi

Nanodiamonds, an attractive class of carbon-based nanomaterials, have gathered considerable attention in recent years due to their unique properties and promising applications in various fields. The most common approach is the detonation method, where high-pressure detonation of carbon-containing explosives yields nanodiamonds in large quantities. This method results in high-energy surface defects and functional groups, making it suitable for various surface modifications. Alternatively, the bottom-up approach of nanodiamond synthesis allows precise control over nanodiamond size and crystallinity, but it is often limited to smaller quantities. Other techniques, such as the high-pressure high-temperature (HPHT) method and the chemical vapor deposition (CVD) method, offer additional possibilities for modifying nanodiamond properties. The characteristics of each method influence the nanodiamonds' properties, making their synthesis crucial in modifying these nanomaterials for specific applications. Nanodiamonds possess remarkable properties that distinguish them from other carbon nanomaterials. Their exceptional mechanical properties, such as high hardness and stiffness, contribute to their potential as reinforcing agents in composite materials. Furthermore, nanodiamonds exhibit superior thermal conductivity and excellent electrical insulation, presenting opportunities in thermal management and electronic applications. Additionally, nanodiamonds' biocompatibility and low toxicity enhance their attractiveness for biomedical applications, including drug delivery systems, bioimaging, and as platforms for targeted therapies. Nanodiamonds represent a versatile class of nanomaterials with exciting potential in various industries and scientific disciplines. Their synthesis methods, tailored properties, and expanding applications are paving the way for innovative technologies and solutions. Continued research and development in this field are likely to present even more exciting applications and open new boundaries for nanodiamond-based technologies in the near future.

doi:10.1088/978-0-7503-5939-9ch8

8.1 Introduction

The brilliance and durability of diamonds have long stimulated human interest. Recent developments in nanotechnology have sparked interest in the investigation of nanodiamonds, a fascinating subset of diamonds. Nanodiamonds are diamond particles that are smaller than a nanometer and have special qualities that set them apart from their larger counterparts. Nanodiamonds are an excellent and attractive material that have appeared in the huge field of nanotechnology. Nanodiamonds, at the crossroads of nanoscience, have attracted the interest of scientists, researchers, and engineers alike. These diamond particles have special qualities that open a wide range of fascinating applications in a variety of fields. This chapter seeks to present a thorough and in-depth review of the synthesis, characteristics, and various uses of nanodiamonds as given in figure 8.1.

The first section of this chapter explores the various synthetic techniques used to produce nanodiamonds. The explosive detonation mechanism, where shock waves and high pressures cause the production of nanodiamonds, will be explained to readers. Additionally, the exact control over size and characteristics possible through gas-phase synthesis will be highlighted by the investigation of CVD techniques. This chapter will also discuss electrochemical synthesis, which enables customized nanodiamond growth and functionalization, and HPHT techniques, which imitate the natural diamond creation process. The advantages, restrictions, and prospective uses of each synthesis technique will be carefully studied, giving readers a comprehensive grasp of the various nanodiamond manufacturing techniques.

After discussing synthesis techniques, this chapter moves on to a thorough examination of the characteristics that set nanodiamonds apart from other nanomaterials. The high surface area, biocompatibility, and functionalization of nanodiamonds will be examined along with their intrinsic hardness, which results from strong covalent connections between carbon atoms. This topic will focus on the fascinating photoluminescence displayed by some nanodiamonds, especially those with nitrogen-vacancy (NV) centers and explain how these materials might be used in quantum technologies and bioimaging. How nanodiamonds behave in terms of thermal conductivity, electrical insulation, chemical stability, and optical

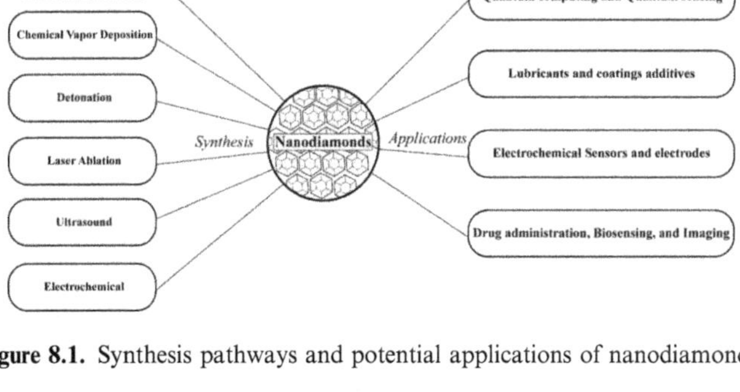

Figure 8.1. Synthesis pathways and potential applications of nanodiamonds.

transparency will also be revealed, giving readers a thorough grasp of their distinctive qualities.

This chapter will also provide a thorough overview of the uses of nanodiamonds across numerous fields in addition to their outstanding features. The fascinating potential of nanodiamonds in biosensing, targeted therapies, and drug delivery systems will be presented, emphasizing their growing importance in the field of biomedicine. The focus will then shift to quantum technology, including details of how nanodiamonds are used in single-photon sources, quantum sensing, and computing. In addition, this chapter will describe how nanodiamonds can be added to lubricants and coatings to increase wear resistance and reduce friction. It will also describe how they can be used in electrodes and electrochemical sensors. Nanodiamonds' unique qualities are considered in relation to each application, highlighting their adaptability and wide range of potential uses.

This is a crucial tool for academics, professionals in the sciences, and engineers who want to tap into the enormous potential of nanodiamonds in developing industries including electronics, healthcare, quantum technology, materials science, and more. This chapter serves as a portal to the exciting world of nanodiamonds, influencing the future of nanotechnology and materials research with its in-depth analysis of production methods, exploration of special features, and thorough overview of applications.

8.2 Synthesis of nanodiamonds

Nanometer-sized diamond particles are synthesized during the production of nanodiamonds. The method chosen for creating nanodiamonds will rely on the desired qualities and the applications for which they will be used. This section will discuss a few common synthesis techniques.

8.2.1 Hydrothermal synthesis

Nanodiamonds are produced using the hydrothermal synthesis process in an aqueous solution at high pressure and high temperature (Boruah and Saikia 2022). Diamond crystals with diameters in the nanometer range, typically from 1 to 100 nm, can be developed under regulated conditions (Auciello and Sumant 2010). Diamond seeds, which act as nucleation sites for the formation of nanodiamonds, are first prepared in the process by either starting with other diamond particles or by finely powdering larger diamonds (Wang *et al* 2004).

The diamond seeds are subsequently mixed with an aqueous solution that contains several chemical additives, such as acids or alkalis, to regulate the pH and promote the growth of nanodiamonds (Boruah and Saikia 2022). Benzenesulfonic-modified nanodiamonds were prepared for the removal of methylene blue in dyestuff sewage. The pH influences the charge on the surface of the adsorbent and organic dyes, and hence the initial concentration was varied. The adsorption effect of the prepared material was found to be better in the alkaline environment than in the acidic environment after analysis (Lei *et al* 2020). A high-pressure reactor that is made to resist the harsh conditions necessary for hydrothermal synthesis has been used to contain the aqueous diamond seed suspension. Stainless steel or other materials that can withstand high pressures and temperatures are frequently used to make this type of reactor (Xie and Liu 2022). In addition, low temperatures and atmospheric pressure plasma reactors have also been utilized to synthesize carbon nanostructures (Chandrakar *et al* 2023). To prevent any solution or gas leaks, the reactor must be completely enclosed. The desired pressure is then achieved by pressurizing it with an inert gas, such as nitrogen or argon. A few hundred megapascals (MPa) to several gigapascals (GPa) of pressure can be applied in hydrothermal synthesis (Bhagabati *et al* 2019).

The reactor is set inside a furnace or other high-temperature heating chamber, often with temperatures between 1000°C and 2000°C (Liu *et al* 2021). Throughout the process, the temperature and pressure should be carefully regulated. As the diamond seed suspension is exposed to the high-pressure and high-temperature conditions, the hydrothermal reaction occurs. The solution's chemical ingredients boost the development of diamond crystals surrounding the seed particles. Molybdenum disulfide (MoS_2)/nanodiamond nanocomposite was effectively created using a hydrothermal synthesis technique, and its humidity sensing properties were examined. It was found that the nanodiamonds in MoS_2 not only act as a binder but also act as active sites and surrounds the nanodiamond (Yu *et al* 2020). Because graphite is the stable form of carbon at ambient conditions, the high pressure serves to stabilize the diamond phase by preventing its conversion to that form (Narayan and Bhaumik 2015). Heating is stopped once the desired reaction time has passed, and the reactor is then allowed to gradually cool. To avoid abrupt changes in pressure that can have unfavorable effects, regulated cooling is necessary. Once the internal pressure reaches ambient levels, the pressure is gradually released, and the reactor is opened (Alzahrani and Alkahtani 2023). The resulting nanodiamonds are extracted from the solution through a variety of separation processes, including centrifugation or filtration (Baker and Baker 2010, Laura Soriano *et al* 2018, Zheng *et al* 2009). To obtain the appropriate level of purity, additional purification processes, such as acid treatments or annealing, may be required because the collected nanodiamonds may contain impurities and some reactants (Hurtado *et al* 2020).

Due to their distinctive surface features, the nanodiamonds produced during hydrothermal synthesis have a wide range of uses, including in the biomedical sector (drug administration, imaging, etc), as additives to improve the mechanical properties of materials, and as catalysts (Bacakova *et al* 2020, Liu and Speranza 2019). The procedure provides superior control over the size and shape of nanodiamonds, making it a useful technique for synthesizing nanomaterials.

8.2.2 Chemical vapor deposition synthesis

CVD is one of the most widely applied processes for creating nanodiamonds. It involves the development of diamond films or nanoparticles on a substrate from a gas-phase precursor at high temperatures and relatively moderate pressures (Das and Roy 2022). An appropriate substrate must first be prepared on which nanodiamonds will be placed (Manawi *et al* 2018). Quartz, silicon carbide, and silicon are often used as substrates, although other materials have also been employed, depending on the application (Shah and Tali 2016).

A CVD chamber is used to place the substrate. Typically, quartz or other materials that can tolerate high temperatures and corrosive gases are used to construct the chamber (Whitlow *et al* 2017). A microwave CVD chamber was prepared with a quartz tube as a reaction chamber, as presented in figure 8.2. It was discovered that the deposition process could only last a few minutes at pressures greater than 25 Torr due to the formation of a conducting coating inside the quartz tube (Manawi *et al* 2018). Then, to get rid of any pollutants and moisture, the chamber is purged and evacuated. Precursor gases are mixed and introduced into the CVD chamber. Typically, hydrogen and a carbon-containing gas (such as methane, acetylene, or ethanol) are utilized as precursor gases when creating nanodiamonds (Kumar and Ando 2010). In hot filament CVD, undoped diamond films were fabricated using gas precursors such as methane and acetone. It was shown that the liquid precursors have higher growth rates than the gas precursors (Lu *et al* 2023). The introduction of other gases, such as oxygen or nitrogen, can sometimes be used to regulate the diamond development process and give the nanodiamonds particular features (Krueger and Lang 2012). The charged carbon clusters can be generated in CVD diamond and are suspended in the gas phase and deposited as diamond films. It was reported that the acetylene to oxygen ratio of 1.04 with the captured clusters of 1.5 nm in a gas mixture were mostly amorphous with few having the diamond lattice (Shenderova *et al* 2002). The precursor gases are activated to initiate the growth of nanodiamonds by thermal activation or plasma activation (Li *et al* 2016).

Diamond seeds (nuclei) are created on the substrate's surface by the activated carbon species. These nuclei aid as a model for developing diamonds in the future. The structures of the nanodiamonds are gradually constructed as carbon atoms continue to deposit and bond to the nuclei. The size, shape, and crystalline structure of the nanodiamonds are affected by the growth conditions, which include temperature, pressure, gas composition, and activation technique. By stopping the flow of precursor gases or turning off the plasma, the growth process is stopped when it reaches the

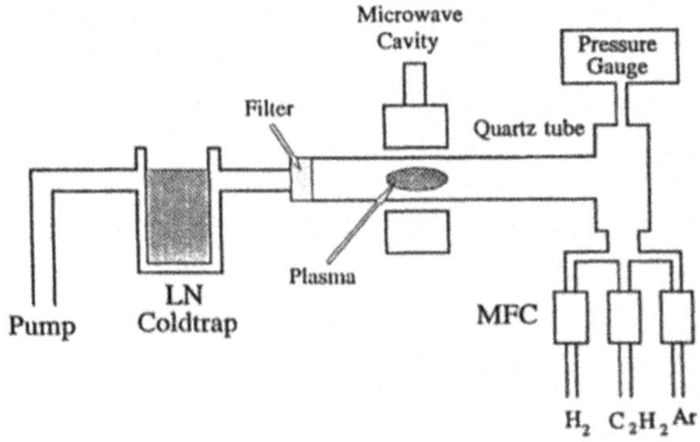

Figure 8.2. Schematic representation of a microwave assisted CVD chamber. Reprinted with permission from Kleckley *et al.* (1997). Copyright (1997) American Chemical Society.

desired stage. The CVD chamber is then allowed to gradually cool to prevent damage to the nanodiamonds and avoid abrupt changes in pressure. The nanodiamond-coated substrate is removed from the CVD chamber once it has reached ambient temperature. It is then subjected to several characterization procedures to assess the caliber and characteristics of the generated nanodiamonds. The nanodiamonds can be extracted from the substrate and used in a variety of ways.

The CVD method offers excellent control over the synthesis parameters, enabling the production of well-defined and uniform nanodiamonds with tailored properties suitable for a wide range of applications, including biomedicine, electronics, and tribology.

8.2.3 Detonation synthesis

Detonation synthesis of nanodiamonds is a technique that uses high-energy explosions to produce extremely small diamond particles that are typically between a few to tens of nanometers (Jain *et al* 2023, Mostofizadeh *et al* 2011). This process is also known as shock-wave synthesis or detonation nanodiamond synthesis (Danilenko 2005). A mixture of an energetic substance (commonly a combination of TNT and RDX) and a carbon-containing precursor, usually graphite, serves as the starting ingredient for detonation synthesis (Yan *et al* 2016). The carbon used to make diamonds comes from graphite. An explosive mixture with a net negative oxygen balance can produce nanodiamonds because it serves as both a source of carbon and energy for the conversion (figure 8.3(A)). Figure 8.3(A) shows the propagation of the explosion wave. Stage A starts with the explosion that creates the shock wave, and Stage B is when the chemical reaction that breaks down the explosive molecules takes place. Stage CJ, when pressure and temperature are equivalent to point A in figure 8.3(B), is the Chapman–Jouguet plane that depicts the conditions when the reaction and energy release are essentially complete. In Stage C, the detonation products expand at a high temperature, and carbon nanoclusters are formed in Stage D. In Stage E, the liquid nanodroplets coagulate. Finally, Stage F is where nanodiamonds crystallize, grow, and aggregate (Hwang *et al* 2020).

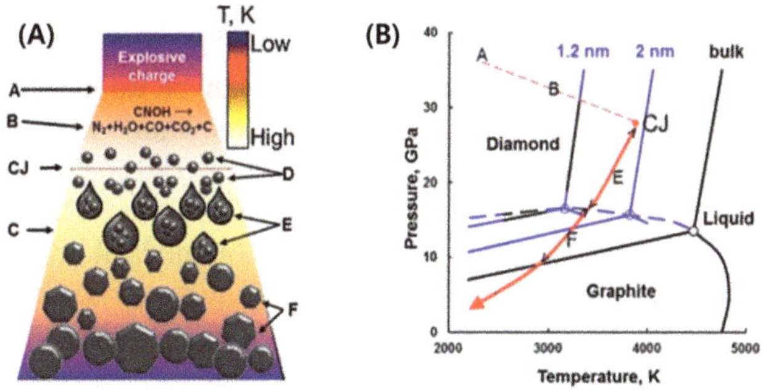

Figure 8.3. (A) Schematic representation of detonation wave propagation. (B) Indication of conditions when the reaction and energy release reach completion. Reprinted from Shenderova and Nunn (2017), Copyright (2017), with permission from Elsevier.

A metal chamber containing the combination is then subjected to a powerful shock wave produced by detonation (Shames *et al* 2012). There are several ways to start the detonation, including electrical discharge or collision with an explosive primer. The nanodiamond and potassium chlorate compositions were considered as green primers, and the reactive formulations of the compositions usually decompose depending on the pressing level of nanodiamonds (Shen *et al* 2013). The pressure, temperature, and velocity of the shock wave are extremely high (Danilenko 2005). The high values of pressure and temperature are obtained under the assumptions that the carbon is in the form temperature conditions in enclosed spaces for long enough periods to partially transform graphite into diamond grains as small as 20 nm, which then compact into micron-sized polycrystalline particles. To quickly dissipate heat at the high temperatures attained during the explosion and prevent the transformation of the diamond back into graphite, copper is combined with graphite in shock-wave synthesis (Shenderova and McGuire 2006).

Detonation can produce pressures of up to a few hundred thousand atmospheres (GPa) and temperatures that temporarily reach over a 1000°C. The precursor experiences a quick and severe compression of the carbon atoms as a result of the shock wave (Tolochko and Zolotarev 2016). As a result, the high-pressure and high-temperature conditions correspond to the conditions needed for the carbon to be in graphite form while lower values of pressure and temperature corresponds to the carbon present in the form of diamond in detonation products. (Danilenko 2005). A high-pressure shock wave is created when the explosive mixture explodes and travels through the carbon-containing precursor. The shock-compressed material rapidly expands and cools down after the detonation, producing a quenching effect (Batsanov 2018). This chilling procedure is essential because it helps stop diamond from turning back into graphite. The diamond structure is frozen into place by the rapid quenching. For instance, graphitic C_3N_4 was compressed at room temperature and desired pressure. It was then heated to 800°C–2000°C for 5–30 min, and quenched and decompressed back to ambient condition (Fang *et al* 2013). Diamond particles of a diameter of a few nanometers, or 'nanodiamonds', are produced as a result of the quick quenching (Boudou *et al* 2009). These particles typically have a cuboid or spherical form and range in size from a few to tens of nanometers. The resultant nanodiamonds from the detonation synthesis are recovered using a variety of techniques, such as filtration, centrifugation, and acid treatments to remove impurities and unreacted carbon (Liang *et al* 2009, Wang 2011).

The properties of the resulting nanodiamonds can be adjusted by varying the parameters of the detonation processes, such as the kind of precursor and the strength of the shock wave (Shenderova *et al* 2002). Detonation synthesis is an effective and scalable approach for creating nanodiamonds. The produced nanodiamonds are used in a variety of industries, such as biomedical imaging, medication delivery, catalysis, and as reinforcements in composite materials (Cha *et al* 2013).

8.2.4 Laser ablation synthesis

High-power lasers are employed in the laser ablation synthesis of nanodiamonds method to create diamonds at the nanoscale. A pulsed laser is used to vaporize a target material that contains carbon, usually graphite, in a controlled setting (Kumar *et al* 2019). A nanodiamond particle is created by the condensation and nucleation of the evaporated carbon atoms. The process takes place in a gaseous environment (Basso *et al* 2020), and the produced nanodiamonds are gathered and subsequently processed for use in a variety of applications.

A high-power pulsed laser, a target substance (such as graphite), and a reaction chamber filled with an appropriate gas environment (such as He or Ar) often make up a laser ablation setup (Liu and Speranza 2019). In the synthesis of carbon nanotubes, the reaction chamber is filled with helium gas and the electrodes are separated at 1–2 mm. A quartz tube that has been heated in a tubular furnace is used for the synthesis process. A high-purity graphite target with a modest amount of metallic catalyst is employed as a carbon source inside the quartz tube. The carbon and metallic catalyst vaporize when a scanning laser beam is focused onto a target at a high temperature and in the presence of an inert gas flow (Liu and Speranza 2019). Furthermore, a regulated pressure is maintained in the reaction chamber. The carbon atoms on the target's surface are quickly heated and turned into vapor when the laser is pulsed onto the substance. A small area of the graphite is changed into a high-energy plasma that contains carbon atoms and ions as a result of the tremendous thermal energy (Yang 2007). Rapid laser-induced plasma expansion produces a high-temperature and high-pressure zone in the reaction chamber's gas environment (Yang 2007). The plasma loses energy and cools down as it enlarges. The plasma's carbon atoms and ions supersaturate the gas environment during its fast growth and cooling (Bundaleska *et al* 2018). This means that nucleation and the production of new particles are encouraged because the concentration of carbon atoms is higher than its equilibrium concentration. Carbon atoms start to condense and form tiny clusters known as nanodiamond nuclei as the plasma cools.

The nanodiamond nuclei serve as starting points for new diamond synthesis (Haque *et al* 2019). These nuclei enlarge as a result of the deposition of carbon atoms from the gas phase. The length of the laser pulse and the particular conditions of the reaction govern how big the nanodiamonds can grow before stopping (Marabotti *et al* 2022). The reaction chamber is cooled down and the gas is expelled following the laser ablation procedure. The gas and any byproducts that might have developed throughout the procedure are then removed from the collected nanodiamond particles (Gorrini *et al* 2016). The nanodiamonds may go through additional purification processes, depending on the application, to get rid of any impurities or undesirable particles. An approach for liquid-phase nanodiamonds purification from the non-diamond phase was proposed, where nanodiamonds are imbedded in the case of graphite ablation. This cleaning method is efficient because it has the benefits of liquid-phase purification procedures, particularly the high density of the oxidizer, without the use of harsh chemicals or a hazardous acidic environment (Basso *et al* 2017).

The exact control over particle size, scalability, and potential for incorporating dopants into the diamond lattice during the synthesis process are the benefits of laser ablation nanodiamond production (Baig *et al* 2021, Zhang *et al* 2021). Due to their distinctive qualities, including their high surface area, chemical stability, and biocompatibility, these nanodiamonds find use in a variety of industries, including biomedicine, electronics, and nanotechnology (Kumar *et al* 2019, Nunn *et al* 2017). To produce high-quality and tightly controlled nanodiamonds, it is important to note that the reaction conditions must be carefully managed during the procedure.

8.2.5 Ultrasound synthesis

Ultrasound synthesis is a process that uses intense ultrasound vibrations to create nanoscale diamonds in a carbon-containing solution (Ullal *et al* 2022). Nanodiamond particles are formed with cavitation, which generates concentrated areas of high pressure and temperature. The process is easy, economical, and environmentally friendly.

High-intensity ultrasound waves are often produced by an ultrasonic bath or an ultrasonic probe in the setup. A carbon source is dissolved in an appropriate solvent to create a precursor solution that contains carbon. Organic carbon components such as sugars, alcohols, or other carbon-rich substances can serve as the source of carbon. Research has found that the carbon dots produced from honey have a quantum yield of 1.6%, whereas the carbon dots produced from bread, jaggery, or sugar have a quantum yield of less than 1.2% (Manzoor *et al* 2023). Hence, the production efficiency is highly dependent on the source of carbon. Using the probe or the ultrasonic bath, the carbon-containing solution is exposed to ultrasound. Cavitation bubbles are small bubbles created by the high-intensity ultrasonic waves alternating cycles of compression and rarefaction in the liquid medium. The pressure inside the bubbles falls during the rarefaction cycle, which causes them to expand. The bubbles develop until they reach a critical size, at which point they can no longer withstand the internal pressure and abruptly burst during the compression cycle. Localized areas of the liquid experience high pressures and temperatures due to the cavitation bubbles collapsing. The bubbles that are collapsing can reach temperatures of several thousand degrees Celsius and pressures of more than one thousand atmospheres. The breaking of carbon–carbon bonds in the carbon-containing precursors and the emergence of extremely reactive carbon species, such as free radicals, are just two of the chemical reactions brought on by these intense conditions (Liu *et al* 2022).

Diamond particles form on the extremely reactive carbon species that cavitation produces as their nucleation sites. On these nucleation sites, the carbon atoms and radicals mix and condense to produce nanodiamond particles. As more carbon species are added to the nanodiamonds surfaces, their size increases. The nanodiamonds are normally extracted from the solution using centrifugation or filtration techniques following ultrasonication production (Turcheniuk *et al* 2016). Furthermore, an efficient method for purifying nanodiamonds by salt assisted air oxidation methods has been reported for removing the surface adsorbed impurities. The impurities were etched by the etching atmosphere introduced by NaCl at high temperature. In addition, this purification method can be employed at commercial scale producing the high-quality nanodiamonds (Zhang *et al* 2021). To get rid of any leftover contaminants or unreacted precursors, the collected nanodiamonds could go through additional purification procedures.

Nanodiamond synthesis by ultrasound has a number of advantages over other diamond synthesis techniques, including simplicity, scalability, and the lack of high-temperature and high-pressure reaction conditions. Due to their distinctive qualities, including their high surface area, biocompatibility, and good optical properties, the resulting nanodiamonds find use in a variety of industries, such as biomedical imaging, drug delivery, and sensors. However, careful tuning of the reaction parameters, such as ultrasonic intensity, duration, and precursor material selection, is necessary to regulate the size and characteristics of the nanodiamonds.

8.2.6 Electrochemical synthesis

The controlled deposition of carbon atoms onto a conductive substrate under electrochemical conditions results in nanoscale diamond formations in the electrochemical synthesis of nanodiamond. The carbon deposition reactions are typically driven by an electric field and an electrolyte that contains carbon (Hongrutai *et al* 2023). As the source of carbon for the formation of nanodiamonds, the first step is to prepare a suitable electrolyte solution that contains carbon precursors. Precursors that include carbon, such as ethanol or methanol, are frequently used (Boruah and Saikia 2022). To regulate the morphology and size of the nanodiamonds, specialized additives may also be added to the electrolyte solution (Dimas Chandra Permana *et al* 2022).

The electrode on which nanodiamond will be placed is a conducting substrate (Puthongkham and Venton 2019). For this, a metal substrate, such platinum or gold, is frequently employed. For instance, Fei *et al* reported that electrochemical reduction of Pt (IV) complex ions on the surface of CNTs resulted in the electrochemical deposit of platinum (Pt) on the activated CNT/graphite electrode (Hwang *et al* 2020). To guarantee a clean and uniform surface for deposition, the substrate needs to be cleaned and prepared. The substrate should be cleansed in an electrochemical cell with a ready-made electrolyte solution. The cell is made to permit the movement of current between a counter electrode and the substrate, which serves as the working electrode. Graphite or another inert substance is typically used for the counter electrode (Wei *et al* 2011). Applying an electric voltage between the working electrode and the counter electrode is a step in the electrochemical deposition process. This potential generates an electric field close to the working electrode's surface. Redox processes are induced at the electrode–electrolyte interface when the electric field is applied.

Due to electrochemical processes, the carbon precursors in the electrolyte solution are broken down, resulting in the creation of carbon radicals or ions. Due to the applied electric field, these carbon species are then drawn to the working electrode's (substrate) surface. They may experience further reactions and reorganizations as they approach the surface, finally creating nanoscale diamond crystallites (Azevedo *et al* 2012). The size, shape, and quality of the nanodiamonds can be adjusted by altering several parameters throughout the electrochemical manufacturing process. The electrolyte temperature applied electric voltage, the concentration and composition of the carbon precursors, and the usage of additives are some of these parameters. The nanodiamonds may undergo further post-synthesis processes to further purify and functionalize them for particular purposes after electrochemical production. These processes could involve washing to get rid of any electrolyte residue, annealing to enhance crystallinity, or surface functionalization to make the material more compatible with other substances.

To improve the control and effectiveness of nanodiamond manufacturing, researchers are always looking at new variants and advances in the electrochemical synthesis of nanodiamonds. Nanodiamonds are a topic of considerable interest in nanotechnology research because they have applications in a variety of sectors, including electronics, medicine, and materials science.

8.3 Properties of nanodiamond

Due to their size and diamond-like structure, nanodiamonds have a variety of special qualities. These characteristics make nanodiamonds very appealing for a wide range of applications in numerous sectors. A full discussion of the properties of nanodiamonds follows.

8.3.1 Mechanical properties

Nanodiamonds, typically between a few and hundreds of nanometers in size, are a special type of carbon nanomaterial because of their nanoscale size and diamond-like structure, and they have remarkable mechanical properties. The crystal structure, size, and presence of flaws and impurities are examples of the variables that affect a material's mechanical properties.

The sp^3-bonded diamond-like structure of nanodiamonds contributes to their extraordinary toughness. Each carbon atom in the sp^3 hybridization is connected to four nearby carbon atoms in a tetrahedral configuration. Nanodiamonds are one of the hardest materials known due to their extraordinary hardness produced by their strong covalent bonding (Vejpravová 2021). In a recent study by Zhang *et al*, spark plasma sintering was used to create dense tungsten carbide (WC) bulks by using a tungsten and nanodiamond self-densification reaction at 1700°C. This novel mechanism of self-densification reaction sintering was demonstrated using nanodiamond, which enables dense WC bulks with a density of 98.1% and Vickers hardness of 21.8 GPa for a volume expansion associated with the change from diamond to graphite, which greatly reduced the porosity of as-prepared WC bulks. In addition, the reaction between tungsten and diamond produced less volume shrinkage (Zhang *et al* 2023). Nanodiamonds have high wear resistance because of their hardness and stiffness. Nanodiamonds can greatly increase the wear resistance of composites or coatings, making them suitable for high-performance uses in sectors such as aerospace and automotive. Yin *et al* demonstrated the friction-reducing and wear-resistant properties by allowing MXene/nanodiamond coating to slide against a PTFE ball. This resulted from the development of a tribolayer, namely a multilayer made up of PTFE, MXene, and nanodiamonds. Significantly less friction and wear were produced due to the shear slip caused by MXene and the rolling effect of the nanodiamond. Figure 8.4 shows that it is beneficial to form dense

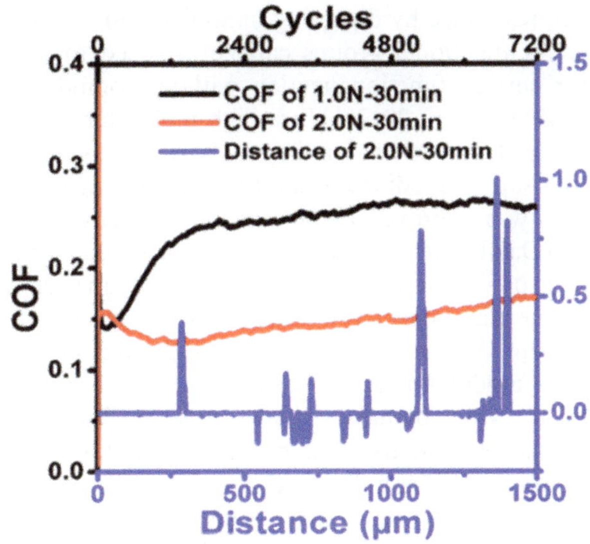

Figure 8.4. Friction curves and cross-sectional profile of the wear track demonstrated by the tribological properties for MXene/nanodiamond coating against a PTFE ball. Reprinted with permission from Yin *et al.* (2019). Copyright (2019) American Chemical Society.

tribofilms to reduce coefficient of friction. In addition, the PTFE's porous nature makes it simple for nanoparticles to penetrate the tribolayer and achieve effective lubrication (Yin *et al* 2019). A material's strength is its capacity to endure external forces without permanent deformation or failure. Because of their robust covalent connections and diamond lattice structure, nanodiamonds also possess exceptional strength (Nunn *et al* 2017).

The high elasticity property of nanodiamonds can deform when exposed to external forces and then recover their original shape after the forces have been eliminated (Ekimov *et al* 2021). For applications where materials are subjected to cyclic loads or impacts, this feature is crucial. When poly-vinyl alcohol reinforced with nanodiamond particles was synthesized with nanodiamond content up to 0.6 wt.%, it was reported that there was significant enhancement in the hardness and elastic modulus of PVA with small addition of nanodiamond in the polymer matrix (Maitra *et al* 2009). Nanodiamonds have an organized diamond lattice structure that contributes to their high thermal conductivity (Kidalov *et al* 2007). They are excellent for thermal management applications in electronics and other heat-sensitive equipment because of this characteristic. The composites prepared from the epoxy polymer binder and nanodiamond showed significant increase in scratch resistance and thermal conductivity, which is due to the direct contacts between the single diamond nanoparticles forming an interconnected network held together by a polymer binder. This property is beneficial in applications ranging from lubricants to electrochemical applications (Neitzel *et al* 2011). Although extremely hard, nanodiamonds can be fragile and their fracture toughness may not be as great as that of some other materials (Protopapa *et al* 2011). However, it is feasible to increase their fracture resistance by designing nanodiamond structures and adding specific flaws. Different functional groups can be attached to the nanodiamond surface, allowing for surface functionalization without compromising the nanodiamond properties (Zhang *et al* 2017). Ultra-high molecular weight polyethylene (UHMPE) was reinforced by various loadings of surface modified nanodiamond. The nanodiamond was surface modified by amine functionalization. It was observed that the fracture stress was improved by 30% at an optimum loading of 0.5 wt.% nanodiamond filler (Dalai and Sreekanth 2021). Nanodiamonds' mechanical characteristics can display size-dependent behavior (Zhang *et al* 2016). The correlation between the mechanical properties and the grain size of the nanodiamond films has been studied. The investigation showed grain size-dependent hardness and Young's modulus that can be associated with the grain boundary volume fraction and the grain boundary related defects (Wiora *et al* 2009). The mechanical performance of nanodiamond particles is increasingly influenced by surface effects and flaws as their size decreases.

High hardness, stiffness, wear resistance, strength, thermal conductivity, and compression strength are just a few of the excellent mechanical qualities that nanodiamonds possess. However, because of their nanoscale structure and potential size dependence, they might have certain limitations in terms of fracture toughness. Nanodiamonds are important for a variety of applications, including nanocomposites, coatings, biomedical devices, and high-performance engineering materials, because of the special combination of their mechanical qualities.

8.3.2 Optical properties

Nanoparticles called nanodiamonds are made of carbon atoms organized in a diamond lattice. Usually, they are between a few nanometers and a few hundred nanometers in size. The remarkable optical characteristics of these distinctive nanoparticles make them stand out from bulk diamonds and other carbon-based materials.

When exposed to specific stimuli, one of the most remarkable optical characteristics of nanodiamonds is their capacity to emit visible light (Hu *et al* 2009, Wu and Weil 2017). Nanodiamond with graphitic carbon nitride (g-C_3N_4) nanosheet heterostructure was synthesized via one step pyrolysis method. There exists a difference between the refractive index of diamond (2.5) and g-C_3N_4 (<1.5). Multiple interfacial reflection and an increase in optical paths were observed when incident light propagated between denser to thinner medium as a result of strong light scatting in the ND-gC_3N_4 interface (Su *et al* 2018). In addition, some carbon vacancies or impurity centers within the diamond lattice are stimulated to higher energy levels when nanodiamonds are exposed to ultraviolet (UV) light or high-energy photons (Kaur and Badea 2013). Vacancies are a result of radiation damage due to irradiation with high-energy electrons (2–10 MeV), protons (2–3 MeV) or low energy (40 keV) particles (Torelli *et al* 2019). They emit visible light in the green to near-infrared range after returning to their ground condition. This fluorescence feature is very beneficial for many applications, including biomedical imaging and sensing, and labeling in biological systems (Reineck and Gibson 2017). Because of their narrow emission spectra, diamonds only emit light at a limited range of wavelengths. Depending on the color of their centers, they emit light at a range of wavelengths, from visible to near-infrared, with either narrow or broad bandwidths (Alkahtani *et al* 2018). The ability to concurrently detect and distinguish between numerous nanodiamonds with various emission wavelengths makes this capability very useful for multiplexed imaging and sensing applications (Torelli *et al* 2019). Because they can keep their fluorescence characteristics even after extended exposure to strong light, nanodiamonds are very photostable (Alkahtani *et al* 2018). Due to the long-term dependability of the fluorescence of nanodiamonds, this trait is helpful in applications requiring long-term imaging or sensing.

For biological applications, nanodiamonds have undergone substantial research, and their optical characteristics are essential (Jung and Neuman 2021). They are excellent prospects for biomarker, drug delivery, and bioimaging applications due to their tiny size, biocompatibility, and fluorescence (Jung and Neuman 2021). Covalent attachment of (3-aminopropyl) trimethoxy silane ND-NH_2 is one method for drug delivery using nanodiamond. In the presence of borane, the carbonyl functionalities were converted to hydroxymethyl groups, and then (3-aminopropyl) trimethoxy silane was used. Once more, the amino group on the surface of nanodiamonds may bind DNA. After surface functionalization via noncovalent 800 Da polyethyleneimine (PEI 800) adsorption and covalent amino conjugation, increased cellular uptake of nanodiamond particles was seen (Chipaux *et al* 2018). Single-photon emission can be seen in some nanodiamond lattice flaws, such as the NV core. In quantum communications, quantum cryptography, and information processing, this characteristic is extremely valuable. Nanodiamonds can aid in the development of quantum technologies by spontaneously producing single photons (Aharonovich *et al* 2016). Nanodiamonds have distinctive Raman scattering characteristics (Osswald *et al* 2009). Due to a process known as surface-enhanced Raman scattering, they can dramatically boost the Raman signals of surrounding molecules or materials. Figure 8.5 displays the average crystallite sizes determined by XRD (green) and the particle size distribution of the examined materials

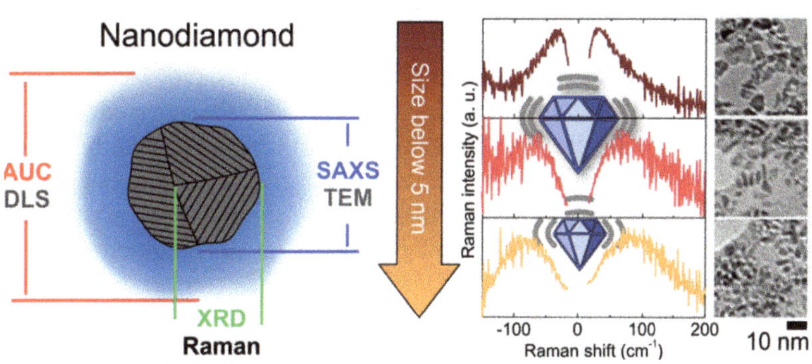

Figure 8.5. Volumetric size distribution of single digit detonation nanodiamonds and Raman spectra shifting. Reprinted with permission from Stehlik *et al.* (2021). Copyright (2021) American Chemical Society.

determined by DLS (black), AUC (red), and SAXS (blue, dashed blue). This figure presents a graphical representation of the particle size as determined by the various methods in use (Stehlik *et al* 2021). This capacity has implications for surface characterization, biosensing, and analytical chemistry (Zang *et al* 2012). Because they scatter light, nanodiamonds can be optically imprisoned using laser beams. Individual nanodiamonds may be precisely controlled and manipulated using optical trapping, which is useful for a variety of scientific and nanotechnology applications. In situ studies of distance and orientation-dependent phenomena, such as energy transfer and force dynamics between various types of single nanoparticle building blocks, will be made possible by contactless trapping and tweezing of individual nanometer-sized nanoparticles in combination with single particle spectroscopy. Using an intensely focused laser beam, optical trapping has been utilized to contain ultracold atoms and nanoparticles (Zhou *et al* 2020). Due to their semiconducting characteristics, nanodiamonds can also produce photoelectrons when exposed to light (Chang *et al* 2008). Devices that use photocatalysis and optoelectronics can take advantage of this property.

Nanodiamonds are flexible and important materials for a variety of applications, including biomedicine, quantum technology, photonics, and nanotechnology, because of their optical capabilities. To create cutting-edge technology and progress numerous scientific domains, researchers continue to investigate and take advantage of these distinctive qualities.

8.3.3 Electrical properties

Due to their distinct structure and size, nanodiamonds, which are carbon-based materials, have remarkable electrical properties. Numerous applications, ranging from electronics and sensors to energy storage and quantum technologies, are relevant to these features.

Bulk diamonds are renowned for their exceptional electrical insulating qualities due to their tightly packed carbon lattice structure and scarcity of free charge carriers (Nan *et al* 2020). However, based on their unique production processes and surface functionalization, nanodiamonds can display variable levels of electrical conductivity (Wang and Cui 2019). By using hot pressing and spark plasma sintering in conjunction with graphitization, it was possible to generate bulk porous electrically conductive materials using nanodiamond powders as a basic material. Nanodiamonds that were compact spark plasma sintered at 1200°C and hot-pressed at 1500°C displayed high electrical conductivities that were 40–50 times lower than that of bulk graphite but one order of magnitude greater than those of the powders made by nanodiamond graphitization. Based on these findings, it was concluded that hot pressing and spark plasma sintering of nanodiamonds might be employed to create porous electrically conductive materials with tailored surface areas (Ukhina *et al* 2015). Nanodiamonds can exhibit semiconducting properties, particularly those with certain flaws, such as the NV center (Schirhagl *et al* 2014). When a nitrogen atom replaces a carbon atom in the lattice, an atomic defect known as the NV center is created. Due to the introduction of new energy levels within the bandgap, this flaw can potentially cause electron transitions and semiconducting behavior. Applications for quantum information processing and sensing are of great interest to the NV center (Andrich *et al* 2017). Nanodiamonds also display intriguing dielectric qualities, which are crucial in capacitor and energy storage applications (Batsanov *et al* 2012). Nanodiamonds can be employed as fillers in polymer composites to improve their breakdown strength and dielectric constant because of their electrical insulating properties. To address the issue of poor thermal conductivity, a novel method of solid-state drawing of ultra-high molecular weight polyethylene (UHMWPE) films doped with a small quantity of nanodiamonds was presented. The composite film (2 wt. % nanodiamonds) exhibits very low dielectric loss at both low and high electric fields, as well as ultra-high thermal conductivity of 60 W m^{-1} K^{-1} in the drawing direction. The creation of innovative polymer-based dielectric and functional materials has a new path after it was shown that the interfaces created by the nanofillers do not always result in an increase in dielectric loss at high electric field and a decrease in thermal conductivity (Yu *et al* 2022). The development of high-performance capacitors and energy storage technologies benefits from this property. The special surface characteristics of nano-diamonds, especially when functionalized with particular groups, can result in quantum capacitance effects. The discrete energy levels of the electrons confined to the surface of the nanodiamond are related to quantum capacitance, which takes place at the nanoscale (Weingarth *et al* 2014). The performance of nanoscale electronic devices, such as nanodiamond-based transistors, may be improved as a result of this effect.

Nanodiamonds have been studied in field emission applications, where electrons are released from a pointed nanodiamond tip under the influence of a strong electric field (Terranova *et al* 2015). Effective electron emission is made possible by the diamond tip's electrical characteristics and nanoscale sharpness. Nanodiamond field emission has been considered for flat panel displays and vacuum electrical devices (Kang *et al* 2007). Nanodiamonds have exceptional electrochemical characteristics in addition to their electrical conductivity. Due to their enormous surface area, high chemical stability, and capacity for charge storage, nanodiamonds can improve the electrochemical performance of batteries and supercapacitors when utilized as an electrode material (Wang and Cui 2019). In addition, composite material based on DND and hydroxypropyl cellulose (HPC) was prepared as a separator and a solid electrolyte (Polino *et al* 2020). The production and surface functionalization of nanodiamonds affects their ability to conduct electric charge (Baranov *et al* 2018). Nanodiamonds are valuable in biological applications, including medication delivery and bioimaging, because their surface charge affects interactions with biomolecules.

The electrical characteristics of nanodiamonds can be controlled and altered using a variety of techniques, including doping, surface functionalization, and size control. As nanodiamond research advances, the electrical properties are being investigated and refined for a variety of applications, opening new potential in numerous scientific and technological domains.

8.3.4 Chemical properties

As carbon-based nanoparticles, nanodiamonds display a variety of fascinating chemical characteristics that make them adaptable for a variety of applications.

These characteristics result from the diamond lattice structure, surface functional groups, and a variety of imperfections that are present.

Nanodiamonds offer numerous potentials for functionalization because of their enormous surface area in relation to their size. Chemical changes to the surface of nanodiamonds can add various functional groups, including carboxyl (–COOH), amino (–NH_2), hydroxyl (–OH), and others (Fraczyk *et al* 2018). To reduce the membrane biofouling in the membrane bioreactor system and improve the surface hydrophobicity of the membrane, a cellulose acetate nanocomposite membrane was prepared by embedding amine functionalized nanodiamond and polyethylene glycol (PEG) grafted nanodiamond. At 0.5 wt.% loading of ND-PEG, the CA membrane's tensile strength and hydrophilicity parameters showed the greatest increase. Additionally, due to the modification of the surface properties, the CA/ND-PEG (0.5 wt.%) nanocomposite membrane was able to achieve increased critical flux and anti-fouling qualities (Etemadi *et al* 2017). These functional groups can be introduced through various chemical processes, permitting interactions with other molecules and materials for custom uses. Nanodiamonds can typically be used in biological systems without posing a serious threat because they are generally biocompatible. Their biocompatibility can be improved by surface functionalization, making them desirable candidates for drug administration, bioimaging, and other biological applications (Uthappa *et al* 2020). The surfaces of a nanodiamond can store an electric charge, which is influenced by the pH levels and functional groups in the material (Mumtaz *et al* 2022, Zhang *et al* 2018). The interactions of nanodiamonds with biomolecules and other substances are influenced by their charged surfaces. For them to perform at their best in particular applications, the surface charge must be controlled. Nanodiamonds are good adsorbents for a variety of compounds and ions due to their high surface area and functional groups (Wei *et al* 2009). The efficiencies of thermally oxidized nanodiamond (Ox-ND) and unmodified nanodiamond for the removal of various heavy metal ions from wastewater have been compared. The adsorption process was exothermic and spontaneous, and at pH 6.9 and 4 g l^{-1} dose, Cd (II) ion elimination was most effective. These results proved the versatility of thermally oxidized nanodiamond (Ox-ND) as an adsorbent for Cd (II) ion removal from wastewater (Ahmadijokani *et al* 2022). As a result of their capacity to adsorb gases, heavy metals, organic chemicals, and biomolecules, they are useful for applications in environmental cleanup, analytical chemistry, and water purification.

Some nanodiamonds, particularly those with surface functional groups or flaws, can show catalytic activity (Duan *et al* 2018). Because they have active sites on their surface, they can act as catalysts for a variety of chemical processes, including hydrogenation, oxidation, and other transformations. Nanodiamond-gold nanocomposites (NDAus) were prepared to examine their oxidative catalytic activity. The oxidation reaction rate was found to be proportional with the substrate concentration with an activation energy of 93.3 kJ mol^{-1}. The association of nanodiamonds with the gold nanoparticles makes them a good carbonic catalyst facilitating substrate adsorption due to the charge transfer at the metal–carbon interface (Kim *et al* 2016). Because of variations in surface charge and functional group protonation, nanodiamonds can be pH sensitive (Nie *et al* 2021). Applications for biosensing and pH-responsive drug delivery benefit from this characteristic. The solubility of nanodiamonds can be controlled by altering the surface chemistry of the crystals (Chauhan *et al* 2020). Hydrophobic functional groups enable nanodiamonds to disperse in non-polar solvents, whereas hydrophilic functional groups improve their dispersibility in water. Nanodiamonds can be added to improve the mechanical, thermal, and electrical properties of diverse composite materials, such polymers, ceramics, and metals (Khan *et al* 2021). The nanodiamond lattice's strong covalent connections increase the composite's mechanical strength, while its high thermal conductivity aids in heat dissipation. The NV center, one of the flaws in the nanodiamond lattice, exhibits special quantum characteristics. Nanodiamonds are desirable for quantum sensing, quantum information processing, and quantum cryptography because of the spin characteristics of the NV center (Andrich *et al* 2017).

The chemical characteristics of nanodiamonds allow for a wide range of applications in industries such as biomedicine, catalysis, environmental cleanup, and materials research. Researchers are still investigating and modifying these characteristics to make nanodiamonds more suitable for particular uses and to find new applications for these extraordinary nanoparticles.

8.4 Overview of nanodiamonds

The versatile class of nanomaterials known as nanodiamonds has special features that result from its nanoscale size and diamond-like structure. Nanodiamonds are diamond particles with a diameter typically between a few and a few hundred nanometers, are made of carbon atoms arranged in a crystal lattice similar to that of macroscopic diamonds, and they exhibit unique properties because of their size and surface features. Nanodiamonds can be created using a variety of techniques, such as detonation, hydrothermal, electrochemical, and laser ablation. The best technique depends on the required features and applications. Nanodiamonds are desirable for a variety of applications due to their assortment of distinctive qualities.

One of the hardest known materials, nanodiamonds inherit the extraordinary hardness of bulk diamonds. It is possible to tune interactions with other substances by functionalizing the surface of nanodiamonds with various chemical groups. Due to their biocompatibility, low toxicity, and simplicity of surface functionalization, nanodiamonds have demonstrated potential in biological applications. When stimulated by an external light source, some nanodiamonds exhibit photoluminescence, which can be beneficial for biological imaging and sensing.

A wide range of prospective applications in numerous sectors have been made possible by the special combination of features in nanodiamonds. Because of their biocompatibility and surface functionalization property, nanodiamonds have uses in drug administration, imaging (including fluorescent tagging of cells), and biosensing. Nanodiamonds with NV centers are employed in single-photon sources, quantum computing, and quantum sensing. Nanodiamonds are helpful as additions in lubricants and coatings because of their exceptional hardness, which increases wear resistance and decreases friction. Furthermore, the high surface area and electrical conductivity of nanodiamonds make them suitable for use in electrochemical sensors and electrodes.

Nanodiamonds face certain difficulties despite their many promising applications. It is still difficult to increase high-quality nanodiamond manufacturing at a reasonable price. In addition, to accomplish specific applications, precise control of the surface chemistry and functionalization of nanodiamonds is necessary. Even while nanodiamonds typically exhibit little toxicity, substantial research is still being conducted to completely comprehend their long-term impacts. In conclusion, nanodiamonds are an intriguing class of nanomaterials with special qualities and uses that have enormous potential in a wide range of industries. Future applications for nanodiamonds are likely to be much more creative as science and technology develop.

References

Aharonovich I, Englund D and Toth M 2016 Solid-state single-photon emitters *Nat. Photon.* **10** 631–41

Ahmadijokani F, Molavi H, Peyghambari A, Shojaei A, Rezakazemi M, Aminabhavi T M and Arjmand M 2022 Efficient removal of heavy metal ions from aqueous media by unmodified and modified nanodiamonds *J. Environ. Manage.* **316** 115214

Alkahtani M H, Alghannam F, Jiang L, Almethen A, Rampersaud A A, Brick R, Gomes C L, Scully M O and Hemmer P R 2018 Fluorescent nanodiamonds: past, present, and future *Nanophotonics* **7** 1423–53

Alzahrani Y A and Alkahtani M H 2023 Growth of sub-10 nm fluorescent nanodiamonds *Opt. Mater. Express* **13** 2192–202

Andrich P, de las Casas C F, Liu X, Bretscher H L, Berman J R, Heremans F J, Nealey P F and Awschalom D D 2017 Long-range spin wave mediated control of defect qubits in nanodiamonds *npj Quantum Inf.* **3** 28

Auciello O and Sumant A V 2010 Status review of the science and technology of ultrananocrystalline diamond (UNCD™) films and application to multifunctional devices *Diam. Relat. Mater.* **19** 699–718

Azevedo A F, Baldan M R and Ferreira N G 2012 Nanodiamond films for applications in electrochemical systems *Int. J. Electrochem.* **2012** 508453

Bacakova L, Pajorova J, Tomkova M, Matejka R, Broz A, Stepanovska J, Prazak S, Skogberg A, Siljander S and Kallio P 2020 Applications of nanocellulose/nanocarbon composites: focus on biotechnology and medicine *Nanomaterials* **10**

Baig N, Kammakakam I and Falath W 2021 Nanomaterials: a review of synthesis methods, properties, recent progress, and challenges *Mater. Adv.* **2** 1821–71

Baker S N and Baker G A 2010 Luminescent carbon nanodots: emergent nanolights *Angew. Chem. Int. Ed.* **49** 6726–44

Baranov O, Levchenko I, Bell J M, Lim J W M, Huang S, Xu L, Wang B, Aussems D U B, Xu S and Bazaka K 2018 From nanometre to millimetre: a range of capabilities for plasma-enabled surface functionalization and nanostructuring *Mater. Horizons* **5** 765–98

Basso L, Cazzanelli M, Orlandi M and Miotello A 2020 Nanodiamonds: synthesis and application in sensing, catalysis, and the possible connection with some processes occurring in space *Appl. Sci.* **10** 4094

Basso L, Gorrini F, Bazzanella N, Cazzanelli M, Dorigoni C, Bifone A and Miotello A 2017 The modeling and synthesis of nanodiamonds by laser ablation of graphite and diamond-like carbon in liquid-confined ambient *Appl. Phys.* A **124** 72

Batsanov S S 2018 Detonation transformation in materials ed S S Batsanov *Shock and Materials* (Singapore: Springer) pp 153–95

Batsanov S S, Gavrilkin S M, Batsanov A S, Poyarkov K B, Kulakova I I, Johnson D W and Mendis B G 2012 Giant dielectric permittivity of detonation-produced nanodiamond is caused by water *J. Mater. Chem.* **22** 11166–72

Bhagabati P *et al* (ed) 2019 *Carbon-Containing Polymer Composites* (Singapore: Springer) pp 1–64

Boruah A and Saikia B K 2022 Synthesis, characterization, properties, and novel applications of fluorescent nanodiamonds *J. Fluoresc.* **32** 863–85

Boudou J-P, Curmi P A, Jelezko F, Wrachtrup J, Aubert P, Sennour M, Balasubramanian G, Reuter R, Thorel A and Gaffet E 2009 High yield fabrication of fluorescent nanodiamonds *Nanotechnology* **20** 235602

Bundaleska N, Tsyganov D, Dias A, Felizardo E, Henriques J, Dias F M, Abrashev M, Kissovski J and Tatarova E 2018 Microwave plasma enabled synthesis of free standing carbon nanostructures at atmospheric pressure conditions *Phys. Chem. Chem. Phys.* **20** 13810–24

Cha C, Shin S R, Annabi N, Dokmeci M R and Khademhosseini A 2013 Carbon-based nanomaterials: multifunctional materials for biomedical engineering *ACS Nano* **7** 2891–7

Chandrakar K, Jassal M and Agrawal A K 2023 In-situ synthesis of carbon nanodiamonds via atmospheric pressure plasma for functionalization of poly (ethylene terephthalate) (PET) fabric surface Available at SSRN 4334703

Chang Y-R *et al* 2008 Mass production and dynamic imaging of fluorescent nanodiamonds *Nat. Nanotechnol.* **3** 284–8

Chauhan S, Jain N and Nagaich U 2020 Nanodiamonds with powerful ability for drug delivery and biomedical applications: recent updates on *in vivo* study and patents *J. Pharm. Anal.* **10** 1–12

Chipaux M, van der Laan K J, Hemelaar S R, Hasani M, Zheng T and Schirhagl R 2018 Nanodiamonds and their applications in cells *Small* **14** 1704263

Dalai N and Sreekanth P S R 2021 UHMWPE/nanodiamond nanocomposites for orthopaedic applications: a novel sandwich configuration based approach *J. Mech. Behav. Biomed. Mater.* **116** 104327

Danilenko V V 2005 Specific features of synthesis of detonation nanodiamonds *Combust., Explos. Shock Waves* **41** 577–88

Das D and Roy A 2022 Growth of nanostructured diamond films on glass substrates by low-temperature microwave plasma-enhanced chemical vapor deposition for applications in nanotribology *ACS Appl. Nano Mater.* **5** 3558–71

Dimas Chandra Permana A, Omar A, Guillermo Gonzalez-Martinez I, Oswald S, Giebeler L, Nielsch K and Mikhailova D 2022 MOF-derived onion-like carbon with superior surface area and porosity for high performance lithium-ion capacitors *Batter. Supercaps* **5** e202100353

Duan X, Ao Z, Zhang H, Saunders M, Sun H, Shao Z and Wang S 2018 Nanodiamonds in sp^2/sp^3 configuration for radical to nonradical oxidation: core–shell layer dependence *Appl. Catal.* B **222** 176–81

Ekimov E A, Lebed Y B and Kondrin M V 2021 Influence of surface reconstruction on elastic properties of nanosized diamond films and nanodiamonds *Carbon* **171** 634–8

Etemadi H, Yegani R and Seyfollahi M 2017 The effect of amino functionalized and polyethylene glycol grafted nanodiamond on anti-biofouling properties of cellulose acetate membrane in membrane bioreactor systems *Sep. Purif. Technol.* **177** 350–62

Fang L, Ohfuji H and Irifune T 2013 A novel technique for the synthesis of nanodiamond powder *J. Nanomater.* **2013** 201845

Fraczyk J, Rosowski A, Kolesinska B, Koperkiewcz A, Sobczyk-Guzenda A, Kaminski Z J and Dudek M 2018 Orthogonal functionalization of nanodiamond particles after laser modification and treatment with aromatic amine derivatives *Nanomaterials* **8** 908

Gorrini F, Cazzanelli M, Bazzanella N, Edla R, Gemmi M, Cappello V, David J, Dorigoni C, Bifone A and Miotello A 2016 On the thermodynamic path enabling a room-temperature, laser-assisted graphite to nanodiamond transformation *Sci. Rep.* **6** 35244

Haque A, Sachan R and Narayan J 2019 Synthesis of diamond nanostructures from carbon nanotube and formation of diamond-CNT hybrid structures *Carbon* **150** 388–95

Hongrutai N, Nganglumpoon R, Watmanee S, Pinthong P, Szilágyi P Á, Titirici M M and Panpranot J 2023 In-situ electrodeposited Ag/Cu for electrochemical reduction of acetic acid to nanodiamond under ambient conditions *Mater. Today Chem.* **30** 101509

Hu S, Tian F, Bai P, Cao S, Sun J and Yang J 2009 Synthesis and luminescence of nanodiamonds from carbon black *Mater. Sci. Eng.* B **157** 11–4

Hurtado C R *et al* 2020 A simple procedure to obtain nanodiamonds from leftover of HFCVD system for biological application *SN Appl. Sci.* **2** 352

Hwang H S, Jeong J W, Kim Y A and Chang M 2020 Carbon nanomaterials as versatile platforms for biosensing applications *Micromachines* **11** 814

Jain G P, Pathan S A, Jadhav S Y and Surana J S 2023 Nanodiamond: insight from introduction to application *Curr. Nanosci.* **19** 817–24

Jung H-S and Neuman K C 2021 Surface modification of fluorescent nanodiamonds for biological applications *Nanomaterials* **11** 153

Kang W P, Davidson J L, Subramanian K, Choi B K and Galloway K F 2007 Nanodiamond lateral VFEM technology for harsh environments *IEEE Trans. Nucl. Sci.* **54** 1061–5

Kaur R and Badea I 2013 Nanodiamonds as novel nanomaterials for biomedical applications: drug delivery and imaging systems *Int. J. Nanomed.* **8** 203–20

Khan M, Tiehu L, Zaidi S B A, Javed E, Hussain A, Hayat A, Zada A, Alei D and Ullah A 2021 Synergistic effect of nanodiamond and titanium oxide nanoparticles on the mechanical, thermal and electrical properties of pitch-derived carbon foam composites *Polym. Int.* **70** 1733–40

Kidalov S V, Shakhov F M and Vul A Y 2007 Thermal conductivity of nanocomposites based on diamonds and nanodiamonds *Diam. Relat. Mater.* **16** 2063–6

Kim M-C, Lee D, Jeong S H, Lee S-Y and Kang E 2016 Nanodiamond–gold nanocomposites with the peroxidase-like oxidative catalytic activity *ACS Appl. Mater. Interfaces* **8** 34317–26

Krueger A and Lang D 2012 Functionality is key: recent progress in the surface modification of nanodiamond *Adv. Funct. Mater.* **22** 890–906

Kumar M and Ando Y 2010 Chemical vapor deposition of carbon nanotubes: a review on growth mechanism and mass production *J. Nanosci. Nanotechnol.* **10** 3739–58

Kumar S, Nehra M, Kedia D, Dilbaghi N, Tankeshwar K and Kim K-H 2019 Nanodiamonds: emerging face of future nanotechnology *Carbon* **143** 678–99

Laura Soriano M, Carrillo-Carrion C, Ruiz-Palomero C and Valcárcel M 2018 Cyclodextrin-modified nanodiamond for the sensitive fluorometric determination of doxorubicin in urine based on its differential affinity towards β/γ-cyclodextrins *Microchim. Acta* **185** 115

Lei Y, Huang Q, Gan D, Huang H, Chen J, Deng F, Liu M, Li X, Zhang X and Wei Y 2020 A novel one-step method for preparation of sulfonate functionalized nanodiamonds and their utilization for ultrafast removal of organic dyes with high efficiency: kinetic and isotherm studies *J. Environ. Chem. Eng.* **8** 103780

Li M, Liu D, Wei D, Song X, Wei D and Wee A T S 2016 Controllable synthesis of graphene by plasma-enhanced chemical vapor deposition and its related applications *Adv. Sci.* **3** 1600003

Kleckley S, Wang H, Oladeji I, Chow L, Daly T K, Buseck P R, Solouki T and Marshall A Synthesis and Characterization of Advanced Materials 1997 51–60

Liang Y, Ozawa M and Krueger A 2009 A general procedure to functionalize agglomerating nanoparticles demonstrated on nanodiamond *ACS Nano* **3** 2288–96

Liu W and Speranza G 2019 Functionalization of carbon nanomaterials for biomedical applications *C* **5**

Liu Y, Ge Z, Li Z and Chen Y 2021 High-power instant-synthesis technology of carbon nanomaterials and nanocomposites *Nano Energy* **80** 105500

Liu Y, Liu X, Cui Y and Yuan W 2022 Ultrasound for microalgal cell disruption and product extraction: a review *Ultrason. Sonochem.* **87** 106054

Lu M, Zhang C, Liu D, Wang X and Sun F 2023 Theoretical and experimental research on growth and doping mechanisms of diamond films fabricated using liquid carbon source precursors *Appl. Surf. Sci.* **629** 157389

Maitra U, Prasad K E, Ramamurty U and Rao C N R 2009 Mechanical properties of nanodiamond-reinforced polymer-matrix composites *Solid State Commun.* **149** 1693–7

Manawi Y M, Ihsanullah , Samara A, Al-Ansari T and Atieh M A 2018 A review of carbon nanomaterials' synthesis via the chemical vapor deposition (CVD) method *Materials* **11** 822

Manzoor S, Dar A H, Dash K K, Pandey V K, Srivastava S, Bashir I and Khan S A 2023 Carbon dots applications for development of sustainable technologies for food safety: a comprehensive review *Appl. Food Res.* **3** 100263

Marabotti P, Peggiani S, Vidale A and Spartaco Casari C 2022 Pulsed laser ablation in liquid of sp-carbon chains: status and recent advances *Chin. Phys.* B **31** 125202

Mostofizadeh A, Li Y, Song B and Huang Y 2011 Synthesis, properties, and applications of low-dimensional carbon-related nanomaterials *J. Nanomater.* **2011** 685081

Mumtaz M, Hussain N, Salam S and Bilal M 2022 Multifunctional nanodiamonds as emerging platforms for cancer treatment, and targeted delivery of genetic factors and protein medications—a review *J. Mater. Sci.* **57** 8064–99

Nan B, Xiao L, Wu K, Xu C-a, Zhang E, Zheng H, Zhan Y, Zhang Q, Shi J and Lu M 2020 Covalently introducing amino-functionalized nanodiamond into waterborne polyurethane via *in situ* polymerization: enhanced thermal conductivity and excellent electrical insulation *Colloids Surf.,* A **596** 124752

Narayan J and Bhaumik A 2015 Novel phase of carbon, ferromagnetism, and conversion into diamond *J. Appl. Phys.* **118** 215303

Neitzel I, Mochalin V, Knoke I, Palmese G R and Gogotsi Y 2011 Mechanical properties of epoxy composites with high contents of nanodiamond *Compos. Sci. Technol.* **71** 710–6

Nie L, Zhang Y, Li L, van Rijn P and Schirhagl R 2021 pH sensitive dextran coated fluorescent nanodiamonds as a biomarker for HeLa cells endocytic pathway and increased cellular uptake *Nanomaterials* **11** 1837

Nunn N, Torelli M, McGuire G and Shenderova O 2017 Nanodiamond: a high impact nanomaterial *Curr. Opin. Solid State Mater. Sci.* **21** 1–9

Osswald S, Mochalin V N, Havel M, Yushin G and Gogotsi Y 2009 Phonon confinement effects in the Raman spectrum of nanodiamond *Phys. Rev.* B **80** 075419

Polino G, Scaramella A, Manca V, Palmieri E, Tamburri E, Orlanducci S and Brunetti F 2020 Nanodiamond-based separators for supercapacitors realized on paper substrates *Energy Technol.* **8** 1901233

Protopapa P, Kontonasaki E, Bikiaris D, Paraskevopoulos K M and Koidis P 2011 Reinforcement of a PMMA resin for fixed interim prostheses with nanodiamonds *Dental Mater. J.* **30** 222–31

Puthongkham P and Venton B J 2019 Nanodiamond coating improves the sensitivity and antifouling properties of carbon fiber microelectrodes *ACS Sens.* **4** 2403–11

Reineck P and Gibson B C 2017 Near-infrared fluorescent nanomaterials for bioimaging and sensing *Adv. Opt. Mater.* **5** 1600446

Schirhagl R, Chang K, Loretz M and Degen C L 2014 Nitrogen-vacancy centers in diamond: nanoscale sensors for physics and biology *Annu. Rev. Phys. Chem.* **65** 83–105

Shah K A and Tali B A 2016 Synthesis of carbon nanotubes by catalytic chemical vapour deposition: a review on carbon sources, catalysts and substrates *Mater. Sci. Semicond. Process.* **41** 67–82

Shames A I, Yu Osipov V, von Bardeleben H J and Ya Vul' A 2012 Spin $S = 1$ centers: a universal type of paramagnetic defects in nanodiamonds of dynamic synthesis *J. Phys. Condens. Matter* **24** 225302

Shen M-Y, Li K-Y, Kuan C-F, Kuan H-C, Chen C-H, Yip M-C, Chou H-W and Chiang C-L 2013 Preparation of expandable graphite via ozone-hydrothermal process and flame-retardant properties of high-density polyethylene composites *High Perform. Polym.* **26** 34–42

Shenderova O A and McGuire G 2006 Types of nanocrystalline diamond ed O A Shenderova and D M Gruen *Ultrananocrystalline Diamond* (Norwich, NY: William Andrew Publishing) ch 3 pp 79–114

Shenderova O and Nunn N 2017 Chapter 2 - Production and purification of nanodiamonds ed J-C Arnault *In Micro and Nano Technologies* (Nanodiamonds: Elsevier) pp 25–56

Shenderova O A, Zhirnov V V and Brenner D W 2002 Carbon nanostructures *Crit. Rev. Solid State Mater. Sci.* **27** 227–356

Stehlik S *et al* 2021 Size effects on surface chemistry and raman spectra of sub-5 nm oxidized high-pressure high-temperature and detonation nanodiamonds *J. Phys. Chem.* C **125** 5647–69

Su L-X *et al* 2018 Effective light scattering and charge separation in nanodiamond@g-C3N4 for enhanced visible-light hydrogen evolution *Carbon* **139** 164–71

Terranova M L, Orlanducci S, Rossi M and Tamburri E 2015 Nanodiamonds for field emission: state of the art *Nanoscale* **7** 5094–114

Tolochko B P and Zolotarev K V 2016 Research on fast solid-state chemical reactions at the Siberian Synchrotron and Terahertz Radiation Center *J. Struct. Chem.* **57** 1288–313

Torelli M D, Nunn N A and Shenderova O A 2019 A perspective on fluorescent nanodiamond bioimaging *Small* **15** 1902151

Turcheniuk K, Trecazzi C, Deeleepojananan C and Mochalin V N 2016 Salt-assisted ultrasonic deaggregation of nanodiamond *ACS Appl. Mater. Interfaces* **8** 25461–8

Ukhina A V, Dudina D V, Anisimov A G, Mali V I, Bulina N V, Bataev I A, Skovorodin I N and Bokhonov B B 2015 Porous electrically conductive materials produced by spark plasma sintering and hot pressing of nanodiamonds *Ceram. Int.* **41** 12459–63

Ullal N, Muthamma K and Sunil D 2022 Carbon dots from eco-friendly precursors for optical sensing application: an up-to-date review *Chem. Pap.* **76** 6097–127

Uthappa U T, Arvind O R, Sriram G, Losic D, Ho Young J, Kigga M and Kurkuri M D 2020 Nanodiamonds and their surface modification strategies for drug delivery applications *J. Drug Delivery Sci. Technol.* **60** 101993

Vejpravová J 2021 Mixed sp^2–sp^3 nanocarbon materials: a status quo review *Nanomaterials* **11** 2469

Wang C X, Yang Y H, Xu N S and Yang G W 2004 Thermodynamics of diamond nucleation on the nanoscale *J. Am. Chem. Soc.* **126** 11303–6

Wang H-D 2011 *Nanodiamond Particles for Biomacromolecule Immobilization and Dye Contaminant Adsorption* (Saskatoon: University of Saskatchewan)

Wang H and Cui Y 2019 Nanodiamonds for energy *Carbon Energy* **1** 13–8

Wei L-M, Shen Q, Lu H-J and Yang P-Y 2009 Pretreatment of low-abundance peptides on detonation nanodiamond for direct analysis by matrix-assisted laser desorption/ionization time-of-flight mass spectrometry *J. Chromatogr.* B **877** 3631–7

Wei Y-S, Jin Q-Q and Ren T-Z 2011 Expanded graphite/pencil-lead as counter electrode for dye-sensitized solar cells *Solid-State Electron.* **63** 76–82

Weingarth D, Zeiger M, Jäckel N, Aslan M, Feng G and Presser V 2014 Graphitization as a universal tool to tailor the potential-dependent capacitance of carbon supercapacitors *Adv. Energy Mater.* **4** 1400316

Whitlow J, Pacelli S and Paul A 2017 Multifunctional nanodiamonds in regenerative medicine: recent advances and future directions *J. Control. Release* **261** 62–86

Wiora M, Brühne K, Flöter A, Gluche P, Willey T M, Kucheyev S O, Van Buuren A W, Hamza A V, Biener J and Fecht H J 2009 Grain size dependent mechanical properties of nanocrystalline diamond films grown by hot-filament CVD *Diam. Relat. Mater.* **18** 927–30

Wu Y and Weil T 2017 Nanodiamonds for biological applications *Phys. Sci. Rev.* **2** 20160104

Xie S-Y and Liu F 2022 The structures, synthesis, properties, and applications of diamond *Functional Carbon Materials* (Bristol, UK: IOP Publishing) 7-1–7-27

Yan Q-L, Gozin M, Zhao F-Q, Cohen A and Pang S-P 2016 Highly energetic compositions based on functionalized carbon nanomaterials *Nanoscale* **8** 4799–851

Yang G W 2007 Laser ablation in liquids: applications in the synthesis of nanocrystals *Prog. Mater Sci.* **52** 648–98

Yin X, Jin J, Chen X, Rosenkranz A and Luo J 2019 Ultra-wear-resistant MXene-based composite coating via *in situ* formed nanostructured tribofilm *ACS Appl. Mater. Interfaces* **11** 32569–76

Yu X *et al* 2022 Dielectric polymer composites with ultra-high thermal conductivity and low dielectric loss *Compos. Sci. Technol.* **229** 109695

Yu X, Chen X, Ding X, Yu X, Zhao X and Chen X 2020 Facile fabrication of flower-like MoS_2/nanodiamond nanocomposite toward high-performance humidity detection *Sensors Actuators* B **317** 128168

Zang J, Wang Y, Bian L, Zhang J, Meng F, Zhao Y, Ren S and Qu X 2012 Surface modification and electrochemical behaviour of undoped nanodiamonds *Electrochim. Acta* **72** 68–73

Zhang D, Gökce B and Barcikowski S 2017 Laser synthesis and processing of colloids: fundamentals and applications *Chem. Rev.* **117** 3990–4103

Zhang P, Wang J, Wang L, Li J and Jiang W 2023 Dense tungsten carbide bulks by self-densification reaction sintering of nanodiamond with tungsten at 1700°C *Ceram. Int.* **49** 9287–97

Zhang S, Liu X, Jiang Y, Ren Y and Li S 2016 Mechanical properties of the interface structure of nanodiamond composite films: first-principles studies *Appl. Surf. Sci.* **363** 522–31

Zhang T, Ma L, Wang L, Xu F, Wei Q, Wang W, Lin Y and Chu Z 2021 Scalable fabrication of clean nanodiamonds via salt-assisted air oxidation: implications for sensing and imaging *ACS Appl. Nano Mater.* **4** 9223–30

Zhang Y, Rhee K Y, Hui D and Park S-J 2018 A critical review of nanodiamond based nanocomposites: synthesis, properties and applications *Compos. Part B: Eng.* **143** 19–27

Zheng W-W, Hsieh Y-H, Chiu Y-C, Cai S-J, Cheng C-L and Chen C 2009 Organic functionalization of ultradispersed nanodiamond: synthesis and applications *J. Mater. Chem.* **19** 8432–41

Zhou J, Chizhik A I, Chu S and Jin D 2020 Single-particle spectroscopy for functional nanomaterials *Nature* **579** 41–50

Chapter 9

DFT studies of diamane and 2D diamonds: properties and applications

Bibhuti B Sahu, Tarun Yadav, Srikanta Moharana and Sunday J Olusegun

The intriguing new area of density functional theory (DFT) studies of diamane and two-dimensional (2D) diamonds is explored in depth in this chapter. Owing to their fascinating structural and electrical characteristics, these hypothetical carbon-based materials have attracted considerable interest, opening up exciting potential for a wide range of materials science and nanotechnology applications. The atomic arrangements and bonding interactions of diamane and 2D diamond are introduced, together with their crystal structures and unit cells. Subsequently, we will explore DFT's foundational ideas, breaking down how this potent computational tool may reveal previously hidden details about a material's characteristics. This chapter demonstrates how DFT enables the evaluation of mechanical stability and structural integrity through the discussion of the prediction of mechanical parameters such as elastic constants and mechanical strength. The optical characteristics of these materials are also investigated in depth; this includes their optical absorption spectra and excitonic effects, which show their promise for use in optoelectronics and light–matter interactions. Diamane and 2D diamond are discussed for their potential as cutting-edge nanoelectronics due to their adaptability in dielectric applications. Even for theoretical materials, practical validation is crucial. This chapter stresses the value of using DFT predictions to direct experimental efforts. Future research goals and problems in DFT are also discussed to help map out where the science is headed next.

9.1 Introduction

The field of materials science has benefited greatly from the use of DFT, which has shed light on the nature and potential uses of a wide range of carbon-based materials [1, 2]. Because of their unique properties and potential for ground-breaking applications, diamane and 2D diamonds have been the subject of much DFT

doi:10.1088/978-0-7503-5939-9ch9

research. Two-dimensional diamonds, or 2D diamond, depict atomically thin diamond-like formations with unique electrical characteristics, whereas three-dimensional (3D) diamond, or diamane, is predicted to have amazing mechanical properties [3–5].

DFT is a powerful computational technique that is widely employed in materials science to investigate the electronic structure and properties of various materials [6, 7]. DFT's foundation lies in quantum physics. It is founded on the idea that a system's electronic density is the only factor that can account for its total energy. DFT greatly decreases the computing complexity of solving the many-body Schrödinger equation by instead characterising the behaviour of the system in terms of the electron density distribution [8]. DFT has many useful and varied material applications. Researchers may use it to learn more about the stability, reactivity, and electrical characteristics of a broad variety of materials, from individual atoms and molecules to complex solids and surfaces [9, 10]. Predicting structural parameters such as lattice constants, bond lengths, and angles is essential for understanding the stability and behaviour of a material, and DFT is a powerful tool for doing so. The study of electronic features, such as band structures and electronic density of states, is essential to comprehending a material's electrical conductance and optical characteristics, and DFT plays a significant part in this [11]. Useful information for experimental validation may be gleaned from the method's predictions and interpretations of a wide range of spectroscopic features, including as infra-red and Raman spectra, x-ray absorption, and photoemission. DFT's ability to forecast whether or not new materials may be created with desired properties is extremely useful in the field of materials design [12]. Researchers can optimise materials for many uses, such as catalysis, energy storage, and semiconductors, by experimenting with different chemical compositions, dopants, and flaws [13, 14]. For a deeper comprehension of events happening in planetary interiors or industrial processes, DFT is also widely employed to explore the behaviour of materials under severe circumstances, such as high pressure or temperature. DFT has established itself as a crucial component of the field of materials science because it provides a robust theoretical framework within which to probe the electronic structure and characteristics of a wide range of materials [15, 16]. Its precision and adaptability make it a vital resource for scientists developing new materials and improving existing materials.

Diamane and 2D diamonds are extremely important materials, and DFT investigations are crucial to understanding their characteristics and discovering new uses. Ultrahigh hardness and elastic stiffness are only two of the projected extraordinary mechanical qualities of diamane, which is a hypothetical 3D carbon structure made of diamond-like tetrahedral units [17]. Diamane's stability and possible real-world uses in cutting tools, abrasives, and protective coatings may be better understood with the help of DFT simulations, which reveal the material's electronic band structure and phonon characteristics. Diamond formations that are just two dimensions thick are known as 2D diamond or graphene-derived diamond [18]. The electrical and mechanical characteristics of these exotic materials are essential to understand, and DFT investigations play a crucial role in doing so. This knowledge might lead to breakthroughs in nanoelectronics and quantum devices. To forecast their performance and direct experimental efforts to synthesise and use them in electronic applications, DFT models of their electronic bandgap, carrier mobility, and responsiveness to external stimuli are essential [19]. DFT research help clarify basic characteristics and investigate the potential of diamane and 2D diamonds in cutting-edge contexts. Because the sp^3 hybridisation of carbon atoms in these structures affords favourable adsorption sites, they may be studied with DFT simulations to see if they can be used as highly efficient hydrogen storage materials. Research into their possible uses in sensors, catalysis, and energy storage devices can be aided by DFT simulations' analyses of their interactions with other molecules and materials [20]. Understanding the electrical, mechanical, and chemical characteristics of diamane and 2D diamonds through DFT investigations is crucial because it provides a more complete picture of their behaviour and their uses. DFT simulations open the door to using these materials' distinctive features and designing them for cutting-edge technological developments in a wide variety of industries by giving crucial insights at the atomic and electronic levels [21].

In this chapter we will examine the exciting field of DFT investigations to learn about the basic characteristics of diamane and 2D diamonds, and their potential uses in nanoelectronics, quantum devices, and other fields.

9.2 Theoretical framework: DFT

9.2.1 Fundamentals of DFT and its key principles

Based on quantum mechanical concepts, DFT is a potent computational tool that can be used to investigate the atomic, molecular, and material electronic structure and behaviour [22]. While many-body wave functions of electrons in a system are complicated, DFT simplifies things by focusing on the electron density, (r), which is the chance of finding an electron at a given place in space. The Hohenberg–Kohn theorem [23] is the cornerstone of DFT, stating that the electronic density in the ground state is the only determinant of the system's ground-state energy. This allows scientists to seek the most stable electronic configuration by minimising the total energy relative to the electron density in their studies of electronic structure. The Kohn–Sham method [24], which was introduced by Walter Kohn and Lu Jeu Sham, is the second essential idea for putting DFT into practise. It recasts the electron–electron interactions via the exchange–correlation functional, transforming the many-electron system into an analogous system of non-interacting electrons travelling in an effective potential that is known as the Kohn–Sham potential [25]. It is difficult to directly quantify the effects of electron exchange and correlation that are captured by the exchange–correlation functional, which is the sole unknown component in DFT. In addition, DFT abides with conservation rules, keeping the overall electron density unchanged throughout the computation process. Different approximations have been established to strike a compromise between computing efficiency and accuracy in DFT findings, which are sensitive to the selection of the exchange–correlation functional. It is common practise to approximate a functional's value using either the generalised gradient approximation (GGA) or the local density approximation (LDA) [26].

DFT is based on the notion of electronic density and its relation to the ground-state energy, which permits the effective and precise investigation of electronic characteristics across a wide range of systems. DFT has become an essential tool in materials science, quantum chemistry, and condensed matter physics by providing a framework for investigating the behaviour of electrons in materials [27]. This has led to the discovery of many useful materials and a deeper understanding of their properties.

9.2.2 Mathematical formulation of DFT

The mathematics behind DFT [28] is made up of a few key parts that, when put together, make it possible to calculate the electronic structure and qualities of materials quickly and accurately. Here are the most important mathematics steps and ideas:

- Electronic density ((*r*)): The electronic density, (*r*), is the most important number in DFT. It shows how likely it is that an electron will be at a certain point in space, *r*. It is the sum of the squares of the wave functions of all the electrons in the system. It depends on where the electrons are in 3D space.
- Hohenberg–Kohn theorems: The Hohenberg–Kohn theories [29] are the basic ideas that DFT is built on. The following is what they say: the ground-state electronic density, (*r*), of a system can only be determined by the external potential, *V*_ext(*r*), that minimises the total energy, *E*[], of that system. The electronic density, (*r*), is a unique function of the ground-state energy, *E*[].
- Kohn–Sham method: The Kohn–Sham method is used by DFT to make the many-body electron–electron interactions easier to understand [30]. It changes the problem by thinking about an extra system of electrons that do not interact with each other and move in an effective potential called the Kohn–Sham potential, *V*_KS(*r*). This system's electronic density, (*r*), is the same as the true electronic density, (*r*), of the system that is engaging with it.
- Kohn–Sham equations: The Kohn–Sham equations [31, 32] are a set of self-consistent equations that figure out the Kohn–Sham orbitals (wave functions) and the Kohn–Sham potential, *V*_KS(*r*). The Kohn–Sham equations are [(−2/2*m*)2 + *V*_eff(*r*)]. *φ*_*i*(*r*) = *ε*_*iφ*_*i*(*r*)
- Here, _*i*(*r*) are the Kohn–Sham wave functions (orbitals), _*i* are the orbital energies, and *V*_eff(*r*) is the effective potential felt by the Kohn–Sham electrons, which includes both the external potential (*V*_ext(*r*)) and the exchange–correlation potential (*V*xc[*r*]).
- Exchange–correlation functional: The exchange–correlation potential, *V*xc[*r*], is the only unknown in the Kohn–Sham equations [33]. It shows how electrons move and interact with each other. Most of the time, the exact shape of *V*xc[*r*] is unknown and has to be guessed. Approximations such as LDA and the GGA are often used to get close to *V*xc[*r*] and make DFT calculations easier.
- Self-consistency loop: A self-consistency loop is used to solve the Kohn–Sham equations [34]. The original density of electrons is guessed, and then the Kohn–Sham equations are solved to get a new density of electrons. This

process is repeated until the electron density settles on an answer that makes sense on its own.

In brief, the Kohn–Sham method is used to describe the ground-state energy of a system in terms of its electronic density (r). This is part of the mathematical formulation of DFT. The Kohn–Sham equations are then solved in a way that makes sense by themselves to get the wave functions and concentrations of the electronic particles [35]. How accurate and fast DFT calculations are depends on which exchange–correlation functional is used and how quickly the self-consistency cycle converges.

9.2.3 Computational methods for DFT calculations

The electrical structure problem for complicated materials requires the use of computational tools for DFT calculations. When performing DFT calculations, the Kohn–Sham equations must be solved consistently to get the electronic wave functions and density, which provide information about the material's energy, structure, electronic band structure, and magnetic behaviour [36]. The main computational techniques for DFT computations are as follows:

(i) Basis sets: The Kohn–Sham orbitals (wave functions) of electrons are represented by basis sets, which are collections of mathematical functions. It is possible to employ a variety of basis sets, including localised atomic orbitals, plane waves, and Gaussian basis functions. Based on the system's periodicity, size, and necessary level of precision, a basis set is selected.

(ii) Pseudopotentials: To considerably lower the computational cost of DFT computations, pseudopotentials are approximations used to replace the electron-core interactions [37]. By more effectively describing the behaviour of the core electrons, pseudopotentials enable the study of the valence electrons, which are vital for understanding the characteristics of materials.

(iii) Fast Fourier transform (FFT): FFT is used with plane wave basis sets and periodic boundary conditions. To calculate the eigenvalues and eigenfunctions, it effectively converts the Kohn–Sham equations from real space to reciprocal (k) space. Calculations involving periodic systems, such crystals and surfaces, are significantly sped up by FFT.

(iv) Exchange–correlation functionals: The exchange–correlation functional, which takes into account both electron exchange and correlation interactions, is a crucial part of DFT computations. The accuracy of the findings is impacted by the functional choice for exchange–correlation. LDA, GGA, hybrid functionals, and meta-GGA functionals are examples of popular functionals.

(v) Iteratively calculating the Kohn–Sham equations is a step in the self-consistency method, which is used in DFT calculations. The Kohn–Sham equations are first solved, and the electron density is first estimated. The process continues until the electron density converges to a self-consistent solution by using the estimated electron density in the subsequent iteration.

(vi) Parallelisation: DFT computations can require a lot of processing power, especially for complex systems. By distributing the calculations across several processors or nodes, parallelisation techniques dramatically reduce the overall computation time [38].

(vii) Methods for linear scaling: Linear scaling by reducing the computational scaling with system size, DFT approaches are intended to make huge systems computationally practical. Localised orbitals are used in techniques like Divide-and-Conquer DFT and density matrix minimisation, which break the system up into smaller pieces that can be solved independently and combined effectively [39].

(viii) Hybrid functionals: Hybrid functionals strike a balance between accuracy and processing efficiency by combining the exchange–correlation functionals from DFT with a portion of Hartree–Fock exchange [40]. Hybrid functionals are especially helpful for precisely anticipating a material's electrical and structural properties.

9.3 Properties of diamane and 2D diamonds

9.3.1 Structural properties

- Crystal structures and unit cells of diamane and 2D diamonds

The 'two-dimensional diamond' or 'ortho-graphene' name is given to the hypothetical 3D carbon allotrope known as diamane [41]. Its crystal structure is similar to that of graphene in that it consists of a hexagonal lattice of carbon atoms arranged in two dimensions. Graphene is a 2D material. Diamane comprises several layers that are layered one on top of the other, in contrast to the structure of graphene, which only consists of a single sheet.

The hexagonal unit cell that makes up diamane contains two atoms of carbon. The structure of the crystal is built on what are known as diatomic units, which are repeated pairs of these two atoms. The formation of sp^2 hybridisation is due to the fact that each carbon atom is covalently bonded to the three carbon atoms that are immediately adjacent to it. The exceptional mechanical capabilities of diamane are a direct result of the in-plane carbon–carbon bonds that characterise the material, which are comparable to those of graphene. Meanwhile, van der Waals forces acting in an out-of-plane orientation are responsible for the majority of interlayer interactions that occur between carbon layers in diamane [41]. When numerous layers are piled one on top of the other, covalent bonds occur within each layer; however, interactions between the layers become weaker, resulting in the formation of a structure that is 3D.

It is hypothesised that the intriguing mechanical properties of diamane will result from the presence of diamond-like sp^3 bonding in the in-plane direction of the material. Although the experimental synthesis of diamane has not yet been completed, the features of diamane continue to be of theoretical interest and enquiry in the field of materials science. The carbon-based polymer known as 2D diamond, often known as 'diamond film,' is a one-of-a-kind substance with only two dimensions [42]. It was proposed in 2014 and is thought to be the variant of diamond that is the thinnest theoretically. In its crystal structure, 2D diamond, like graphene and diamane, has carbon atoms arranged in a hexagonal pattern, just like diamane and graphene.

In contrast to graphene's sp^2 hybridisation, the carbon atoms in 2D diamond have a sp^3 hybridisation, which causes them to generate tetrahedral connections in a 2D plane that resemble diamonds. Each carbon atom establishes a powerful covalent relationship with its three nearest neighbours, similar to the way that sp3 bonds are formed in bulk diamond. In a 2D diamond atom, the unit cell is shaped like a rhombus, which groups together four carbon atoms. This particular unit cell is iterated across the crystal lattice, resulting in a continuous, 2D diamond structure. Researchers have looked at 2D diamond because of the unique electrical and mechanical properties it possesses, and it is believed to be quite stable. Because of the material's anticipated high bandgap, it is a prospective candidate for usage in the electrical and optoelectronic industries [43]. Even though 2D diamond, like diamane, has not yet been synthesised in a laboratory setting, scientists are continuing to investigate the substance's potential use and work on creating methods for synthesising it. Moreover, the crystal structures of both diamane and 2D diamonds (2D diamond) are fascinating, but in different ways. In contrast to 2D diamond, which is composed of a single layer of carbon atoms arranged in a 2D hexagonal lattice with sp^3 bonding, diamante is made up of multiple stacked layers of a hexagonal lattice with sp^2 bonding in the in-plane direction. Meanwhile, 2D diamond is composed of a single layer of carbon atoms arranged in a 2D hexagonal lattice with sp^3 bonding. Both materials are still in the theoretical stage, but they show enormous potential for the development of nanotechnology and applications of nanomaterials in the future.

- DFT predictions of lattice parameters and bond lengths

DFT has been shown to be an effective technique for estimating the lattice characteristics and bond lengths of diamane and 2D diamonds (2D diamond), none of which have been synthesised experimentally. None of these compounds has been called 2D diamond. Quantum mechanical density functional theory, also known as DFT, is a useful tool for gaining knowledge on the atomic structures and electrical characteristics of various materials. In this section, we will describe the DFT predictions for the lattice parameters and bond lengths of diamane and 2D diamond [44].

- Diamane

Each carbon atom in the crystal structure of diamane creates three powerful covalent connections in the in-plane direction, which results in a 2D hexagonal lattice. Lattice parameters, in particular the lattice constant '*a*' and the interlayer gap '*c*,' have a significant role in determining the unit cell and the stacking pattern of these layers. Simulations based on DFT have allowed researchers to generate predictions regarding the lattice characteristics of diamane [17]. Due to the fact that both materials have a hexagonal lattice structure, the predicted lattice constant '*a*' is comparable to that of graphene. In contrast to graphite, which is a stacked arrangement of graphene layers, the interlayer spacing 'c' in diamane may be different from that in graphite as a result of the many bonding interactions that take place between carbon layers. With the assistance of DFT, the bond lengths and lattice characteristics of the diamane unit cell may be comprehended on a deeper level. Because of the presence of sp^2 hybridisation in the in-plane direction in both graphene and fullerene, the bond lengths that are measured between carbon atoms should be uniform and equivalent to those seen in graphene. Each carbon atom in 2D diamonds makes four strong covalent connections in a tetrahedral arrangement, precisely like in bulk diamond. This arrangement is similar to the way diamond is structured [45, 46]. Lattice characteristics, such as the lattice constant '*a*' and bond lengths within the unit cell, play an important role in defining the structure of 2D diamond.

Calculations based on DFT have been utilised to make predictions regarding the lattice properties and bond lengths of 2D diamond. It is anticipated that 2D diamond, with its sp^3 hybridisation and tetrahedral bonding structure, will have a different estimated lattice constant '*a*' than graphene and diamane. Because diamond and 2D diamond share the property of sp^3 hybridisation, which results in strong covalent bonding between carbon atoms, it is reasonable to anticipate that the bond lengths within the 2D diamond unit cell will be homogeneous and equivalent to the bond lengths in bulk diamond. This happens because sp^3 hybridization leads to strong covalent bonding between carbon atoms in both materials. Keep in mind that the accuracy of DFT predictions for lattice parameters and bond lengths is dependent on the exchange–correlation functionals and theory level that was used in the computations [47]. This is something that you should keep in mind. When creating predictions using functionals that produce varying outputs, it is possible to improve the precision of those predictions by utilising cutting-edge computational approaches such as hybrid functionals or dispersion corrections because these methods consider the fact that different functionals produce varied outcomes.

- Stacking configurations and interlayer interactions

It is only possible to discuss the stacking arrangements of diamane and 2D diamonds based on theoretical expectations and computer simulations because diamane and 2D diamond are both theoretical materials that have not been produced in the laboratory. The manner in which these components are arranged determines a significant portion of how they seem in three dimensions, as well as the degree to which they are stable as a whole. Let us next have a look at the several ways in which diamane and 2D diamonds can be stacked, as well as the different ways in which their layers can interact with one another.

The structure of diamane consists of numerous layers of a hexagonal mesh that are piled one on top of the other in a 2D plane. This gives the impression that the structure is 'diamond-like' in appearance. However, because diamane is only a theoretical substance, there are an infinite number of ways to stack it, and each of these methods results in a unique reaction between the layers. The following is a list of some of the various stacking configurations for diamane:

(a) AB stacking: The layers of diamane are stacked in this arrangement in an ABAB order, which is analogous to the well-known AB stacking in graphene [48]. This configuration is called the AB Stacking. Each layer sits directly over the one that came before it, and the carbon atoms that are in layers that are adjacent to one another sit directly atop one another. The van der Waals forces are responsible for the majority of the interactions that occur between layers when AB stacking is applied.

(b) AA stacking: In AA stacking, the layers of diamane are exactly aligned, with carbon atoms in one layer just above carbon atoms in the layer below. AA stacking is the most common type of diamane stacking [49]. This pattern is similar to the way the layers of graphene are layered. The van der Waals forces that hold the layers together are also significantly weakened when AA stacking is applied.

(c) Other stacking designs: Depending on how the layers interact with one another, further stacking designs may also be conceivable. Some examples of these designs include those that have a relative shift between levels, as well as those that have more complex stacking patterns.

The van der Waals forces, which are far weaker than the powerful covalent bonds that hold the layers of diamane together, are responsible for the majority of the interactions that occur between the layers of diamane. The manner in which a material is layered can have an impact on its stability as well as its electrical properties.

The 2D material known as 2D diamonds has been hypothesised to have a fundamentally distinct stacking pattern from that of diamane. Meanwhile, 2D diamond is made up of a single layer, as opposed to the numerous stacked layers that make up bulk diamond, which is structured like a 3D lattice [50]. Due to the fact that 2D diamond only consists of a single layer, the concept of interactions between layers is not applicable. Instead, the 2D diamond's stability and properties are determined by the strength of the covalent bonds that are contained inside the single layer of 2D diamond. It should be underlined that both diamane and 2D diamond are theoretical materials; the actual stacking patterns of these two substances, as well as the interlayer interactions between them, will not be understood with any degree of accuracy until they are successfully synthesised [51]. Chernozatonskii *et al* have studied diamane, a diamond-Like C_2H nanolayer. They have taken into account compounds of carbon atoms placed opposite each other in neighbouring layers, as well as a new C_2H nanostructure based on bilayer graphene (BLG) changed under the covalent bond of hydrogen atoms adsorbing on its exterior surface. As a whole, they make up a diamane 'film' of $\langle 111 \rangle$ diamond thinner than 1 nm. The authors used DFT, and compared the electron spectra and energy levels of diamond, graphene, and diamane. Figure 9.1 shows that the basic unit of graphane (figure 9.1(a)) and diamane (figures 9.1(b) and (c)) with the chair-like C6 rings

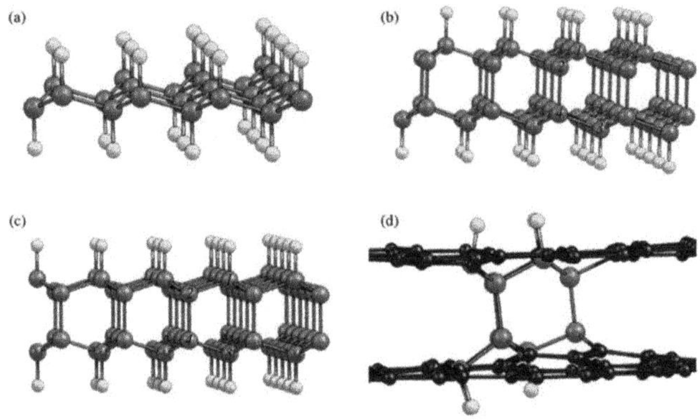

Figure 9.1. Atomic structures of (a) graphane, (b) AB stacking diamane, (c) AA stacking diamane, and (d) the method for forming interlayer C–C bonds on double-layer graphene: hydrogen atoms settle on both sides of the graphene, causing carbon atoms (dark colour) between adjacent layers to form bonds. Reprinted from [52], Copyright (2009), with permission from Springer Nature.

creates little puckered carbon sheets as a result of the pyramidalization of hydrogenated (fluorinated) carbon atoms, in contrast to the flat honeycomb pattern in graphene [52].

9.3.2 Mechanical and thermal properties

- DFT-based predictions of mechanical properties (elastic constants, strength, etc)

Through the use of theoretical computations, DFT-based predictions of their mechanical properties, such as elastic constants and strength, have been made [53]. These predictions should be taken as theoretical estimates rather than as validated experimental data because it is crucial to highlight that there are a number of uncertainties and assumptions that could affect their accuracy.

- Diamane:

The extraordinary crystal structure of diamane, which consists of stacked layers of a 2D hexagonal lattice, is what gives rise to the material's fascinating mechanical capabilities. Calculations based on the DFT have been utilised to make predictions regarding its elastic constants and mechanical strength.

Elastic constants: A material's ability to resist deformation when subjected to mechanical stress can be inferred by examining the material's elastic constants. Calculations using DFT are able to make predictions about the elastic constants for diamane in both the in-plane (C11 and C12) and out-of-plane (C33) directions. These elastic constants are what define the stiffness of the material, as well as how it reacts to outside forces [54].

Mechanical strength: Calculations that are based on DFT can also be used to determine the mechanical strength of diamane. This includes the material's tensile strength and shear strength. Shear strength is a measure of a material's resistance to shear deformation, while tensile strength is the highest amount of stress it can endure before cracking under tension.

- 2D diamonds:

DFT computations can be used to investigate how the crystal structure of a 2D diamond, which is a single layer of diamond, affects its mechanical characteristics. A 2D diamond's crystal structure is a single layer of diamond. Calculations using DFT can make predictions about the elastic constants of 2D diamonds, which can provide information on the material's stiffness and mechanical response [55]. In addition to the in-plane constants C11 and C12, these elastic constants also include the out-of-plane constant C33. The in-plane and out-of-plane elastic constants of materials can be predicted by DFT calculations. The elasticity of a substance is described by a set of parameters called elastic constants. They tell us a lot about the material's stiffness, deformation resistance, and general mechanical stability, among other mechanical qualities.

The elastic constants are determined within the context of DFT calculations by deriving the total energy of the material as a function of strain. Strain is the stress-induced lattice distortion of a crystal. The elastic constants can be calculated using DFT by applying strains to the crystal lattice in different directions and then seeing how the energy of the material changes. Diamane and 2D diamonds are anisotropic materials because their elastic constants vary along distinct crystallographic directions [56]. Two elastic constants are of primary interest in the in-plane direction:C11 represents the stiffness of a material along one in-plane direction (such as the x-direction) under tension while leaving the other in-plane direction (such as the y-direction) unrestrained during the stretch. When the material is stretched in one direction (say, the x-direction) while permitting some deformation in the perpendicular direction (say, the y-direction), C12 represents the coupling between the two in-plane directions. One elastic constant is typically found in the direction perpendicular to the plane of motion. C33 represents the stiffness of the material along an out-of-plane direction (such as the z-direction) under a stretch where the in-plane directions are unrestricted. The mechanical properties of materials, such as Young's modulus, shear modulus, bulk modulus, and Poisson's ratio, can be gleaned from the elastic constants found in DFT calculations. These characteristics are essential for figuring out how a material reacts to external forces and how it acts in various mechanical settings [57]. But keep in mind that DFT predictions are theoretical, and can be affected by things like exchange–correlation functional choice, theory level, and calculational assumptions. The anticipated elastic constants must be experimentally validated to ensure their accuracy and to achieve a complete understanding of the mechanical behaviour of diamane and 2D diamonds. A 2D diamond's mechanical strength, which includes both its tensile and shear strengths, may be approximated using DFT-based computations [58]. This is the case for both of these strengths. However, the prospective uses of 2D diamond in nanomechanical devices cannot be properly evaluated without first gaining an understanding of these features. It is vital to keep in mind that the predictions provided by DFT are very susceptible to a variety of factors, including the level of theory, the choice of exchange–correlation functionals, and the assumptions used while performing the computations. In addition, theoretical predictions need to be validated through experimental synthesis and characterisation to achieve a full understanding of the mechanical properties of diamane and 2D diamond. Wu and his coworkers have reported diamante's mechanical behaviour in tension and bending with a single layer. Using atomistic simulations, they looked into the tensile and bending mechanical properties of a single-layer diamond-diamane. We find that the mechanical properties of diamane are unaffected by the order in which its layers are stacked. The Young's modulus is the same in both the zigzag and armchair directions, but the fracture strain/strength is significantly higher in the former. Fracture of diamane is dominated by crack propagation in zigzag directions, which is not related to the tensile directions, as shown by the atomic configurations. The nanosheet structure of diamane is depicted in figure 9.2, which is composed of 106 272 atoms and measures 307 Å in the armchair direction and 303 Å in the zigzag direction. Similar to the AB- and AA-stacked forms of graphene, both the Bernal (AB) stacked and the AA-stacked configurations (hereinafter abbreviated as D-AB and D-AA, respectively) were investigated [59].

The stress–strain curves for D-AB and D-AA in the zigzag and armchair directions are shown side by side in figure 9.3. These findings are comparable with those obtained in graphene and show that the crystallographic orientation has a major impact on the tensile behaviours of diamane, whereas the stacking sequence has only a little effect. D-AB has a brittleness at fracture of 64 GPa (armchair) and 77 GPa (zigzag), respectively.

As research and computational tools continue to progress, future studies may offer more precise predictions of the mechanical properties of these intriguing carbon-based materials. This could be the case if carbon-based materials are studied further. However, in the absence of experimental data, DFT-based predictions can provide useful insights into the possible mechanical behaviour of diamane and 2D diamonds. Figure 9.4 demonstrates that HD has higher thermal conductivity than

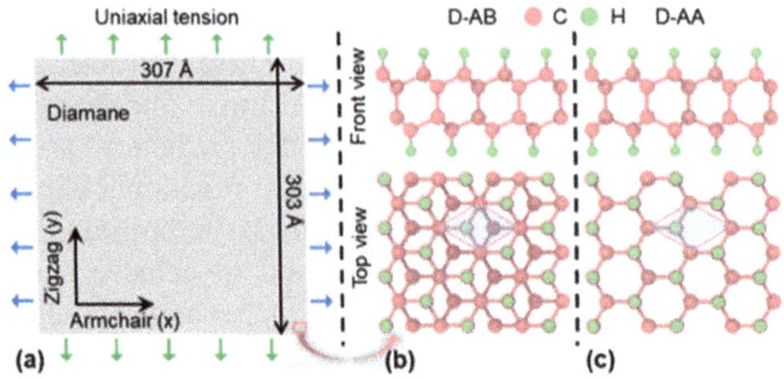

Figure 9.2. The atomic structure of diamanes. (a) Schematic representation of a diamane nanosheet used in the uniaxial tensile test. The two lateral tensile deformation directions are indicated by blue and green arrows. (b) AB-stacked diamane (D-AB) and (c) AA-stacked diamane (D-AA) atomic configurations. The C and H atoms are represented by the red and pink atoms, respectively. Reprinted with permission from [59]. Copyright (2021) American Chemical Society.

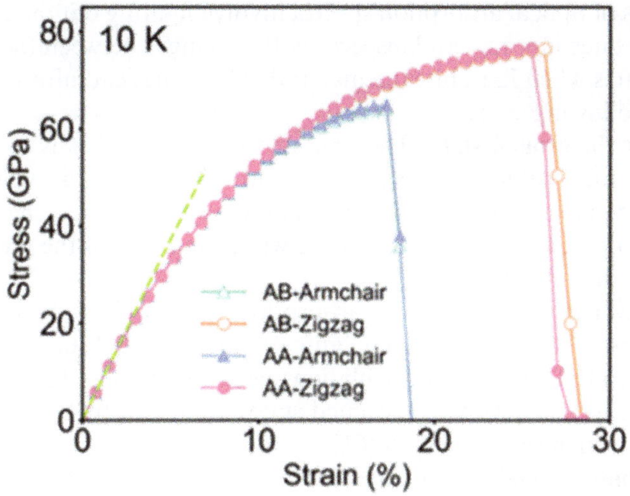

Figure 9.3. At 10 K, stress–strain curves of D-AB and D-AA samples. The green dashed line represents the linear fitting curve for strains up to 2%. Reprinted with permission from [59]. Copyright (2021) American Chemical Society.

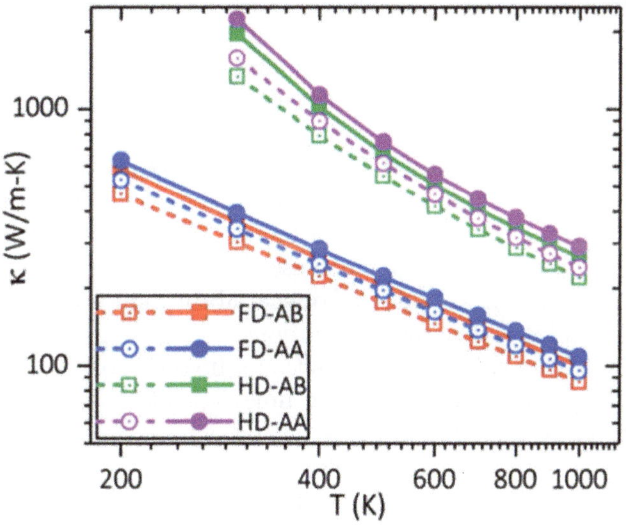

Figure 9.4. Thermal conductivities of several stacked diamanes calculated using the RTA (dashed lines) and CGP (solid lines) methods. Reprinted from [60], with the permission of AIP Publishing.

FD. The AA stacking sequence has better thermal conductivity than the AB stacking sequence.

9.3.3 Optical and vibrational properties

- DFT calculations of optical absorption spectra

DFT studies of optical absorption spectra involve figuring out the energy levels of the electronic states in a material, especially the changes between the occupied and unoccupied states when light hits the material. Mathematical information might be split into the following steps:

Figuring out the ground state: The first step in DFT studies is to figure out the structure of the material's electrons in its ground state. To do this, you have to solve the Kohn–Sham equations, which are based on the system's total energy functional. Self-consistent iterations are used to get the wave functions of the ground state and the energies of the filled states.

Excited state calculation: Optical absorption spectra show how the electronic states change from filled to empty. To figure out these excited states, linear-response time-dependent DFT (TDDFT) calculations need to be done. In TDDFT, a small change is made to the system. The excited states are found by figuring out how the electron density changes in reaction [61].

Linear-response formalism: In linear-response formalism, a change is usually shown by a time-dependent electric field and the reaction of the system is shown by how the electron density changes. The reaction function, which is also called the polarizability, shows how the change in the number of the electrons is caused by the disturbance.

Optical absorption spectrum: To figure out the optical absorption spectrum, the response function can be used to figure out the imaginary part of the dielectric function ($\varepsilon(\omega)$) or complex refractive index ($n(\omega)$) [17]. $\text{Im}[\varepsilon(\omega)] = n(\omega)^2$ describes the relationship between the absorption spectrum and the imaginary part of the dielectric function.

Computational considerations: DFT calculations of optical absorption spectra can be difficult, especially for large systems. In practice, a small number of empty states (usually called 'virtual states') are added to the formula to simulate the exciting states. In addition, it may be necessary to use mixed functionals or range-separated functionals to improve the accuracy of optical absorption forecasts, especially for systems that have a lot of electronic excitations.

DFT studies of optical absorption spectra involve solving the Kohn–Sham equations for the ground state and doing time-dependent linear-response calculations to find the excited states and figure out the optical absorption spectrum. The resulting spectrum gives important information about the material's electronic structure and its ability to capture light at different frequencies. This makes it an important tool for studying the optical features of different materials.

- Study of excitonic activities and optical characteristics

The investigation of excitonic phenomena and their corresponding optical characteristics is a major component of the field of condensed matter physics and materials research [62]. Excitonic effects pertain to the interplay between electrons and holes, specifically electron vacancies, inside a given material. This interaction gives rise to the creation of excitons, which are composite entities consisting of bound electron–hole pairs that exhibit Coulombic interactions. Comprehending these effects is crucial to characterise the electrical and optical properties of materials, and it carries substantial consequences for a range of applications, such as optoelectronics, photonics, and energy harvesting systems. Various theoretical and experimental methodologies are utilised in the investigation of excitonic phenomena and optical characteristics. The combination of DFT and many-body perturbation theory (MBPT) is frequently employed in the computation of the electronic band structure and fundamental characteristics of many materials. Nevertheless, DFT does not explicitly consider the influence of excitonic effects. It is possible to utilise MBPT methodologies to account for excitonic effects, such as the GW approximation and the Bethe–Salpeter equation (BSE). The GW approximations offer a more precise characterisation of the electronic states, whereas the BSE computes the optical absorption spectra, encompassing the influence of excitonic processes. TDDFT is a computational method that extends the capabilities of DFT to calculate excited states and optical characteristics in real-time [64]. The use of TDDFT enables the accurate prediction of optical absorption spectra and excitonic characteristics, such as exciton energies and binding energies, through the incorporation of electron–hole interactions. Exciton Hamiltonian models are frequently employed to characterise excitonic phenomena, particularly in the context of low-dimensional systems such as semiconductor quantum dots or nanowires. These models offer a simplified representation of exciton dynamics. The aforementioned models consider excitons as quantum systems with two energy levels. Their optical characteristics may be examined by analytical methods. Mortazavi and his research group have investigated the mechanical, electrical, and optical properties of hydrogen, fluorine, and chlorine diamane from first principles. In this study, the authors focused on the effect thickness has on the characteristics of functionalized diamane nanomembranes. The results from ab-initio molecular dynamics and phonon dispersions corroborated the thermal and dynamical stability of the systems under study. The results showed that by boosts in elastic modulus and tensile strength can be achieved by increasing the number of carbon atom layers in diamane. Meanwhile, the electronic band structures of H- and F-diamane, unlike Si and Ge, which are indirect bandgap semiconductors and insulators, Cl-diamanes gap semiconductors, hold a great deal of promise for use in nanoelectronics. Calculations in optics evidence that Cl-diamane, unlike its H and F analogues, is able to absorb visible light in the UV spectrum. Finally, the authors used the HSE06 functional to calculate the frequency-dependent absorption coefficients of 2L to 4L diamane nanosheets to investigate their optical features. Figure 9.5 depicts the absorption spectra of these systems for incident light that is both planarly polarised and linearly polarised along the z-axis [65]. Spectroscopic methods, such as photoluminescence, absorption spectroscopy, and optical reflectance, are utilised in experimental settings to investigate the optical characteristics and excitonic phenomena in various materials. These studies yield significant empirical evidence on the energies, lifetimes, and binding energies of excitons. Time-resolved spectroscopy encompasses a range of techniques, including transient absorption spectroscopy and pump–probe spectroscopy [66].

These methods facilitate the study of exciton dynamics and exciton recombination lifetimes, hence providing valuable insights into the behaviour of excitons under

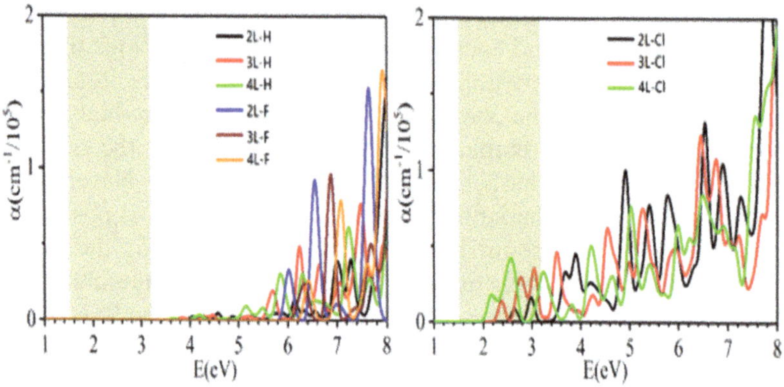

Figure 9.5. HSE06 prediction of optical absorption in diamane nanosheets. Reprinted from [65], Copyright (2020), with permission from Elsevier.

varying circumstances. The investigation of excitonic phenomena and optical characteristics has significant importance in comprehending the interactions between light and matter in various materials. This field of study yields useful insights into the electronic structure of materials and their reactions to external stimuli. By integrating excitonic phenomena into theoretical frameworks and undertaking advanced experimental investigations, scholars may get more profound understandings of the underlying physics of materials and develop innovative optoelectronic devices with improved performance and efficiency.

- Study of the interactions between light and matter and optical implications

The investigation of light–matter interactions and the exploration of optical applications pertaining to diamane and 2D diamonds, commonly referred to as 2D diamond, are captivating fields of research with significant prospects for diverse technological progression. The optical characteristics of theoretical materials are generally examined via DFT computations and theoretical simulations. This discourse aims to explore the interplay between light and matter, specifically focusing on diamane and 2D diamonds, and their possible uses in the field of optics.

The phenomenon of light–matter interactions is a fundamental aspect of physics and plays a crucial role in several scientific disciplines. The interaction of incoming light with diamane and 2D diamond can result in the absorption and reflection of particular wavelengths. DFT investigations have the capability to forecast the optical absorption spectra of materials, therefore providing insights into the energy levels and transitions occurring between electronic states [41]. The remarkable optical features of these materials can be attributed to their exceptional electrical band structures. The phenomenon of excitonic effects is observed in the interaction between light and matter in some materials. These effects arise from the Coulombic interactions between electrons and holes, resulting in the formation of bound electron–hole pairs known as excitons. The use of DFT computations in conjunction with MBPT or TDDFT offers valuable insights into excitonic states, including characteristics such as exciton binding energies and lifetimes. The comprehension of photoluminescence and emission characteristics has significant importance in the context of prospective applications in optoelectronics and light-emitting devices [67]. Theoretical investigations have the capability to forecast photoluminescence spectra and emission properties, therefore aiding researchers in the development of materials that exhibit efficient light emission.

Optical applications refer to the practical uses and implementations of optical technology. The distinctive electrical characteristics and adjustable bandgaps of diamane and 2D diamond render them viable contenders for deployment in photodetectors. The capacity of organisms to capture and utilise certain light wavelengths may be effectively utilised in applications involving the detection and imaging of light. Optical modulators leverage the significant interactions between light and matter, as well as the effects of excitons in these materials, to facilitate the development of optical modulator designs [68]. These designs enable the manipulation and regulation of light signals inside photonic circuits and communication systems. Transparent conductors, such as diamane and 2D diamond, possess distinctive structural characteristics and electronic properties that render them suitable for utilisation in optoelectronic devices. These materials exhibit a desirable amalgamation of electrical conductivity and transparency to visible light. Quantum emitters are materials that have the potential to serve as candidates for the generation of stable excitons [69]. These excitons play a crucial role in facilitating quantum information processing and quantum communication technologies. Photocatalysis refers to the technique of using light absorption and photocarrier formation in diamane and 2D diamond for potential benefits in photocatalytic applications. These applications involve the utilisation of light to facilitate chemical reactions in various environmental and energy-related activities. The capacity to effectively capture a broad spectrum of light wavelengths and transform them into excitonic states renders these materials very appropriate for light harvesting purposes, including use in solar cells and photovoltaic systems. It is imperative to acknowledge that although DFT investigations offer helpful insights, experimental validation via the synthesis and characterisation of diamane and 2D diamond is vital in affirming their optical features and suitability for diverse applications [70]. Moreover, the theoretical investigation of heterostructures including these materials has the potential to increase their functions for particular optical applications. As the advancement of research in this particular domain continues, it is evident that diamane and 2D diamonds has the capability to make significant contributions to the development of next-generation optoelectronic and photonic devices. These materials exhibit distinctive and customised optical characteristics, hence enhancing their potential in this sector.

9.4 Applications of diamane and 2D diamonds

9.4.1 Electronic and optoelectronic devices

Diamane and 2D diamonds, sometimes known as diamondene, are materials that exist in theory but have not yet been successfully synthesised in practical settings. Notwithstanding this constraint, the investigation of prospective electronic and optoelectronic devices utilising theoretical forecasts derived from DFT simulations provides significant elucidation about the distinctive characteristics of these materials.

According to theoretical expectations, diamane and diamondene have captivating electrical characteristics, rendering them potentially viable for a range of electronic applications [71]. The aforementioned materials possess potential for application in the production of high-performance transistors due to their anticipated high carrier mobility and adjustable bandgaps. Field-effect transistors of this nature possess the capability to facilitate electron transport with high efficiency, hence presenting potential implications for the miniaturisation of electronic devices. In addition, the theoretical existence of 2D diamonds presents intriguing prospects for quantum devices. The anticipated capacity to generate stable excitons and distinctive electronic states holds promise for the development of quantum emitters and components tailored for quantum information processing. Moreover, the inherent high carrier mobility and the ability to adjust bandgaps render diamane and diamondene as promising contenders for transparent conductors in electronic devices, thereby merging electrical conductivity and optical transparency [72]. Moreover, the anticipated capacity of these entities to selectively capture particular light wavelengths implies possible utility in the development of photodetectors for imaging and sensing apparatus.

Optoelectronic devices refer to electronic devices that can interact with and manipulate light, hence enabling the conversion of electrical signals into optical. The potential optoelectronic uses of diamane and diamondene are highly promising, given their theoretical optical characteristics. The materials under consideration exhibit excitonic phenomena, rendering them compelling contenders for optoelectronic applications such as light-emitting diodes and lasers [73]. These materials possess an adjustable bandgap, which presents the opportunity to regulate the wavelengths of emission, hence potentially impacting the efficacy of light emission. Additionally, the anticipated capacity to effectively capture light and produce charge carriers renders them appealing for use in solar systems. Solar cells have the potential to utilise them for the purpose of converting sunlight into electrical energy. Moreover, the pronounced interactions between light and matter, coupled with the adjustable bandgap properties, have the potential to facilitate the utilisation of diamane and diamondene in the field of optical modulation. This capability would enable the dynamic manipulation of light signals inside photonic circuits and communication systems. In addition, the excellent absorption of various light wavelengths by these entities presents potential prospects for the use of light harvesting in many applications. These materials might potentially be utilised to enhance the efficiency of energy collecting and conversion in solar cells and other photonic devices. It is crucial to emphasise that experimental synthesis and validation play a vital role in verifying the precision of DFT predictions and uncovering the complete capabilities of diamane and 2D diamonds in the realm of electrical and optoelectronic applications.

It is found that the electronic structure of diamane is highly sensitive to changes in the functional group and nanosheet thickness. To explain this finding, we calculated the PDOS of 2L-H-, 2L-F-, and 2L-Cl-diamane monolayers (figure 9.6) and their valence and conduction band maximum and minimum charge densities (figure 9.7) [65].

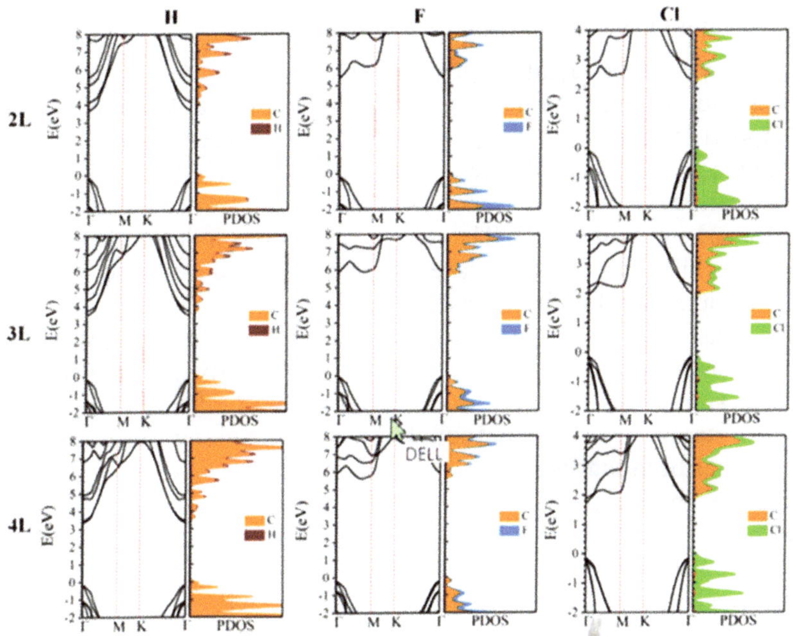

Figure 9.6. The HSE06 functional predicts the electronic band structures and overall electronic density of states of diamane nanosheets. Reprinted from [65], Copyright (2020), with permission from Elsevier.

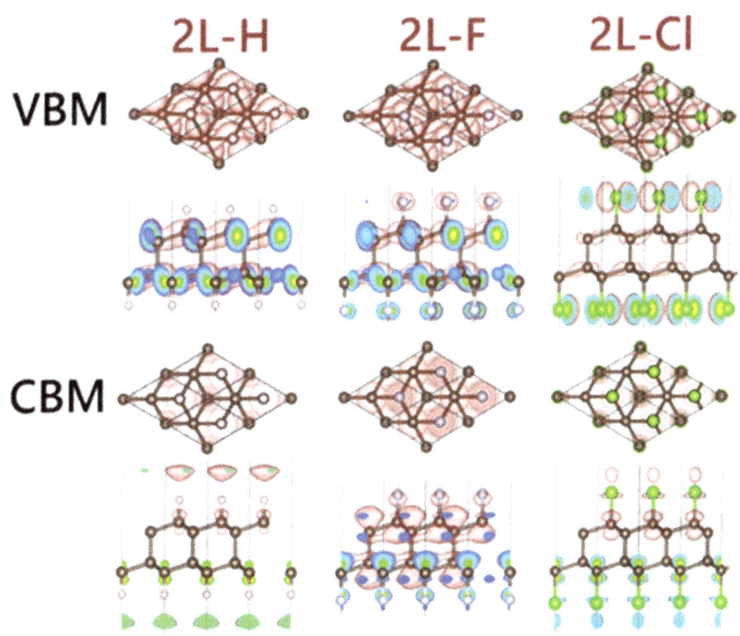

Figure 9.7. The predicted charge density distributions of the H-, F-, and Cl-diamane nanosheets formed of BLG are shown in top and side views. The value of the iso-surface is set at 0.01 e $Å^{-3}$. Reprinted from [65], Copyright (2020), with permission from Elsevier.

9.4.2 Energy storage and conversion

The study of specific possibilities for energy storage and conversion utilising these materials is entirely conjectural and grounded in theoretical forecasts derived from DFT computations [74, 75]. However, it is worth investigating the potential energy storage and conversion capabilities of diamane and 2D diamonds due to their distinctive features.

Batteries and supercapacitors: Theoretical projections indicate that diamane and diamondene exhibit substantial surface areas and electronic conductivity, rendering them viable contenders for electrode materials in batteries and supercapacitors. The ability to adjust their bandgaps and their possession of distinct electrical features have the potential to augment the energy storage capacity and rate capabilities of these energy storage devices.

Hydrogen storage: The potential suitability of 2D diamonds for hydrogen storage applications has been suggested based on theoretical investigations, which show that they may have a high capacity for hydrogen adsorption. The capacity to incorporate hydrogen molecules into their lattice structure has the potential to enhance the efficacy and reversibility of hydrogen storage.

Lithium-ion batteries: The anticipated structural stability and electrical characteristics of diamane and diamondene suggest their potential use as materials for anodes or cathodes in lithium-ion batteries. These materials have the potential to provide improved diffusion and charge storage capacities for lithium-ion batteries.

Energy conversion:

Photovoltaics: The potential suitability of diamane and diamondene as materials for photovoltaic systems is indicated by their theoretical efficiency in light absorption and carrier production. These materials have the potential to be utilised in solar cells for the purpose of converting sunlight into electricity, hence presenting promising prospects for developments in renewable energy technology [76].

Thermoelectric devices: The anticipated superior thermal conductivity and electrical characteristics of diamane and diamondene present potential avenues for their use in thermoelectric devices. These materials have the potential to be utilised for the purpose of converting waste heat into electrical energy that may be effectively utilised, hence making a significant contribution towards improving energy efficiency [80].

Theoretical investigations suggest that diamane and 2D diamonds have the potential to function as catalysts for the process of water splitting. The photocatalytic process may be enhanced by the adjustable bandgaps and distinctive electronic states shown by the material, enabling the effective separation of water into hydrogen and oxygen gases. It is imperative to underscore the need of good experimental synthesis and material characterisation to achieve the implementation of energy storage and conversion applications utilising diamane and 2D diamonds. Furthermore, it is imperative to substantiate the reliability of DFT predictions by conducting experimental validation following the successful synthesis of these materials. Although the aforementioned theoretical investigations provide useful insights, it is imperative to conduct more research and development endeavours to fully exploit the potential of diamane and 2D diamonds in energy-related domains.

9.4.3 Catalysis and chemical sensing

The examination of particular applications in catalysis and chemical sensing utilising these materials is entirely conjectural and grounded in theoretical forecasts derived from DFT computations [78]. However, it is worth examining the possible uses of diamane and 2D diamonds in the fields of catalysis and chemical sensing, taking into consideration their distinctive characteristics.

Heterogeneous catalysis is a branch of catalysis that involves the use of catalysts in a different phase from the reactants. According to theoretical expectations, diamane and 2D diamonds have promising potential as effective catalysts owing to their adjustable electronic characteristics and substantial surface areas. The catalytic activity of these theoretical materials may be investigated in several heterogeneous catalysis processes, including the conversion of tiny molecules, such as CO_2 reduction, as well as other ecologically significant reactions.

Electrocatalysis: The anticipated electrical conductivity and stability of diamane and diamondene render them potentially viable for use in electrocatalytic applications. These materials have the potential to serve as electrodes for a range of electrochemical processes, including the oxygen evolution reaction and the hydrogen evolution reaction (HER) [79].

Electrochemical water splitting: The materials' notable light absorption capabilities indicate promising prospects for use in the field of photocatalysis. It is conceivable that solar energy might be utilised to facilitate catalytic reactions, specifically in the context of photoelectrochemical processes, such as water splitting to produce hydrogen [80].

Gas sensing: The potential large surface area and anticipated gas molecule interactions suggest that diamane and diamondene exhibit attractive characteristics for utilisation in gas sensing applications. These nanosensors have the potential to be utilised for the detection of particular gas species at low concentrations, rendering them highly beneficial for applications in environmental monitoring and industrial safety.

Chemical sensor platforms provide theoretical qualities that have the potential to facilitate the development of highly sensitive and selective sensing systems. By incorporating particular chemical receptors onto their surfaces, these materials may be customised to detect a diverse array of analytes in a variety of chemical sensing applications [81]. Emphasising their importance, it is imperative to underscore the indispensability of conducting experimental synthesis and validation procedures to corroborate the precision of DFT predictions and to fully investigate the extensive capabilities of diamane and 2D diamonds in the realms of catalysis and chemical sensing applications. Although the theoretical studies presented in this study provide useful insights, more research and development endeavours are necessary to comprehensively comprehend and effectively utilise these materials in actual real-world applications. Upon synthesis, the use of diamane and 2D diamonds holds the potential to unlock a plethora of captivating opportunities in the realm of ground-breaking catalytic processes and cutting-edge chemical sensing technologies.

9.4.4 Biomedical and environmental applications

The use of diamane and 2D diamonds in biomedical and environmental applications, as per theoretical predictions derived from DFT computations, provides significant contributions towards understanding the distinctive characteristics of these materials [41].

Biomedical imaging: Theoretical investigations propose that diamane and diamondene possess potential utility as contrast agents within biomedical imaging modalities, including magnetic resonance imaging and fluorescence imaging. The materials under consideration include distinct electrical and magnetic characteristics that have been projected to potentially provide enhanced imaging contrast and signal amplification. Consequently, these qualities might potentially lead to greater visualisation of tissues and structures, thereby enhancing their diagnostic utility [57, 82].

Drug delivery: The potential of utilising diamane and 2D diamonds as drug delivery carriers has been suggested through theoretical analyses. The substantial surface areas and adjustable characteristics of these entities may provide prospects for precise and directed medication delivery. By modifying the surfaces of these materials with specific biomolecules, it is possible to engineer them in such a way that they may effectively transport therapeutic chemicals to target cells or tissues. This targeted delivery approach has the potential to enhance the effectiveness of drugs while minimising any adverse effects.

Environmental sensing: The potential of diamane and diamondene as biosensor platforms is suggested by their projected large surface area and electrical characteristics. These materials provide the potential for customisation by functionalization using specific biomolecules or receptors, enabling the detection of biomarkers and analytes with a notable degree of sensitivity and specificity. In the realm of biosensors, these devices have the potential to assume a pivotal function in the realm of disease diagnostics, the real-time monitoring of health parameters, and environmental sensing [83].

Water purification: Theoretical investigations suggest that diamane and diamondene possess promising prospects for use in the field of water purification. The considerable surface area possessed by these materials, together with their anticipated interactions with water molecules and pollutants, may facilitate the effective elimination of contaminants. These materials have the potential to be employed in improved filtration and adsorption processes; therefore, improving water quality and ultimately ensuring the provision of clean and safe drinking water.

The potential of diamane and 2D diamonds as environmental sensors is suggested by their theoretical features. These devices have the potential to be employed for the purpose of continuously monitoring pollutants and environmental factors in real-time; therefore, offering significant data that can be utilised for the evaluation and management of the environment. The capacity of these sensors to identify certain molecules and analytes may facilitate the timely identification of environmental concerns and contribute to endeavours aimed at safeguarding the environment.

Photocatalysis: The notable light absorption characteristics exhibited by diamane and diamondene indicate their prospective utility in the field of photocatalysis for the purpose of environmental cleanup [84]. These materials have the potential to utilise solar energy for the purpose of driving catalytic processes, namely the degradation of organic pollutants. The use of photocatalytic processes involving diamane and 2D diamonds has the potential to make significant contributions to sustainable and environmentally friendly technology. These processes have the ability to provide clean and renewable energy-powered remediation procedures.

It is important to emphasise that experimental synthesis and validation are required to verify the precision of DFT predictions and to fully investigate the potential of diamane and 2D diamonds for applications in the fields of biomedicine and the environment [85]. Although the theoretical studies presented in this study provide useful insights, more research and development endeavours are necessary to comprehensively comprehend and effectively utilise these materials in practical applications within the realms of biomedicine and the environment. Upon synthesis, diamane and 2D diamonds have the potential to present novel resolutions to a multitude of issues in the realms of healthcare and environmental preservation; hence, establishing a fundamental basis for the advancement of biomedical and environmental technologies in the future.

9.5 Challenges and future directions

Significant insights into the characteristics and prospective applications of diamane and 2D diamonds (2D diamond) can be gained by DFT investigations; however, there are unique problems and promising future paths that must be overcome. Due to the theoretical character of these materials, a lack of experimental evidence makes it difficult to fully validate DFT predictions. Given the one-of-a-kind bonding and electrical properties of diamane and 2D diamond, it is still vitally important to correctly select exchange–correlation functionals to achieve valid findings. In addition, their intricate crystal structures create computational challenges, particularly for large systems, necessitating a trade-off between speed and precision. The use of linear scaling methods and hybrid functional computations has the potential to enhance precision and open the door to more expansive studies in the future. Better understanding of these materials and the doors they can open in nanotechnology, optoelectronics, and beyond can be gained by investigating their electrical and optical properties, investigating heir interface interactions, and characterising their defects and functionalization. Diamane and 2D diamond will be realised, and their incorporation into novel technologies and materials will be sped up thanks in large part to experimental synthesis and characterisation corrating DFT predictions. Theoretical and experimental researchers should work together to accelerate these frontier investigations.

9.6 Conclusion

Key findings and insights into the characteristics and prospective applications of these interesting carbon-based materials have been uncovered through DFT studies of diamane and 2D diamonds. Diamane's crystal structure is several layers of a 2D hexagonal lattice, while 2D diamond's crystal structure is a single layer with sp^3

tetrahedral bonding, both of which were predicted theoretically. Studies have shown that 2D diamond has a strength comparable to bulk diamond and that diamane has extraordinary rigidity and mechanical stability due to its unusually high elastic constants. Electronic properties have also been investigated using DFT simulations, with promising results for large bandgaps and optical properties that point towards optoelectronic applications. Diamane and 2D diamond are attractive possibilities for cutting-edge nanoelectronics and magnetic devices, and these studies have provided predictions for their magnetic and dielectric properties. However, experimental data is still lacking, thus further experimental synthesis and characterisation are needed to verify the results. Despite difficulties, DFT investigations have paved the way for the investigation of novel materials and interfaces, as well as hybrid functional computations and larger-scale simulations. Advances are anticipated to be driven by collaborative efforts between theory and experiment, bringing these materials closer to practical applications in cutting-edge technologies. The characteristics and potential uses of diamane and 2D diamonds (2D diamond) are best understood with the help of DFT. The practical synthesis of theoretical materials such as diamond and 2D diamond is difficult. DFT allows us to theorise about their physical characteristics including structure, electronics, mechanics, magnetism, and dielectrics. DFT helps researchers to precisely anticipate qualities by modelling their electronic structure and bonding interactions, providing direction for future experiments. The crystal structures and bonding patterns in diamane and 2D diamond are revealed by DFT calculations. They all have different structures, and therefore knowing how those structures affect their stability, mechanical behaviour, and electrical properties is crucial. Mechanical characteristics, such elastic constants and mechanical strength, may be computed with DFT. The electrical band structure, energy levels, and optical characteristics of diamane and 2D diamond are revealed using DFT. Although it is important for gauging their usefulness in fields like nanoelectronics, optoelectronics, and quantum technologies, this data is often overlooked. Magnetic and dielectric devices may benefit from the use of diamane and 2D diamond. Magnetic moments, magnetic anisotropy, and dielectric constants may all be predicted with DFT calculations, shedding light on whether or not they can be used in these technologies. The experimental attempts to synthesise and characterise diamane and 2D diamond are greatly aided by the DFT predictions. This is useful for refining experimental setups and confirming expected features. Diamane, 2D diamond, and other material interactions can be studied by DFT, leading to the development of heterostructures with desirable characteristics. The electrical, optical, and magnetic characteristics of heterostructures may be significantly altered. DFT enables researchers to investigate a plethora of possible materials with varying compositions and structures, such as those introduced by dopants and defects, in an effort to optimise the qualities required for a given application. When compared to other, more involved quantum mechanical approaches, DFT is computationally efficient. It is well-suited for investigating the characteristics of large and complex systems because it strikes a good balance between accuracy and processing expense. Diamane and 2D diamond research relies heavily on DFT due to the theoretical insights it provides into the characteristics of these materials, its ability to guide experimental efforts, and its ability to identify new applications in a wide range of sectors. Due to its accuracy in prediction and speed in calculation, it is a useful tool for expanding our knowledge of and application to these special carbon-based materials.

References

[1] Neugebauer J and Hickel T 2013 Density functional theory in materials science *Wiley Interdiscip. Rev. Comput. Mol. Sci.* **3** 438–48

[2] Jia W, Cao Z, Wang L, Fu J, Chi X, Gao W and Wang L W 2013 The analysis of a plane wave pseudopotential density functional theory code on a GPU machine *Comput. Phys. Commun.* **184** 9–18

[3] Chernozatonskii L A, Demin V A and Kvashnin D G 2021 Fully hydrogenated and fluorinated bigraphenes–diamanes: theoretical and experimental studies *C* **7** 17

[4] Bharadwaj V, Jedrkiewicz O, Hadden J P, Sotillo B, Vázquez M R, Dentella P and Eaton S M 2019 Femtosecond laser written photonic and microfluidic circuits in diamond *J. Phys.: Photonics* **1** 022001

[5] Rahman S and Lu Y 2022 Nano-engineering and nano-manufacturing in 2D materials: marvels of nanotechnology *Nanoscale Horiz.* **7** 849–72

[6] He Q, Yu B, Li Z and Zhao Y 2019 Density functional theory for battery materials *Energy Environ. Mater.* **2** 264–79

[7] Tang Q, Zhou Z and Chen Z 2015 Innovation and discovery of graphene-like materials via density-functional theory computations *Wiley Interdiscip. Rev. Comput. Mol. Sci.* **5** 360–79

[8] Segall M D, Lindan P J, Probert M A, Pickard C J, Hasnip P J, Clark S J and Payne M C 2002 First-principles simulation: ideas, illustrations and the CASTEP code *J. Phys.: Condens. Matter* **14** 2717

[9] Hantanasirisakul K and Gogotsi Y 2018 Electronic and optical properties of 2D transition metal carbides and nitrides (MXenes) *Adv. Mater.* **30** 1804779

[10] Sun Z, Zeng Z and Wu J 2014 Zethrenes, extended p-quinodimethanes, and periacenes with a singlet biradical ground state *Acc. Chem. Res.* **47** 2582–91

[11] Rahman M A, Rahaman M Z and Rahman M A 2016 The structural, elastic, electronic and optical properties of MgCu under pressure: a first-principles study *Int. J. Mod. Phys.* **B 30** 1650199

[12] Ward L, Agrawal A, Choudhary A and Wolverton C 2016 A general-purpose machine learning framework for predicting properties of inorganic materials *npj Comput. Mater.* **2** 1–7

[13] Liu W, Li M, Jiang G, Li G, Zhu J, Xiao M and Chen Z 2020 Graphene quantum dots-based advanced electrode materials: design, synthesis and their applications in electrochemical energy storage and electrocatalysis *Adv. Energy Mater.* **10** 2001275

[14] Li G, Blake G R and Palstra T T 2017 Vacancies in functional materials for clean energy storage and harvesting: the perfect imperfection *Chem. Soc. Rev.* **46** 1693–706

[15] Hochbaum A I and Yang P 2010 Semiconductor nanowires for energy conversion *Chem. Rev.* **110** 527–46

[16] Zhang L, Wang Y, Lv J and Ma Y 2017 Materials discovery at high pressures *Nat. Rev. Mater.* **2** 1–16

[17] Bouzidi S, Barhoumi M and Said M 2021 Optical properties of Janus and non-Janus diamanes monolayers using ab-initio calculations *Optik* **235** 166642

[18] Chen L Q, Chen L D, Kalinin S V, Klimeck G, Kumar S K, Neugebauer J and Terasaki I 2015 Design and discovery of materials guided by theory and computation *npj Comput. Mater.* **1** 1–2

[19] Tang X, Li S, Ma Y, Du A, Liao T, Gu Y and Kou L 2018 Distorted Janus transition metal dichalcogenides: stable two-dimensional materials with sizable band gap and ultrahigh carrier mobility *J. Phys. Chem.* C **122** 19153–60

[20] Heinz H, Pramanik C, Heinz O, Ding Y, Mishra R K, Marchon D and Ziolo R F 2017 Nanoparticle decoration with surfactants: molecular interactions, assembly, and applications *Surf. Sci. Rep.* **72** 1–58

[21] Mahor A, Singh P P, Bharadwaj P, Sharma N, Yadav S, Rosenholm J M and Bansal K K 2021 Carbon-based nanomaterials for delivery of biologicals and therapeutics: a cutting-edge technology *C* **7** 19

[22] Hafner J 2008 Ab-initio simulations of materials using VASP: density-functional theory and beyond *J. Comput. Chem.* **29** 2044–78

[23] Teale A M, Helgaker T, Savin A, Adamo C, Aradi B, Arbuznikov A V and Yang W 2022 DFT exchange: sharing perspectives on the workhorse of quantum chemistry and materials science *Phys. Chem. Chem. Phys.* **24** 28700–81

[24] Joseph K S, Dabhi S and Chakraborty B 2023 Importance of 2D materials for electrochemical sensors: theoretical perspectives *2D Materials-Based Electrochemical Sensors* (Amsterdam: Elsevier) pp 133–58

[25] Kurzweil Y, Lawler K V and Head-Gordon M 2009 Analysis of multi-configuration density functional theory methods: theory and model application to bond-breaking *Mol. Phys.* **107** 2103–10

[26] Gritsenko O V, Schipper P R T and Baerends E J 1997 Exchange and correlation energy in density functional theory: comparison of accurate density functional theory quantities with traditional Hartree–Fock based ones and generalized gradient approximations for the molecules Li_2, N_2, F_2 *J. Chem. Phys.* **107** 5007–15

[27] Hasnip P J, Refson K, Probert M I, Yates J R, Clark S J and Pickard C J 2014 Density functional theory in the solid state *Phil. Trans. R. Soc.* A **372** 20130270

[28] Marom N, Tkatchenko A, Scheffler M and Kronik L 2010 Describing both dispersion interactions and electronic structure using density functional theory: the case of metal–phthalocyanine dimers *J. Chem. Theory Comput.* **6** 81–90

[29] Kryachko E S and Ludena E V 2014 Density functional theory: foundations reviewed *Phys. Rep.* **544** 123–239

[30] Gulans A, Kontur S, Meisenbichler C, Nabok D, Pavone P, Rigamonti S and Draxl C 2014 Exciting: a full-potential all-electron package implementing density-functional theory and many-body perturbation theory *J. Phys.: Condens. Matter* **26** 363202

[31] Castro A, Marques M A and Rubio A 2004 Propagators for the time-dependent Kohn–Sham equations *J. Chem. Phys.* **121** 3425–33

[32] Mendl C B, Malet F and Gori-Giorgi P 2014 Wigner localization in quantum dots from Kohn–Sham density functional theory without symmetry breaking *Phys. Rev.* B **89** 125106

[33] Li L, Hoyer S, Pederson R, Sun R, Cubuk E D, Riley P and Burke K 2021 Kohn–Sham equations as regularizer: building prior knowledge into machine-learned physics *Phys. Rev. Lett.* **126** 036401

[34] Zhou Y, Chelikowsky J R and Saad Y 2014 Chebyshev-filtered subspace iteration method free of sparse diagonalization for solving the Kohn–Sham equation *J. Comput. Phys.* **274** 770–82

[35] Posvyanskii D V and Shul'man A Y 2009 Self-consistent solution of Kohn–Sham equations for infinitely extended systems with inhomogeneous electron gas *J. Exp. Theor. Phys.* **109** 145–68

[36] Milman V, Winkler B, White J A, Pickard C J, Payne M C, Akhmatskaya E V and Nobes R H 2000 Electronic structure, properties, and phase stability of inorganic crystals: a pseudopotential plane-wave study *Int. J. Quantum Chem.* **77** 895–910

[37] Valencia F, Romero A H, Ancilotto F and Silvestrelli P L 2006 Lithium adsorption on graphite from density functional theory calculations *J. Phys. Chem.* B **110** 14832–41

[38] Lagardère L, Jolly L H, Lipparini F, Aviat F, Stamm B, Jing Z F and Piquemal J P 2018 Tinker-HP: a massively parallel molecular dynamics package for multiscale simulations of large complex systems with advanced point dipole polarizable force fields *Chem. Sci.* **9** 956–72

[39] Prentice J C, Aarons J, Womack J C, Allen A E, Andrinopoulos L, Anton L and Skylaris C K 2020 The ONETEP linear-scaling density functional theory program *J. Chem. Phys.* **152**

[40] Harvey J N 2006 On the accuracy of density functional theory in transition metal chemistry *Annu. Rep. Sect.' C'(Phys. Chem.)* **102** 203–26

[41] Tiwari S K, Pandey R, Wang N, Kumar V, Sunday O J, Bystrzejewski M and Mishra Y K 2022 Progress in diamanes and diamanoidsnanosystems for emerging technologies *Adv. Sci.* **9** 2105770

[42] Varga M, Vretenar V, Kotlar M, Skakalova V and Kromka A 2014 Fabrication of free-standing pure carbon-based composite material with the combination of sp^2–sp^3 hybridizations *Appl. Surf. Sci.* **308** 211–5

[43] Samantaray M R, Mondal A K, Murugadoss G, Pitchaimuthu S, Das S, Bahru R and Mohamed M A 2020 Synergetic effects of hybrid carbon nanostructured counter electrodes for dye-sensitized solar cells: a review *Materials* **13** 2779

[44] Chermette H 1998 Density functional theory: a powerful tool for theoretical studies in coordination chemistry *Coord. Chem. Rev.* **178** 699–721

[45] Yakobson B I and Smalley R E 1997 Fullerene nanotubes: C 1,000,000 and beyond: some unusual new molecules—long, hollow fibers with tantalizing electronic and mechanical properties—have joined diamonds and graphite in the carbon family *Am. Sci.* **85** 324–37

[46] Georgakilas V, Perman J A, Tucek J and Zboril R 2015 Broad family of carbon nanoallotropes: classification, chemistry, and applications of fullerenes, carbon dots, nanotubes, graphene, nanodiamonds, and combined superstructures *Chem. Rev.* **115** 4744–822

[47] Uzunova E L, Göltl F, Kresse G and Hafner J 2009 Application of hybrid functionals to the modeling of NO adsorption on Cu–SAPO-34 and Co–SAPO-34: a periodic DFT study *J. Phys. Chem.* C **113** 5274–91

[48] Machado A S, Maroudas D and Muniz A R 2013 Tunable mechanical properties of diamond superlattices generated by interlayer bonding in twisted bilayer graphene *Appl. Phys. Lett.* **103** 013113

[49] Javvaji B, Mortazavi B, Zhuang X and Rabczuk T 2021 Exploring tensile piezoelectricity and bending flexoelectricity of diamane monolayers by machine learning *Carbon* **185** 558–67

[50] Sorokin P B and Yakobson B I 2021 Two-dimensional diamond—diamane: current state and further prospects *Nano Lett.* **21** 5475–84

[51] Qin G, Wu L and Gou H 2022 Diamane: design, synthesis, properties, and challenges *Funct. Diam.* **1** 83–92

[52] Chernozatonskii L A, Sorokin P B, Kvashnin A G and Kvashnin D G E 2009 Diamond-like C_2 H nanolayer, diamane: simulation of the structure and properties *JETP Lett.* **90** 134–8

[53] Barhoumi M, Sfina N, Said M and Znaidia S 2021 Elastic and mechanical properties of aluminium and silicon carbide using density functional theory and beyond *Solid State Commun.* **334** 114369

[54] Pakornchote T, Ektarawong A, Busayaporn W, Pinsook U and Bovornratanaraks T 2020 Roles of nitrogen substitution and surface reconstruction in stabilizing nonpassivated single-layer diamond *Phys. Rev.* B **102** 075418

[55] Jensen B D, Wise K E and Odegard G M 2015 Simulation of the elastic and ultimate tensile properties of diamond, graphene, carbon nanotubes, and amorphous carbon using a revised ReaxFF parametrization *J. Phys. Chem.* A **119** 9710–21

[56] Wang B, Zhao J, Hu Y, He Y, Rodionov N, Han J and Zhu J 2022 Anomalous thermal response of bulk diamond to uniaxial (100) strain: a first-principles prediction *Phys. Rev.* B **106** 184303

[57] Hu Y, Li D, Feng C, Li S, Chen B, Li D and Zhang G 2022 Nanostructure engineering of two-dimensional diamonds toward high thermal conductivity and approaching zero Poisson's ratio *Phys. Chem. Chem. Phys.* **24** 15340–8

[58] John R and Merlin B 2016 Theoretical investigation of structural, electronic, and mechanical properties of two-dimensional C, Si, Ge, Sn *Cryst. Struct. Theory Appl.* **5** 43–55

[59] Wu Y C, Shao J L, Zheng Z and Zhan H 2020 Mechanical properties of a single-layer diamane under tension and bending *J. Phys. Chem.* C **125** 915–22

[60] Zhu L and Zhang T. 2019 Suppressed thermal conductivity in fluorinated diamane: Optical phonon dominant thermal transport *Applied Physics Letters* **115**(15)

[61] Yabana K, Nakatsukasa T, Iwata J I and Bertsch G F 2006 Real-time, real-space implementation of the linear response time-dependent density-functional theory *Phys. Status Solidi (b)* **243** 1121–38

[62] Kumar S, Pratap S, Kumar V, Mishra R K, Gwag J S and Chakraborty B 2022 Electronic, transport, magnetic, and optical properties of graphene nanoribbons and their optical sensing applications: a comprehensive review *Luminescence* **38** 909–53

[63] Zahedifar M and Kratzer P 2018 Band structure and thermoelectric properties of half-Heusler semiconductors from many-body perturbation theory *Phys. Rev.* B **97** 035204

[64] Burke K, Werschnik J and Gross E K U 2005 Time-dependent density functional theory: past, present, and future *J. Chem. Phys.* **123** 062206

[65] Mortazavi B, Shojaei F, Javvaji B, Azizi M, Zhan H, Rabczuk T and Zhuang X 2020 First-principles investigation of mechanical, electronic and optical properties of H-, F- and Cl-diamane *Appl. Surf. Sci.* **528** 147035

[66] Ulbricht R, Hendry E, Shan J, Heinz T F and Bonn M 2011 Carrier dynamics in semiconductors studied with time-resolved terahertz spectroscopy *Rev. Mod. Phys.* **83** 543

[67] Wu Z, Wang Y, Dou Y, Zhou L and Zhu J 2023 Property modulations of two-dimensional materials under compression *Nano Res. Energy* **2** e9120080

[68] An J, Zhao X, Zhang Y, Liu M, Yuan J, Sun X and Li D 2022 Perspectives of 2D materials for optoelectronic integration *Adv. Funct. Mater.* **32** 2110119

[69] Ren S, Tan Q and Zhang J 2019 Review on the quantum emitters in two-dimensional materials *J. Semicond.* **40** 071903

[70] Molina-Sánchez A, Palummo M, Marini A and Wirtz L 2016 Temperature-dependent excitonic effects in the optical properties of single-layer MoS_2 *Phys. Rev.* B **93** 155435

[71] Bianco A, Chen Y, Chen Y, Ghoshal D, Hurt R H, Kim Y A and Terrones M 2018 A carbon science perspective in 2018: current achievements and future challenges *Carbon* **132** 785–801

[72] Liu B, Obot E E, Liang T and Wang B 2023 Advancements in theoretical and experimental investigations on diamane materials *Nanoscale* **15** 10498–512

[73] Chen Q, De Marco N, Yang Y M, Song T B, Chen C C, Zhao H and Yang Y 2015 Under the spotlight: the organic–inorganic hybrid halide perovskite for optoelectronic applications *Nano Today* **10** 355–96

[74] Lei X and Medford A J 2019 Design and analysis of machine learning exchange-correlation functionals via rotationally invariant convolutional descriptors *Phys. Rev. Mater.* **3** 063801

[75] Decin L 2021 Evolution and mass loss of cool aging stars: a daedalean story *Annu. Rev. Astron. Astrophys.* **59** 337–89

[76] Tyagi P K and Singh V 2022 N-type diamane: an effective emitter layer in crystalline silicon heterojunction solar cell *Carbon Trends* **9** 100209

[77] Hu Y, Li D, Yin Y, Li S, Ding G, Zhou H and Zhang G 2020 The important role of strain on phonon hydrodynamics in diamond-like bi-layer graphene *Nanotechnology* **31** 335711

[78] Cepitis R, Kongi N, Grozovski V, Ivaništšev V and Lust E 2021 Multifunctional electrocatalysis on single-site metal catalysts: a computational perspective *Catalysts* **11** 1165

[79] dos Reis Souza A C 2022 Ab initio investigations in two-dimensional nanomaterials: electronic and structural properties

[80] Zhang Y, Shen Y, Liu J, Lv L, Gao X, Zhou M and Zhou Z 2022 Novel 2D HfTeS4 for water splitting with high visible-light absorption *Appl. Surf. Sci.* **578** 151992

[81] Misawa N, Mitsuno H, Kanzaki R and Takeuchi S 2010 Highly sensitive and selective odorant sensor using living cells expressing insect olfactory receptors *Proc. Natl Acad. Sci.* **107** 15340–4

[82] Zhou X J 2019 Biomedical imaging magnetic resonance imaging *Engineering-Medicine* (Boca Raton, FL: CRC Press) pp 219–39

[83] Brothers M C, DeBrosse M, Grigsby C C, Naik R R, Hussain S M, Heikenfeld J and Kim S S 2019 Achievements and challenges for real-time sensing of analytes in sweat within wearable platforms *Acc. Chem. Res.* **52** 297–306

[84] Vinu R and Madras G 2010 Environmental remediation by photocatalysis *J. Indian Inst. Sci.* **90** 189–230

[85] Paul S and Momeni K 2019 Mechanochemistry of stable diamane and atomically thin diamond films synthesis from bi-and multilayer graphene: a computational study *J. Phys. Chem.* C **123** 15751–60

IOP Publishing

Diamane
Fabrication, properties and new advances in 2D diamond
Santosh K Tiwari and Arpan Kumar Nayak

Chapter 10

Fine-tuning fabrication parameters for high-grade diamane and derivatives

Bibhuti B Sahu, Tarun Yadav, Pradeep Kumar Gautam and Srikanta Moharana

The fabrication of diamane, a theoretical carbon-based material closely related to diamond, and its derivatives presents exciting possibilities for engineering novel materials with exceptional properties. This paper explores the critical role of temperature and pressure in the synthesis and formation of diamane and its derivatives. Theoretical studies have predicted that diamane could possess a unique three-dimensional (3D) diamond-like structure, composed of two-dimensional (2D) layers of carbon atoms arranged in a tetrahedral lattice. To realize such a structure, extreme conditions of temperature and pressure are essential. High-pressure conditions promote the rearrangement of carbon atoms into a tetrahedral arrangement, representing the sp^3-hybridized structure found in diamonds. The impact of pressure and temperature on the thermodynamic stability of diamane is analyzed, providing insights into the regions of phase stability under varying conditions. The synthesis of diamane and its derivatives requires specialized techniques that can create and sustain high-pressure environments, often in the gigapascal range and elevated temperatures reaching thousands of degrees Celsius. Methods such as diamond-anvil cells, multi-anvil presses, and chemical vapor deposition under pressure have been theoretically explored as potential avenues for diamane fabrication. Additionally, this study delves into the significance of temperature and pressure in the functionalization and derivatization of diamane. By introducing specific chemical groups and altering the synthesis conditions, researchers aim to tailor the properties of diamane derivatives for specific applications, ranging from electronics and optics to superhard materials. Despite the immense theoretical potential, the fabrication of diamane and its derivatives remains a significant challenge due to the extreme conditions required and the complexity of controlling the synthesis process. Nonetheless, the insights gained from this exploration shed light on the fundamental understanding of carbon materials and pave the way for future research and potential breakthroughs in the synthesis of this elusive carbon allotrope and its derivatives.

doi:10.1088/978-0-7503-5939-9ch10

10.1 Introduction

Diamane is a theoretical carbon-based material and is considered to have close similarities with diamonds. It is believed to consist of 2D layers of carbon atoms arranged in a similar hexagonal lattice as graphene but with a distinct 3D diamond-like structure [1–3]. Diamane is expected to have a unique tetrahedral structure formed by sp^3-hybridized carbon atoms, similar to diamond [4]. This tetrahedral arrangement gives the diamond its exceptional hardness and strength. Diamane is a theoretical material had not been synthesized or observed experimentally. Its existence was only predicted based on theoretical calculations and simulations [5–7]. Diamane is believed to exhibit extraordinary properties, such as exceptional hardness, thermal conductivity, and optical transparency. Its unique structure could potentially make it suitable for various applications, ranging from electronics to optics and even as a superhard material [8]. Researchers have also explored the possibility of creating derivatives of diamane by functionalizing its surface with different chemical groups. These derivatized diamane structures could potentially exhibit tailored properties for specific applications [9]. It is important to note that much of the information on diamane is speculative and theoretical. The synthesis and characterization of novel materials such as diamane can be a complex and challenging process. Its unique structure could potentially make it suitable for various applications, ranging from electronics to optics and even as a superhard material [10–12]. Researchers have also explored the possibility of creating derivatives of diamane by functionalizing its surface with different chemical groups. These derivatized diamane structures could potentially exhibit tailored properties for specific applications. It is important to note that much of the information on diamane is speculative and theoretical. The synthesis and characterization of novel materials such as diamane can be a complex and challenging process.

Based on theoretical studies, understanding the role of temperature and pressure is crucial for potential fabrication attempts of diamane [13]. Diamane's unique tetrahedral structure, akin to diamond, requires a specific arrangement of carbon atoms. To achieve this, it is believed that high pressure is needed to encourage the carbon atoms to bond in a tetrahedral manner. High pressure pushes the carbon atoms together, promoting the formation of the desired sp^3-hybridized carbon bonds, which are characteristic of diamond-like materials [1, 4]. The transition from a typical sp^2-hybridized carbon structure (like in graphene) to a sp^3-hybridized carbon structure (like in diamane) requires an energy barrier to be overcome. This barrier represents the energy difference between the two configurations. High pressure can lower this energy barrier and facilitate the transformation from graphene-like structures to diamond-like structures [14]. At ambient conditions, the formation of diamane is highly improbable due to kinetic limitations, meaning that the carbon atoms will not spontaneously rearrange themselves into the desired diamond-like structure. By increasing the temperature, the mobility of carbon atoms is enhanced, making it more likely for them to find the correct positions and form the tetrahedral structure characteristic of diamane [1, 15]. At certain combinations of temperature and pressure, the thermodynamic stability of diamane may be achieved, meaning that the formation of diamane is energetically favored over other carbon structures [16]. It is essential to point out that the precise temperature and pressure conditions required for the successful synthesis of diamane would depend on the specific fabrication techniques and the starting materials used. Theoretical predictions can provide valuable guidance, but practical challenges in creating such extreme conditions and controlling the synthesis process make the fabrication of diamane a complex task [17].

10.2 High-pressure synthesis techniques

10.2.1 Theoretical high-pressure synthesis methods

High-pressure synthesis methods for diamane were still in the realm of theoretical predictions, and diamane had not been successfully synthesized or observed experimentally [18]. However, researchers have explored various high-pressure methods in theory to potentially create diamane. An overview of some of these theoretical high-pressure synthesis methods follows:

Diamond-anvil cell method: The diamond-anvil cell (DAC) is a widely used device for creating high pressures in a small sample volume. In this method, a small amount of carbonaceous material (e.g., graphite) is placed between two diamond anvils, and pressure is applied by compressing the anvils together [19]. At extremely high pressures and temperatures, it is believed that the carbon atoms could rearrange into a tetrahedral structure, potentially leading to the formation of diamane.

Shock compression: In shock compression methods, intense shockwaves are generated, which create extreme pressures and temperatures for a brief moment. These shockwaves can be generated by different means, such as laser-induced shock compression or impact from high-velocity projectiles. Under such conditions, it is theorized that carbon atoms might be forced into the desired tetrahedral arrangement [20].

Chemical vapor deposition under pressure: Chemical vapor deposition (CVD) is a common method for depositing thin films of materials. In the high-pressure version of CVD, carbon-containing gases are introduced into a high-pressure chamber, and elevated temperatures promote chemical reactions that could potentially lead to the formation of diamane [21, 22].

Synthesis inside carbon nanotubes: Carbon nanotubes can act as nanoscale reaction chambers. It has been suggested that encapsulating carbon atoms or small carbon clusters within carbon nanotubes and subjecting them to high-pressure conditions could favor the formation of diamane due to spatial confinement effects. It is important to note that high-pressure synthesis methods, especially at the extreme conditions required for diamane formation, present significant experimental challenges. The high pressures and temperatures involved often require specialized equipment and expertize. Additionally, high-pressure synthesis can lead to the formation of various carbon allotropes and structures other than diamane, making characterization and identification complex.

10.2.2 Diamond anvil cell technique

DAC technique is a powerful experimental method used in high-pressure research to create extreme pressure conditions on small samples [23]. It was invented in the early-1960s by researchers Alvin Van Valkenburg and Percy Bridgman. DAC has since become an indispensable tool for studying the properties and behaviors of materials under very high pressures, which are otherwise challenging to achieve. The basic principle of the DAC involves compressing a sample between two small, flat-tipped diamonds, which act as anvils [24]. These anvils are mounted on opposite sides of a metal gasket, typically made from materials like stainless steel or rhenium, which serve as the pressure-transmitting medium. A description of the DAC technique and its key components follows:

Diamond anvils: The most crucial elements of the DAC are the diamonds themselves. Natural or synthetic diamonds are used due to their exceptional hardness and ability to withstand high pressures. The diamond anvils have a small, flat surface, typically with a diameter of a few hundred micrometers to a few millimeters [25].

Sample chamber: The sample to be studied is placed between the culets (flat tips) of the diamond anvils. The size of the sample is often on the order of tens of micrometers or smaller, making the DAC suitable for studying tiny amounts of materials [26].

Gasket: The sample and the pressure-transmitting medium are enclosed within a metal gasket, which acts as a cell to contain the sample and the pressure medium. The gasket material must be strong enough to withstand the pressure generated by the diamonds [27].

Pressure medium: The pressure-transmitting medium, often a non-reactive material such as a mixture of alcohol and water, surrounds the sample inside the gasket. This medium helps distribute the pressure from the diamond tips evenly onto the sample.

Pressure generation: The pressure is generated by turning a screw or applying an external force to compress the diamond anvils together [28]. As the anvils are brought closer, the sample and the pressure medium between them are subjected to higher pressures. The extreme pressures achieved with the DAC can reach values on the order of hundreds of gigapascals (GPa) or even higher. These high-pressure conditions can simulate the interior of planets, explore the behavior of materials under extreme compression, and lead to the discovery of novel phases and properties.

10.2.3 High-pressure high-temperature techniques

High-pressure high-temperature (HPHT) techniques [29] refer to a broad category of experimental methods used to subject materials to both elevated pressure and temperature simultaneously. These techniques are invaluable for studying the behavior, phase transitions, and properties of materials under extreme conditions, which can mimic the high-pressure and high-temperature environments found deep within the Earth's crust or even within planets.

Here are some other common HPHT techniques apart from the DAC that researchers use:

Multi-anvil press: The multi-anvil press is another popular high-pressure device used in materials science and geosciences research. It consists of a series of tungsten carbide cubes that apply pressure to a sample when compressed together [30]. The pressure is transmitted through the sample contained within a ceramic or graphite capsule. Heating is achieved by electrical resistance or laser heating.

Large volume press: Large volume presses, also known as Belt or Toroid presses, are used to generate high pressures and moderate temperatures over larger sample volumes. These presses use shaped anvils or toroids to create pressure, and they can reach pressures in the range of tens of gigapascals (GPa). Heating is typically achieved using resistive or laser heating [31, 32].

Shock compression: In shock compression techniques, intense shockwaves are generated using methods like laser-induced shock or impact from high-velocity projectiles. The shockwaves propagate through the sample, leading to extreme pressures and temperatures for very short durations [33]. These high-pressure shock conditions can simulate meteorite impacts and are also used in high-energy physics and inertial confinement fusion research.

Hydrothermal synthesis: Hydrothermal synthesis is a method used primarily in materials chemistry. It involves placing reactants in a sealed container and subjecting them to high pressure and temperature in an aqueous or hydrothermal environment [34]. This technique is commonly used to synthesize various materials, including crystals and nanoparticles, under high-pressure conditions.

Piston-cylinder apparatus: The piston-cylinder apparatus is another technique used for high-pressure experiments [35]. It consists of a sample enclosed within a cylinder and pressure is generated by compressing a piston inside the cylinder. The apparatus can reach pressures on the order of several gigapascals (GPa).

Floating zone method: The floating zone method is a crystal growth technique that can be adapted for high-pressure conditions [36]. In this method, a small region of a sample is melted and recrystallized as a floating zone is moved through the material. High pressures and temperatures can be achieved depending on the design and materials used. Each of these HPHT techniques offers unique advantages and limitations, making them suitable for different types of experiments and applications. Researchers choose the most appropriate method based on the specific research goals and the properties of the materials they want to study under high-pressure and high-temperature conditions.

10.3 Temperature effects on fabrication

10.3.1 Impact of temperature on diamond formation

Temperature plays a crucial role in diamond formation because it determines the stability and kinetics of the process. The formation of diamonds occurs deep within the Earth's mantle, where intense pressure and high temperatures create the ideal conditions for its synthesis [37]. Some of the key impacts of temperature on diamond formation follow:

Thermodynamic stability: The formation of diamonds is thermodynamically favored at high temperatures and pressures deep within the Earth's mantle [38]. At depths of about 150 km or more, temperatures exceed 900°C (1652°F), which is above the threshold for the thermodynamic stability of diamonds. Under these conditions, carbon atoms tend to bond in a tetrahedral arrangement, forming the strong and stable sp^3-hybridized structure characteristic of a diamond.

Kinetic barrier: While thermodynamically stable at high temperatures, the formation of diamonds from carbon can be hindered by kinetic barriers at lower temperatures. At Earth's surface temperatures (25°C or 77°F), the conversion of carbon into diamond is an extremely slow process [39]. This is why natural diamond formation deep within the Earth takes millions to billions of years. However, processes such as meteorite impacts or volcanic eruptions can provide the sudden and intense temperature and pressure needed to overcome these kinetic barriers, leading to the formation of impact diamonds or 'popigaiite.'

Synthetic diamond formation: In the laboratory, diamond synthesis can be achieved through HPHT techniques or CVD [40]. HPHT methods require temperatures on the order of 1400°C–2500°C (2552°F –4532°F) to induce the transformation of carbon into diamond at high pressures. In contrast, CVD typically operates at lower temperatures around 700°C –1000°C (1292°F –1832°F), and relies on chemical reactions in a gas phase to deposit carbon atoms on a substrate, facilitating diamond growth.

Diamond quality: The temperature at which diamond forms can also influence its quality and characteristics. Higher temperatures during the synthesis process can lead to the formation of larger and higher-quality diamonds [41]. Additionally, temperature fluctuations and growth conditions can affect the presence of impurities or defects in the diamond lattice, which can impact its color and transparency. Temperature significantly impacts the formation and properties of diamond. Natural diamond formation occurs over long geological timescales and is driven by high temperatures and pressures in the Earth's mantle [42]. In the laboratory, diamond synthesis requires high temperatures to create the conditions for diamond growth. Understanding the relationship between temperature, pressure, and diamond formation is essential for both scientific research and industrial diamond production. The formation of diamond is highly dependent on temperature, and temperature plays a crucial role in determining the conditions under which diamond can be created or converted from other forms of carbon.

Some key impacts of temperature on diamond formation follow:

Stability and phase diagram: The phase diagram of carbon shows the different phases of carbon as a function of pressure and temperature. At ambient conditions (room temperature and low pressure), graphite is the most thermodynamically stable form of carbon [43]. However, as temperature increases, the stability of diamond relative to graphite also increases, especially at high pressures. At elevated temperatures and pressures, diamond becomes the stable form of carbon.

Conversion of graphite to diamond: Although diamond is thermodynamically stable at high pressure and high temperature, its formation from graphite is kinetically challenging at standard atmospheric pressure and temperature [44]. However, under very high pressure and temperature conditions, carbon atoms can rearrange to form the tetrahedral structure of diamond, converting graphite into diamond.

Natural diamond formation: Natural diamonds are believed to have formed deep within the Earth's mantle under extremely high pressure and temperature conditions [45].

The exact depths and temperatures at which diamonds form naturally are subject to ongoing research, but they are likely to occur at depths of around 150–250 km (93–155 miles) and temperatures of over 1000°C (1832°F).

High-pressure synthesis: In laboratory settings, diamonds can be synthesized from various forms of carbon, such as graphite or amorphous carbon, using HPHT methods [46]. The high temperature is necessary to overcome the energy barrier for the transformation of carbon into diamond, while the high pressure is crucial for stabilizing the diamond structure.

Formation of other carbon allotropes: Besides diamond and graphite, the temperature can also influence the formation of other carbon allotropes, such as amorphous carbon, carbon nanotubes, and fullerenes [47]. These materials have distinct structures and properties and can be formed under specific temperature.

10.3.2 Thermodynamics of carbon conversion

The thermodynamics of carbon conversion, specifically the transformation of carbon into different allotropes such as diamond, graphite, and amorphous carbon, is a fundamental aspect of materials science and is governed by the principles of chemical thermodynamics [48, 49]. These transformations are influenced by factors like temperature, pressure, and the free energy of the involved phases. The thermodynamics of carbon conversion and its implications are as follows:

Phase diagram: The thermodynamic behavior of carbon can be visualized using a phase diagram, which shows the stability regions of different carbon allotropes as a function of temperature and pressure. The phase diagram of carbon is complex due to the presence of various allotropes, including diamond, graphite, amorphous carbon, and others [50]. At low temperatures and low pressures, graphite is the thermodynamically stable form of carbon. As pressure increases, diamond becomes more stable than graphite at higher temperatures, and at even higher pressures, other carbon allotropes may become favored.

Gibbs free energy: The transformation of carbon between different allotropes is driven by minimizing the Gibbs free energy of the system. The Gibbs free energy is a measure of the system's total energy (enthalpy) and the degree of disorder (entropy) [51]. For a transformation to occur spontaneously, the change in Gibbs free energy (ΔG) must be negative. The allotrope with the lowest Gibbs free energy under specific temperature and pressure conditions is the most stable form of carbon.

Enthalpy and entropy: The formation of different carbon allotropes involves changes in enthalpy (heat transfer) and entropy (disorder) [52]. The transition from graphite to diamond, for example, is accompanied by an increase in enthalpy (endothermic) but also an increase in entropy (more disorder). The overall stability depends on the balance between these enthalpy and entropy changes.

Kinetic considerations: Thermodynamics provides information about the stability of carbon allotropes under certain conditions, but it does not directly address the rate at which the transformations occur [53]. The kinetics of carbon conversion is also influenced by factors such as activation energy barriers and nucleation processes. These kinetic factors can significantly impact the transformation rates, which is why diamond formation at Earth's surface is exceedingly slow despite being thermodynamically stable.

Metastability: Metastable states are another important aspect of carbon conversion. Some carbon allotropes may be kinetically stabilized, meaning they persist even when thermodynamically less stable than other phases [54]. For example, amorphous carbon is a metastable form that can exist even at conditions where graphite or diamond is more thermodynamically stable.

Overall, the thermodynamics of carbon conversion provide insights into the conditions under which different carbon allotropes are favored and the energy changes associated with these transformations. Combining thermodynamics with considerations of kinetics and activation energy helps us understand the processes that drive the formation, stability, and persistence of various carbon structures in different environments.

10.3.3 Temperature ranges for successful synthesis of diamane

Theoretical studies and simulations have suggested that diamane could potentially form at extremely high pressures and temperatures, akin to the conditions found deep within the Earth's mantle. The temperature required for the synthesis of diamane would likely need to be in the range of thousands of degrees Celsius, possibly exceeding 4000°C or more, to overcome kinetic barriers and allow the carbon atoms to rearrange into the desired tetrahedral structure [55].

However, it is essential to emphasize that the actual temperature range for the successful synthesis of diamane would depend on various factors, including the specific fabrication method, the starting materials used, and the experimental conditions. Additionally, diamane is a highly speculative material, and its synthesis might present significant challenges given the extreme conditions required.

10.4 Pressure effects on fabrication

10.4.1 Role of pressure in carbon rearrangement

To understand how this hypothetical carbon-based material could be made, you have to know how pressure affects the rearrangement of carbon atoms. Diamane is thought to have a unique diamond-like 3D structure, with 2D layers of carbon atoms grouped in a diamond-like tetrahedral lattice. For this structure to happen, the carbon atoms must rearrange themselves into a tetrahedral shape, which is similar to the sp^3-hybridized structure of a diamond [56]. This is made possible by high pressure. A description of the most important things to know about how pressure changes the way carbon is arranged in diamane follows:

Favoring sp^3 hybridization: When there is a lot of pressure, the electron clouds of carbon atoms meet more. This makes the bonds between the atoms stronger and helps sp^3 hybridization happen. Unlike the sp^2-hybridized carbon in graphite and graphene, sp^3-hybridized carbon makes a tetrahedral structure, with each carbon atom linked to four neighboring carbon atoms at equal bond angles [57].

Overcoming kinetic barriers: Under normal conditions, carbon atoms tend to make stable sp^2-hybridized structures, such as those in graphite and graphene. This happens because carbon rearrangement is limited by lower energy states and kinetic barriers. High pressure helps get around these kinetic barriers by giving the energy needed to change sp^2-hybridized carbon into sp^3-hybridized carbon in the right tetrahedral shape [58].

Enhanced carbon mobility: Pressure makes carbon atoms move faster by bringing them closer together, which makes them more mobile [59]. As carbon atoms get closer to each other, they are more likely to switch places and make new bonds. High pressure also makes it more likely that carbon–carbon bonds will form in the sp^3-hybridized form, which helps the tetrahedral structure of diamane form.

Factors for phase stability: Pressure also affects how thermodynamically stable different carbon phases are. Under high pressure, the phase diagram of carbon changes, making structures that look like diamonds more common than other forms of carbon, such as graphite [60, 61]. At certain pressure–temperature ranges, the energy environment changes, and structures that look like diamonds, like diamane, are more likely to form.

Exploration of phase changes: Pressure-induced phase changes in carbon need to be studied to understand how carbon rearranges itself [62]. As the pressure goes up, carbon changes from graphite to diamond or from amorphous carbon to structures that look like diamonds. These changes are necessary for the formation of diamane and its products. Extreme pressures are used in experiments to make diamane and study its qualities, which is a scientific challenge. For carbon rearrangement, high-pressure tools such as DACs or multi-anvil presses are needed to make and keep the right pressure conditions. Understanding the role of pressure in the rearrangement of carbon that leads to the creation of diamane is important for understanding how this theoretical material forms and stays stable. Theoretical models and experiments in high-pressure synthesis tell us a lot about the conditions needed to make a diamane have a structure like a diamond. This could open new ways to make carbon-based materials with unique qualities. Raman spectroscopy of double-layer graphene under high pressure has been performed by Martin *et al* [63] providing spectroscopic evidence for the creation of diamondene. Because of the limited size of the surviving graphene sites surrounded by the diamondene matrix, the observed results are interpreted as a breakdown in the Kohn anomaly. Diamondene is formed when two or more layers of graphene are exposed to high pressures in the presence of particular chemical groups such as hydroxyl groups or hydrogens. This process is clarified using ab initio calculations and molecular dynamics simulations [63].

The two distinct measurement sessions (first and second, respectively) are performed on two distinct samples to obtain the data. Each pressure level is represented by two spectra, one obtained with an excitation laser energy of 2.33 eV (green symbols) and the other with an excitation laser energy of 2.54 eV (blue symbols). Voigt fitting of the experimental data yields the solid lines. The experimental setup is depicted in figure 10.1(a). A DAC with a pressure range of 15 GPa was used to apply the necessary force to the sample. The experimental setup is described in depth in the 'Methods' section 10.4.2. Pressure–temperature–pressure (*P–T–P*) curves for water are shown in figure 10.1(b), illustrating how the first-order Raman-allowed bond-stretching G band changes with increasing pressure (up to 14 GPa). In high-pressure studies,

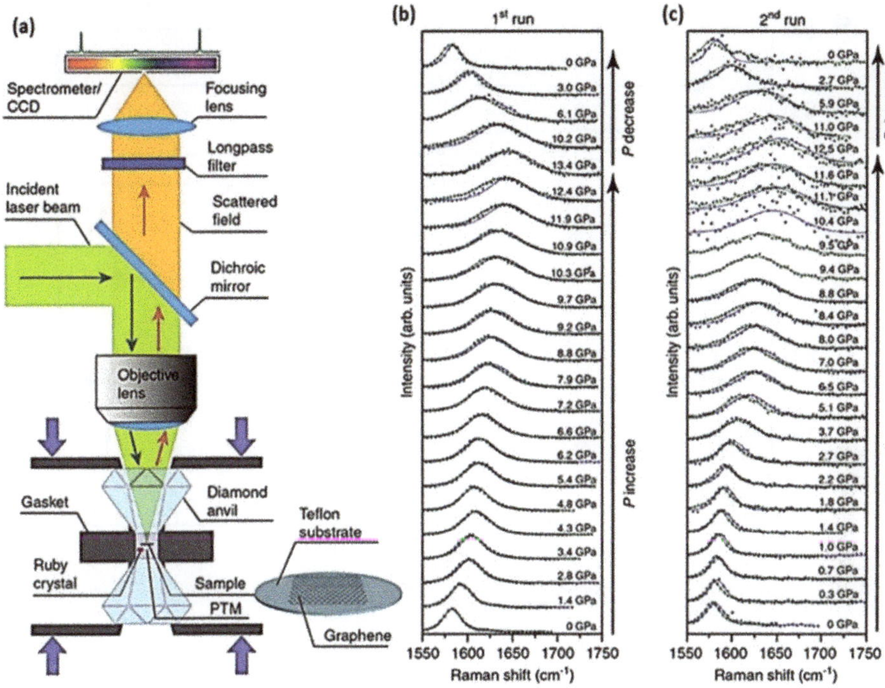

Figure 10.1. (a) A diagram of the experimental apparatus. The sample is pressurized to 15 GPa using water as the pressure transmission medium (PTM) in the DAC. (b) and (c) Raman spectra illustrate the evolution of the G band with increasing pressure. Reproduced from [63]. CC BY 4.0.

the only clearly detectable Raman characteristic from graphene was the G band. This happened because the D (1350 cm^{-1}) and 2D (2700 cm^{-1}) bands were superimposed with the first- and second-order bond-stretching peaks of diamond, respectively.

Figure 10.2(a) depicts a pressure versus time (t) plot that summarizes the MD simulation results. The procedure was broken down into five phases, as shown in figure 10.2(a), and described as follows. In stage (i), thermal equilibrium is attained. The apparatus operates for 100 ps at 300 K with fluctuating pressure close to zero. The loading procedure occurs during phase (ii) 100 ps during the linear approach of the upper piston. The prior stage's burden is held constant for 2 ns. By fixing the ultimate position of the upper piston in stage (iii), the system is maintained in pressure equilibrium. The unloading is conducted in stage (iv), when the piston is released and returns linearly to its original position and an initial location within 100 ps. The final equilibrium of the structure is attained in stage (v), which adds 100 ps to the overall time. The piston is then released. Figure 10.2(b) is a magnified variant of figure 10.2(a), emphasizing its details. Pressure levels near the diamondene transition (between 4 and 5 GPa) are shown. The red curve in figure 10.2 illustrates the evolution of the system in instantaneous (non-equilibrium) peak compression. The pressure indicated by the projectile in panel (b) is 4.92 GPa. As in the stage of pressure equilibration, the system undergoes change [63].

The pressure decreases marginally; 0.95 ns later, a transition to diamondene occurs, which begins at 4.57 GPa (circle in figure 10.2(b)) when there is a rapid

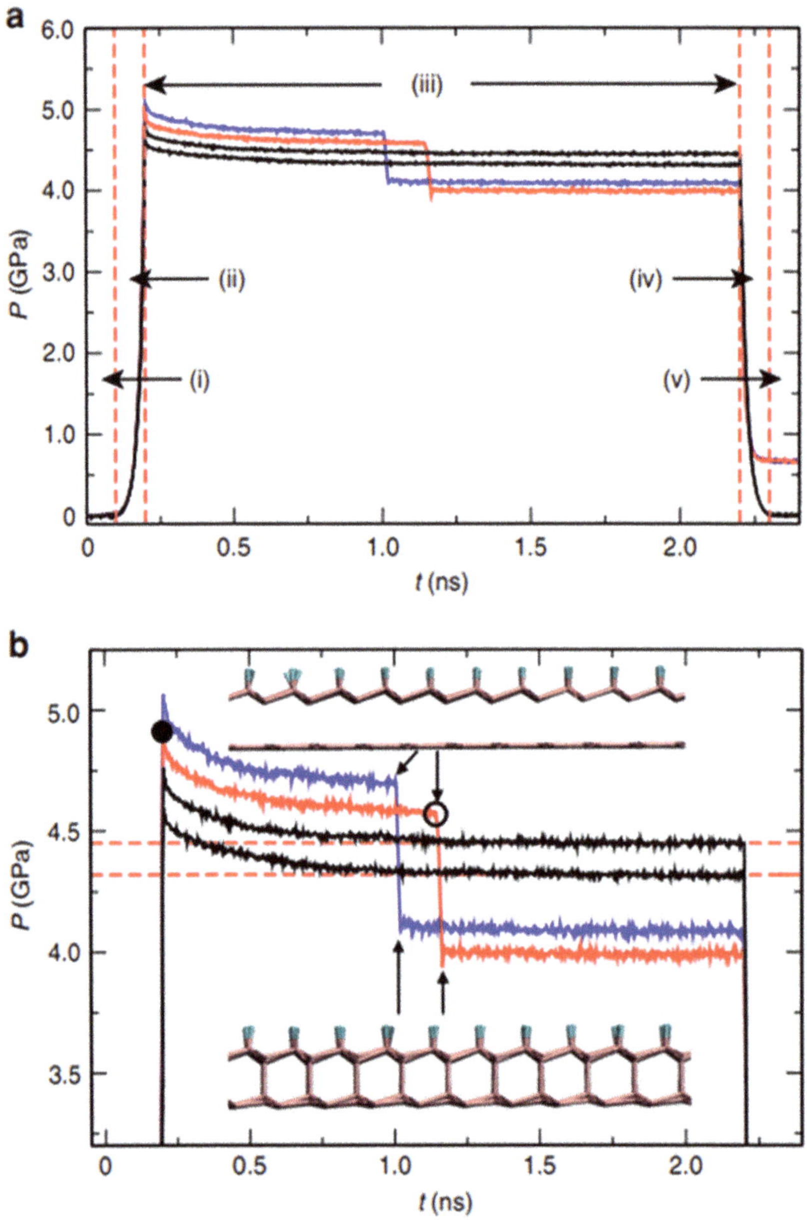

Figure 10.2. Simulations of molecular dynamics. (a) The simulations imposed the following phases on the system: (i) thermal equilibration, (ii) loading, (iii) pressure equilibration, (iv) discharge, and (v) final structure relaxation (see text for details). Cases in which diamondene transition is observed are depicted by blue and red curves, while black curves indicate that it is not observed. (b) Zoomed version of (a), emphasizing pressure levels near the diamondene transition (4–5 GPa). The red curve's red dot indicates the essential peak pressure required for the sp^3 transition to occur (4.92 GPa). The circle indicates the critical pressure (4.57 GPa) at which the transition occurs. When no sp^2-to-sp^3 transition is observed, equilibrium pressures for the graphene bilayer under piston load are depicted by dashed lines. The illustrations depict the characteristic geometries just before (top) and just after (bottom) the transition to diamondene occurs. Reproduced from [63]. CC BY 4.0.

increase in pressure There is then a pressure decrease (to about 4.0 GPa). Figure 10.2(b) (top and bottom panels) depicts the transitional characteristics geometry right before and after it occurs. No significant changes in the diamondene structure are seen, even after the unloading stage and the final equilibration interval, and the pressure value remains constant until the conclusion of this stage. The blue lines in figures 10.2(a) and (b) represent a second experiment that was conducted at a slightly higher peak pressure of 5.06 GPa. Diamondene production occurred in a shorter time frame, as predicted, confirming the phenomenology seen in the previous experiment (red curves). Following the structural transition, the simulation box is not rescaled along the periodic directions, which results in the artifact of residual pressure being detected in both cases (first and second runs) following pressure release (stage (v)) [64]. Similar experiments were carried out at peak pressures lower than 4.92 GPa (black curves in figure 10.2).

10.4.2 Pressure requirements for diamane synthesis

Very specific pressure conditions must be met to successfully synthesize diamane, a theoretical carbon-based substance with a diamond-like structure. Like a diamond, diamane is thought to be composed of layers of carbon atoms organized in a tetrahedral lattice but on a smaller scale [16]. This state can only be achieved by subjecting the material to pressure in the gigapascal (GPa) range or higher. Extreme pressures such as these are required to overcome kinetic barriers that prevent carbon from being converted into the necessary tetrahedral sp^3-hybridized form under normal circumstances. The thermodynamic stability of carbon phases is also modified by meeting the required pressure conditions, shifting the balance towards diamond-like formations. Meeting the pressure requirements for diamane synthesis is an exciting challenge that might lead to novel discoveries in carbon materials and exciting new uses in many different industries [65].

10.5 Applications of diamane and derivatives

10.5.1 Overview of potential applications

Diamane's theoretical features imply that it might have several applications. This subsection gives a rundown of some of the possible uses for diamane, including its use in superhard materials.

It is anticipated that diamane's hardness and mechanical strength would rival or perhaps exceed those of diamond. Cutting tools, drill bits, abrasives, and wear-resistant components for severe environments might all benefit from the synthesis of diamane because it would be a very hard material.

Optoelectronics: Diamane's potential semiconducting characteristics and optical transparency make it a promising material for use in optoelectronics. Light-emitting diodes (LEDs), photodetectors, and lasers are just some of the high-performance optical devices that potentially benefit from its use [18]

Fast computers and electronics: High-speed electronics and nanoelectronics are only two of the possible uses of diamane, which theoretical studies have revealed may have outstanding electron transport capabilities [8]. Diamane has potential as a high-frequency device material or in the next generation of transistors. Diamane's strong thermal conductivity might be used for heat spreaders inefficient electronic devices, power electronics, and microprocessors, among other thermal management applications. Diamane's distinctive porous nature may make it useful as a medium for gas storage and separation. It has potential applications in gas separation procedures and as a molecular sieve for gas storage systems [66]. Because of its stability under high pressure, diamane may be useful in high-pressure experiments and studies in high-energy physics and material science if it is found to be biocompatible. However, these possible uses are very theoretical due to the theoretical nature of diamane and its uncertain synthesis. Diamane's potential uses can only be unlocked by its successful synthesis and characterization. Extensive study is needed to properly understand and use diamane's characteristics, and its qualities may deviate from theoretical expectations. Diamane has the potential to become a fascinating material with many uses in science, engineering, and technology as research and development continue to advance.

10.5.2 Electronics and optoelectronics

Due to its one-of-a-kind tetrahedral sp^3-hybridized structure, diamane has the potential to have high electron mobility, which might lead to the creation of high-speed transistors [67]. Diamane transistors have the potential to improve the performance of electrical circuits by allowing for quicker switching rates. The realization of diamane's quantum features, which might be useful in quantum computing, would make it a promising material for this emerging field. Due to its quantum nature, diamane might pave the way for the development of qubits with improved coherence and entanglement characteristics, which would be a boon for the field of quantum information processing [8]. For this reason, diamane may be used in high-frequency devices such as high-frequency transistors and amplifiers to improve the performance of communication networks and radar systems. Diamane's high thermal conductivity might make it a useful component in power electronic devices, which are known to produce significant amounts of heat [68]. Power devices based on diamane may have better heat dissipation, leading to more stable and effective power electronics. More efficient and powerful integrated circuits may be possible if diamane can be synthesized and incorporated into semiconductor technology. It is possible that novel device designs will emerge from the integration of diamane's unusual features with those of more common semiconductor materials [69].

Diamane is a promising material for LEDs due to its theoretical optical transparency and possible semiconducting characteristics. High-efficiency diamane-based LEDs might revolutionize the lighting and electronics industries. Due to its unusual electrical characteristics, diamane may find use in high-performance photodetectors [70]. Possible benefits to imaging and sensing from diamane-based photodetectors include increased sensitivity and faster reaction times. Theoretical investigations have shown that diamane has the ability to display optical gain features, suggesting that it might be used as a laser medium. Communications, laser spectroscopy, and perhaps medicine might benefit from diamagnetic lasers. Diamane has the potential to be used in solar cells due to its semiconducting characteristics and its theoretical optical transparency. It is possible that diamane-based solar cells will allow for more effective energy conversion and more solar energy gathering. Diamane may find application in nonlinear optical systems due to its unusual electrical and optical characteristics [71]. Harmonic frequencies may be generated using these devices, and signal processing functions can be enabled. The theoretical basis of diamane and the lack of knowledge surrounding its synthesis severely restrict its present practical applicability.

10.5.3 Energy storage and conversion

The theoretical properties of this substance suggest promising applications in these disciplines. This subsection will give a comprehensive discussion of diamane's applicability in energy storage and conversion. The theoretically high thermal conductivity and mechanical robustness of diamane could make it a promising material for lithium-ion batteries [72]. As an anode material, diamane could provide enhanced thermal management and structural stability, resulting in improved battery performance and safety. Diamane's elevated surface area, which is a result of its unique tetrahedral structure, may be advantageous for supercapacitor electrodes. Possible characteristics of supercapacitors based on diamane include high energy and power densities, quicker charging, and extended cycle life. Diamane's porous structure could make it an effective material for hydrogen storage. Diamane could contribute to the development of hydrogen storage systems for fuel cells and renewable energy applications if successfully synthesized [73]. Diamane may find use in solid-state batteries, which promise greater energy density and enhanced safety compared to conventional lithium-ion batteries. Its distinctive properties could improve the overall efficacy and durability of solid-state battery materials. Diamane is a candidate for use in photovoltaic cells due to its potential as a semiconducting material. Compared to conventional solar cell materials, diamane-based solar cells may offer greater efficiency and enhanced energy conversion [56]. Diamane's exceptional thermal conductivity and potential as a semiconducting material make it an attractive material for thermoelectric applications. Thermoelectric devices manufactured from diamane could convert waste heat into electricity in an efficient manner. Diamane's theoretical properties may allow it to serve as a catalyst for water splitting [74], a crucial step in the production of hydrogen for renewable energy applications. Catalysts based on diamane may increase the efficiency of water-splitting reactions. Diamane, if effectively synthesized, could exhibit photocatalytic properties, thereby facilitating light-driven chemical reactions. Photocatalysts based on diamane may have applications in environmental remediation and solar fuel production. Diamane's thermal conductivity and potential semiconducting properties could make it a suitable material for thermophotovoltaic devices [75]. These thermal radiation-to-electricity converters have applications in residual heat recovery and power generation.

10.5.4 Biomedical and sensing applications

The porous structure of diamane, if realized, could be applied to the development of drug delivery systems. Due to its high surface area and ability to encapsulate molecules, diamane is a potential candidate for controlled and targeted drug requires a thorough understanding of, and the capacity to regulate, its stability and possible interactions with other materials. Scalability issues and high prices are further barriers to wider diamane use. Large-scale manufacture may be hindered by the high-pressure methods and specialized equipment needed for its synthesis. Furthermore, the price of these synthesis processes may be too high for widespread commercial use. Research and development into high-pressure synthesis methods, material characterization, and cost-effective manufacturing methods are crucial to finding solutions to these issues.

Despite the obstacles, there are intriguing opportunities in the field of diamane research. The experimental synthesis of diamane is a potential future avenue. High-pressure procedures will be further developed, and new methods of fabricating diamane will be investigated. The experimental production of this compound will pave the way for in-depth research into its characteristics and prospective uses. Diamane characteristics will be an important area of study in the future. Once it is synthesized, diamane will undergo extensive characterization through experimental research and theoretical inquiries into its mechanical, electrical, thermal, and optical characteristics. These investigations will shed light on the properties of diamane and help direct its use in domains including electronics, optoelectronics, energy storage, biomedicine, and sensing. The development and implementation of diamane's functionalities will also be crucial moving forward. To improve diamane's compatibility with other materials, scientists will investigate ways to 'functionalize' it by altering its surface chemistry. By incorporating diamane into devices and systems, its qualities may be modified to meet the needs of various uses. The problems with stability and security will be dealt with immediately. The goal of this research is to stabilize and preserve diamane so that it may be safely handled and used in a wide variety of applications. Even after an experimental synthesis of diamane has been accomplished, theoretical research will continue to investigate its characteristics and prospective behaviors. These investigations will help us to better grasp diamane's properties and direct future experimental endeavors.

In general, experimental synthesis, detailed property characterization, functionalization, and integration are the next steps in diamane research. To fully realize the promise of this exciting carbon-based material, scientists and engineers from different fields will need to work together. Diamane has the potential to pave the way for novel solutions across a range of technical fields in the future, which would be good for businesses and the economy.

10.7 Conclusion

To fully comprehend the synthesis and construction of these hypothetical carbon-based materials, it is essential to have a grasp of the function that temperature and pressure play in the production of diamane and its derivatives. Conditions of high pressure are required to realize the tetrahedral sp^3-hybridized structure that is wanted; this structure is similar to that of diamond. Carbon atoms go through a rearrangement when they are subjected to extremely high pressures. This allows them to overcome kinetic barriers and create the distinctive tetrahedral structure that is seen in diamane. Diamond-like structures are more likely to form as a result of this change in the phase diagram of carbon than other allotropes. To successfully create diamane through experimentation, one needs specialized high-pressure equipment as well as precise control during the creation process. Pressure is a crucial factor in the thermodynamic stability of diamane and its derivatives. Pressure supplies the energy necessary for the rearrangement of carbon, which is required for the reaction. If these materials can be successfully manufactured, then it may be possible to make strides in the fields of high-speed electronics, optoelectronics, thermal management, and very hard materials. However, to make efficient use of their prospective applications, problems such as as stability, responsiveness, scalability, and cost must be solved. This is an essential step towards unlocking the potential of these one-of-a-kind carbon-based materials and their potential contributions to a wide variety of industries and technology to get an understanding of the function that temperature and pressure play in the synthesis of diamane.

References

[1] Lavini F, Rejhon M and Riedo E 2022 Two-dimensional diamonds from sp^2-to-sp^3 phase transitions *Nat. Rev. Mater.* **7** 814–32

[2] Fyta M 2013 Nitrogen-vacancy centers and dopants in ultrathin diamond films: electronic structure *J. Phys. Chem.* C **117** 21376–81

[3] Chernozatonskii L A, Sorokin P B, Kuzubov A A, Sorokin B P, Kvashnin A G, Kvashnin D G and Yakobson B I 2011 Influence of size effect on the electronic and elastic properties of diamond films with nanometer thickness *J. Phys. Chem.* C **115** 132–6

[4] Piazza F, Monthioux M and Puech P 2022 2D (<10 nm) sp^3-C-rich carbon materials, possibly hydrogenated: a review *Carbon Trends* **9** 100219

[5] Mélinon P, Masenelli B, Tournus F and Perez A 2007 Playing with carbon and silicon at the nanoscale *Nat. Mater.* **6** 479–90

[6] Tahenti M, Gatfaoui S, Issaoui N, Roisnel T and Marouani H 2020 A tetrachlorocobaltate (II) salt with 2-amino-5-picolinium: synthesis, theoretical and experimental characterization *J. Mol. Struct.* **1207** 127781

[7] Field J E 2012 The mechanical and strength properties of diamond *Rep. Prog. Phys.* **75** 126505

[8] Tiwari S K, Pandey R, Wang N, Kumar V, Sunday O J, Bystrzejewski M and Mishra Y K 2022 Progress in diamanes and diamanoids nanosystems for emerging technologies *Adv. Sci.* **9** 2105770

[9] Capek I 2009 Dispersions, novel nanomaterial sensors and nanoconjugates based on carbon nanotubes *Adv. Colloid Interface Sci.* **150** 63–89

[10] Baptista F R, Belhout S A, Giordani S and Quinn S J 2015 Recent developments in carbon nanomaterial sensors *Chem. Soc. Rev.* **44** 4433–53

[11] Rjeibi I, Feriani A, Hentati F, Hfaiedh N, Michaud P and Pierre G 2019 Structural characterization of water-soluble polysaccharides from *Nitraria retusa* fruits and their antioxidant and hypolipidemic activities *Int. J. Biol. Macromol.* **129** 422–32

[12] Addoun N, Boual Z, Delattre C, Ursu A V, Desbrieres J, Le Cerf D and Pierre G 2020 Structural features and rheological behavior of a water-soluble polysaccharide extracted from the seeds of *Plantago ciliata* Desf. *Int. J. Biol. Macromol.* **155** 1333–41

[13] Chen S, Liu J, Zhang Q, Teng F and McLellan B C 2022 A critical review on deployment planning and risk analysis of carbon capture, utilization, and storage (CCUS) toward carbon neutrality *Renew. Sustain. Energy Rev.* **167** 112537

[14] Park S, Abate I I, Liu J, Wang C, Dahl J E, Carlson R M and Lin Y 2020 Facile diamond synthesis from lower diamondoids *Sci. Adv.* **6** eaay9405

[15] Berman D and Erdemir A 2021 Achieving ultralow friction and wear by tribocatalysis: enabled by in-operando formation of nanocarbon films *ACS Nano* **15** 18865–79

[16] Sorokin P B and Yakobson B I 2021 Two-dimensional diamond—diamane: current state and further prospects *Nano Lett.* **21** 5475–84

[17] Penev E S, Marzari N and Yakobson B I 2021 Theoretical prediction of two-dimensional materials, behavior, and properties *ACS Nano* **15** 5959–76

[18] Kumar S, Pratap S, Kumar V, Mishra R K, Gwag J S and Chakraborty B 2023 Electronic, transport, magnetic, and optical properties of graphene nanoribbons and their optical sensing applications: a comprehensive review *Luminescence* **38** 909–53

[19] Giriat G, Wang W, Attfield J P, Huxley A D and Kamenev K V 2010 Turnbuckle diamond anvil cell for high-pressure measurements in a superconducting quantum interference device magnetometer *Rev. Sci. Instrum.* **81** 073901–5

[20] Tahir N A, Hoffmann D H H, Kozyreva A, Shutov A, Maruhn J A, Neuner U and Bock R 2000 Shock compression of condensed matter using intense beams of energetic heavy ions *Phys. Rev.* E **61** 1975

[21] May P W 2000 Diamond thin films: a 21st-century material *Phil. Trans. R. Soc.* A **358** 473–95

[22] Zhang Y, Zhao J, Fang Y, Liu Y and Zhao X 2018 Preparation of long linear carbon chain inside multi-walled carbon nanotubes by cooling enhanced hydrogen arc discharge method *Nanoscale* **10** 17824–33

[23] Hirao N, Kawaguchi S I, Hirose K, Shimizu K, Ohtani E and Ohishi Y 2020 New developments in high-pressure X-ray diffraction beamline for diamond anvil cell at SPring-8 *Matter Radiat. at Extremes* **5** 018403–10

[24] Lorenzana H E 1990 *Phase Transitions in Solid Hydrogen at Megabar Pressures* (Cambridge, MA: Harvard University)

[25] Shen G, Rivers M L, Wang Y and Sutton S R 2001 Laser heated diamond cell system at the advanced photon source for *in situ* X-ray measurements at high pressure and temperature *Rev. Sci. Instrum.* **72** 1273–82

[26] Jayaraman A 1983 Diamond anvil cell and high-pressure physical investigations *Rev. Mod. Phys.* **55** 65

[27] Yagi T, Yusa H and Yamakata M A 1996 An apparatus to load gaseous materials to the diamond-anvil cell *Rev. Sci. Instrum.* **67** 2981–4

[28] Shen G and Mao H K 2016 High-pressure studies with x-rays using diamond anvil cells *Rep. Prog. Phys.* **80** 016101

[29] Vejpravová J 2021 Mixed sp^2–sp^3 nanocarbon materials: a status quo review *Nanomaterials* **11** 2469

[30] Stoyanov E, Häussermann U and Leinenweber K 2010 Large-volume multianvil cells designed for chemical synthesis at high pressures *High Press. Res.* **30** 175–89

[31] Miletich R, Allan D R and Kuhs W F 2000 High-pressure single-crystal techniques *Rev. Mineral. Geochem.* **41** 445–519

[32] D'Haenens-Johansson U F, Butler J E and Katrusha A N 2022 Synthesis of diamonds and their identification *Rev. Mineral. Geochem.* **88** 689–753

[33] Thevamaran R, Lawal O, Yazdi S, Jeon S J, Lee J H and Thomas E L 2016 Dynamic creation and evolution of gradient nanostructure in single-crystal metallic microcubes *Science* **354** 312–6

[34] Shandilya M, Rai R and Singh J 2016 Hydrothermal technology for smart materials *Adv. Appl. Ceram.* **115** 354–76

[35] Ayers J C, Brenan J B, Watson E B, Wark D A and Minarik W G 1992 A new capsule technique for hydrothermal experiments using the piston-cylinder apparatus *Am. Mineral.* **77** 1080–6

[36] Dabkowska H A and Dabkowski A B 2010 Crystal growth of oxides by optical floating zone technique *Springer Handbook of Crystal Growth* (Berlin: Springer) 367–91

[37] Martin P, Guizard S, Daguzan P, Petite G, D'oliveira P, Meynadier P and Perdrix M 1997 Subpicosecond study of carrier trapping dynamics in wide-band-gap crystals *Phys. Rev.* B **55** 5799

[38] Frenklach M 1989 The role of hydrogen in vapor deposition of diamond *J. Appl. Phys.* **65** 5142–9

[39] Frezzotti M L, Huizenga J M, Compagnoni R and Selverstone J 2014 Diamond formation by carbon saturation in C–O–H fluids during cold subduction of oceanic lithosphere *Geochim. Cosmochim. Acta* **143** 68–86

[40] Sauer F H D S R 1999 Synthetic diamond—basic research and applications *Cryst. Res. Technol.* **34** 227–41

[41] Strong H M and Chrenko R M 1971 Diamond growth rates and physical properties of laboratory-made diamond *J. Phys. Chem.* **75** 1838–43

[42] Shirey S B, Cartigny P, Frost D J, Keshav S, Nestola F, Nimis P and Walter M J 2013 Diamonds and the geology of mantle carbon *Rev. Mineral. Geochem.* **75** 355–421

[43] Bundy F P, Bassett W A, Weathers M S, Hemley R J, Mao H U and Goncharov A F 1996 The pressure–temperature phase and transformation diagram for carbon; updated through 1994 *Carbon* **34** 141–53

[44] Esmeryan K D, Castano C E, Bressler A H, Abolghasemibizaki M, Fergusson C P, Roberts A and Mohammadi R 2017 Kinetically driven graphite-like to diamond-like carbon transformation in low temperature laminar diffusion flames *Diam. Relat. Mater.* **75** 58–68

[45] Sumiya H, Yusa H, Inoue T, Ofuji H and Irifune T 2006 Conditions and mechanism of formation of nano-polycrystalline diamonds on direct transformation from graphite and non-graphitic carbon at high pressure and temperature *High Press. Res.* **26** 63–9

[46] Guignard J, Prakasam M and Largeteau A 2022 A review of binderless polycrystalline diamonds: focus on the high-pressure–high-temperature sintering process *Materials* **15** 2198

[47] Dai D, Li Y and Fan J 2021 Room-temperature synthesis of various allotropes of carbon nanostructures (graphene, graphene polyhedra, carbon nanotubes and nano-onions, n-diamond nanocrystals) with aid of ultrasonic shock using ethanol and potassium hydroxide *Carbon* **179** 133–41

[48] Liang C, Chen Y, Wu M, Wang K, Zhang W, Gan Y and Pan H 2021 Green synthesis of graphite from CO_2 without graphitization process of amorphous carbon *Nat. Commun.* **12** 119

[49] Balandin A A 2011 Thermal properties of graphene and nanostructured carbon materials *Nat. Mater.* **10** 569–81

[50] Marchant G A, Caro M A, Karasulu B and Pártay L B 2023 Exploring the configuration space of elemental carbon with empirical and machine learned interatomic potentials *npj Comput. Mater.* **9** 131

[51] Ng S F, Chen X, Foo J J, Xiong M and Ong W J 2023 2D carbon nitrides: regulating non-metal boron-doped C3N5 for elucidating the mechanism of wide pH range photocatalytic hydrogen evolution reaction *Chin. J. Catal.* **47** 150–60

[52] Jin Y, Cheng J, Varma-Nair M, Liang G, Fu Y, Wunderlich B and Zettl A K 1992 Thermodynamic characterization of fullerene (C60) by differential scanning calorimetry *J. Phys. Chem.* **96** 5151–6

[53] Xu J, Wan Q, Anpo M and Lin S 2020 Bandgap opening of graphdiyne monolayer via B, N-codoping for photocatalytic overall water splitting: design strategy from DFT studies *J. Phys. Chem.* C **124** 6624–33

[54] Yarbrough W A 1989 Thermodynamics and the CVD of diamond *MRS Online Proc. Lib. (OPL)* **162** 75

[55] Vejpravová J 2021 Mixed sp2–sp3 nanocarbon materials: A status quo review *Nanomater* **11** 2469

[56] Chernozatonskii L A, Demin V A, Kvashnin A G and Kvashnin D G 2022 Diamane quasicrystals *Appl. Surf. Sci.* **572** 151362

[57] Chernozatonskii L A, Katin K P, Kochaev A I and Maslov M M 2022 Moiré and non-twisted sp^3-hybridized structures based on hexagonal boron nitride bilayers: *ab initio* insight into infrared and Raman spectra, bands structures and mechanical properties *Appl. Surf. Sci.* **606** 154909

[58] Baidakova M 2007 New prospects and frontiers of nanodiamond clusters *J. Phys.* D **40** 6300

[59] Schwartz G, Tee B C K, Mei J, Appleton A L, Kim D H, Wang H and Bao Z 2013 Flexible polymer transistors with high pressure sensitivity for application in electronic skin and health monitoring *Nat. Commun.* **4** 1859

[60] Blank V D, Churkin V D, Kulnitskiy B A, Perezhogin I A, Kirichenko A N, Denisov V N and Popov M Y 2018 Phase diagram of carbon and the factors limiting the quantity and size of natural diamonds *Nanotechnology* **29** 115603

[61] Kaur N, Gupta S, Jindal V K and Dharamvir K 2010 Pressure induced transformations in condensed and molecular phases of C60 *Carbon* **48** 744–55

[62] Erohin S V, Ruan Q, Sorokin P B and Yakobson B 2020 Nano-thermodynamics of chemically induced graphene–diamond transformation *Small* **16** 2004782

[63] Martins L G P, Matos M J, Paschoal A R, Freire P T, Andrade N F, Aguiar A L and Cançado L G 2017 Raman evidence for pressure-induced formation of diamondene *Nat. Commun.* **8** 96

[64] Wang Y, Panzik J E, Kiefer B and Lee K K 2012 Crystal structure of graphite under room-temperature compression and decompression *Sci. Rep.* **2** 520

[65] Paoletti A and Tucciarone A (ed) 1997 *The Physics of Diamond* **vol 135** (Amsterdam: IOS Press)

[66] Ahmad Y, Batisse N, Chen X and Dubois M 2021 Preparation and applications of fluorinated graphenes *C* **7** 20

[67] Piazza F, Monthioux M, Puech P and Gerber I C 2020 Towards a better understanding of the structure of diamanoïds and diamanoïd/graphene hybrids *Carbon* **156** 234–41

[68] Chowdhury S, Demin V A, Chernozatonskii L A and Kvashnin A G 2022 Ultra-low thermal conductivity of Moiré diamanes *Membranes* **12** 925

[69] Sorokin P B and Chernozatonskii L A 2013 Graphene-based semiconductor nanostructures *Phys.-Usp.* **56** 105

[70] Tantardini C, Kvashnin A G, Azizi M, Gonze X, Gatti C, Altalhi T and Yakobson B I 2023 Electronic properties of functionalized diamanes for field-emission displays *ACS Appl. Mater. Interfaces* **15** 16317–26

[71] Ghosal S and Jana D 2022 Beyond T-graphene: two-dimensional tetragonal allotropes and their potential applications *Appl. Phys. Rev.* **9** 021314

delivery, which would result in enhanced drug efficacy and decreased adverse effects. Theoretical investigations imply that diamane may be biocompatible and transparent to visible light [76]. Consequently, it could be investigated as a contrast agent in bioimaging techniques such as optical imaging and potentially magnetic resonance imaging or computed tomography scanning, which provide detailed images of biological tissues and structures. Diamane's unique electronic properties could be utilized in biosensors for the detection of biological molecules or biomarkers. Applications for biosensors based on diamane include disease diagnostics and healthcare monitoring [29]. Diamane's potential mechanical strength and biocompatibility may qualify it as a scaffold material for tissue engineering. Diamane-based scaffolds may provide mechanical support for tissue regeneration and facilitate cell adhesion and growth. Emerging in the field of medicine, theranostics refers to the integration of imaging and therapeutic capabilities on a single platform. If effectively synthesized, diamane could be incorporated into theranostic platforms, providing diagnostic imaging and therapeutic functions in a single system.

The porous structure of diamane could allow it to function as an extremely sensitive gas sensor. Gas sensors based on diamane may detect numerous gases with exceptional sensitivity and selectivity [73], making them useful for environmental monitoring, industrial safety, and healthcare applications. Due to its theoretical mechanical strength, diamane could be a candidate for strain sensors. Useful for structural health monitoring and wearable devices, strain sensors based on diamane could measure mechanical deformation with high accuracy. Due to its theoretically exceptional mechanical properties, diamane may be investigated as a material for high-performance pressure sensors [77]. Pressure sensors based on diamane may provide precise measurements over a broad pressure range, making them suitable for a variety of applications, including robotics and healthcare. Theoretical properties of diamane, such as its high surface area and potential biocompatibility, may make it useful for biosensing applications. Biosensors based on diamane could detect biomolecules with high specificity and sensitivity, enabling their use in medical diagnostics and life sciences research [78].

10.6 Challenges and future directions

As a speculative carbon-based substance with unusual characteristics, diamane faces a number of obstacles that limit its usefulness. The most difficult part of working with diamane is getting it synthesized and manufactured. There has been no progress in synthesizing or observing diamane in experiments as of my previous report in September 2021. Its experimental synthesis is difficult because of the high pressure and high-temperature conditions needed for its creation. Adding complexity to the experimental efforts is the need to regulate the synthesis process to produce the required tetrahedral diamond-like structure. The instability and reactivity of diamane can provide considerable difficulties. Diamane's unusual tetrahedral sp^3-hybridized structure and high-energy state have led to theoretical predictions that it may be extremely reactive. Because of its reactivity, diamane may be difficult to manipulate and keep stable in real-world settings. Harnessing its particular features

[72] Chen X, Fan K, Liu Y, Li Y, Liu X, Feng W and Wang X 2022 Recent advances in fluorinated graphene from synthesis to applications: critical review on functional chemistry and structure engineering *Adv. Mater.* **34** 2101665

[73] Liu Y, Gao L, Fu S, Cheng S, Gao N and Li H 2023 Highly efficient VOC gas sensors based on Li-doped diamane *Appl. Surf. Sci.* **611** 155694

[74] Singh D, Khossossi N, Luo W, Ainane A and Ahuja R 2022 2D Janus and non-Janus diamanes with an in-plane negative Poisson's ratio for energy applications *Mater. Today Adv.* **14** 100225

[75] Shah N, Shah A A, Leung P K, Khan S, Sun K, Zhu X and Liao Q 2023 A review of third generation solar cells *Processes* **11** 1852

[76] Liu B, Obot E E, Liang T and Wang B 2023 Advancements in theoretical and experimental investigations on diamane materials *Nanoscale* **15** 10498–512

[77] Qin G, Wu L and Gou H 2022 Diamane: design, synthesis, properties, and challenges *Funct. Diam.* **1** 83–92

[78] Chernozatonskii L A, Katin K P, Demin V A and Maslov M M 2021 Moiré diamanes based on the hydrogenated or fluorinated twisted bigraphene: the features of atomic and electronic structures, Raman and infrared spectra *Appl. Surf. Sci.* **537** 148011

www.ingramcontent.com/pod-product-compliance
Ingram Content Group UK Ltd.
Pitfield, Milton Keynes, MK11 3LW, UK
UKHW051041120926
8556IPUK00030B/677

9 780750 359375